Retinoids in Normal Development and Teratogenesis

# Retinoids in Normal Development and Teratogenesis

Edited by
GILLIAN MORRISS-KAY
*Department of Human Anatomy,*
*University of Oxford*

OXFORD UNIVERSITY PRESS
Oxford   New York   Tokyo
1992

*Oxford University Press, Walton Street, Oxford OX2 6DP*
*Oxford New York Toronto*
*Delhi Bombay Calcutta Madras Karachi*
*Petaling Jaya Singapore Hong Kong Tokyo*
*Nairobi Dar es Salaam Cape Town*
*Melbourne Auckland*
*and associated companies in*
*Berlin Ibadan*

*Oxford is a trade mark of Oxford University Press*

*Published in the United States*
*by Oxford University Press, New York*



*A catalogue record for this book is available from the British Library*

*Library of Congress Cataloging in Publication Data*
*Retinoids in normal development and teratogenesis/edited by Gillian*
*Morriss-Kay.*
*Based on a conference held in Oxford in March 1991.*
*Includes index.*
*1. Retinoids—Physiological effect—Congresses.  2. Teratogenesis—*
*Congresses. 3. Retinoids—Toxicology—Congresses.  4. Retinoids—*
*Receptors—Congresses.  5. Chemical embryology—Congresses.*
*I. Morriss-Kay, Gillian.*
*[DNLM: 1. Abnormalities, Drug Induced—etiology—congresses.*
*2. Retinoids—pharmacology—congresses.  3. Retinoids—toxicity—*
*congresses.  4. Teratogens—congresses.  QU 167 R4389 1991]*
*QP772.V5R48      1992      612.6'4—dc20      91–45147*
*ISBN 0 19 854770 6*

*Typeset by Cotswold Typesetting Ltd, Gloucester*
*Printed in Great Britain by*
*St Edmundsbury Press,*
*Bury St Edmunds*

To the memory of Dame Honor Fell

# Preface

All of the articles in this book arose from a conference held in Oxford in March 1991. Interaction between the participants was considerable, and is reflected in the amount of cross-referencing between the chapters. I wish to thank Hoffmann-La Roche for providing the funds for bringing us together for such stimulating presentations and discussions, and New College for providing an appropriate ambience. All of those present contributed to this book through their active participation at the meeting, but I am particularly grateful to the authors of the articles presented here: they not only provided manuscripts within the allotted time, but were enormously tolerant of my editorial interventions.

*Oxford*
July 1991

G.M.M-K.

# Contents

# Contributors

*Barbara D. Abbott* Developmental Toxicology Division, Perinatal Toxicology Branch, US EPA, Health Effects Research Laboratory, Research Triangle Park, North Carolina 27711, USA

*Dario Acampora* International Institute of Genetics and Biophysics of CNR, Via Marconi 10, 80125 Naples, Italy

*Rosemary J. Akhurst* Duncan Guthrie Institute of Medical Genetics, University of Glasgow, Yorkhill, Glasgow G3 8SJ, UK

*Christian Apfel* Department of Pharmaceutical Research, F. Hoffmann-La Roche Ltd, Basel, Switzerland

*Dawna L. Armstrong* Department of Pathology, Baylor College of Medicine, Houston, Texas, USA

*James P. Basilion* Department of Pharmacology, The Medical School, University of Texas Health Sciences Center at Houston, Houston, Texas, USA

*Edoardo Boncinelli* International Institute of Genetics and Biophysics of CNR, Via Marconi 10, 80125 Naples, Italy

*Paul Brickell* Medical Molecular Biology Unit, Department of Biochemistry and Molecular Biology, University College and Middlesex School of Medicine, The Windeyer Building, Cleveland Street, London W1P 6DB, UK

*Jeremy Brockes* Ludwig Institute for Cancer Research, 91 Riding House Street, London W1P 8BT, UK

*Pierre Chambon* Laboratoire de Génétique Moléculaire des Eucaryotes du CNRS, Unité 184 de Biologie Moléculaire et de Génie Génétique de l'INSERM, Institut de Chimie Biologique, Faculté de Médecine, 11 rue Humann, 67085 Strasbourg Cédex, France

*E. Antonio Chiocca* Department of Pharmacology, The Medical School, University of Texas Health Sciences Center at Houston, Houston, Texas, USA

*Frank Chytil* Department of Biochemistry, Vanderbilt University School of Medicine, Nashville, Tennessee 37232, USA

*Joan Creech Kraft* Department of Pharmacology, University of Washington, Seattle, Washington, USA

*Marco Crettaz* Zyma SA, Nyon, Switzerland

*Maurizio D'Esposito* International Institute of Genetics and Biophysics of CNR, Via Marconi 10, 80125 Naples, Italy

*Duncan R. Davidson* MRC Human Genetics Unit, Western General Hospital, Crewe Road, Edinburgh EH4 2XU, UK

*Peter J. A. Davies* Department of Pharmacology, The Medical School, University of Texas Health Sciences Center at Houston, Houston, Texas, USA

*Danielle Dhouailly* Jeune Equipe Biologie de la Différenciation Epithéliale, Centre d'Etudes et de Recherches sur les Macromolécules Organisées, Université Joseph Fourier, 38041 Grenoble, France

*Pascal Dollé* EMBL, Meyerhofstrasse 1, 6900 Heidelberg, Germany

*Denis Duboule* EMBL, Meyerhofstrasse 1, 6900 Heidelberg, Germany

*Nicholas S. C. Eager* Medical Molecular Biology Unit, Department of Biochemistry and Molecular Biology, University College and Middlesex School of Medicine, The Windeyer Building, Cleveland Street, London W1P 6DB, UK

*Ronald M. Evans* Howard Hughes Medical Institute, The Salk Institute for Biological Studies, La Jolla, California, USA

*Antonio Faiella* International Institute of Genetics and Biophysics of CNR, Via Marconi 10, 80125 Naples, Italy

*Laszlo Fesus* Department of Biochemistry, The Medical School, University of Debrecen, Debrecen, Hungary

*David R. FitzPatrick* Duncan Guthrie Institute of Medical Genetics, University of Glasgow, Yorkhill, Glasgow G3 8SJ, UK

*Emily Gale* Anatomy and Human Biology Group, Division of Biomedical Sciences, Kings College, The Strand, London WC2R 2LS, UK

*Vittorio Gentile* Department of Pharmacology, The Medical School, University of Texas Health Sciences Center at Houston, Houston, Texas, USA

*Adam B. Glick* Laboratory of Cellular Carcinogenesis and Tumor Promotion, National Cancer Institute, Bethesda, Maryland 20892, USA

*Gerhart Graupner* La Jolla Cancer Research Foundation, 10901 North Torrey Pines Road, La Jolla, California 92037, USA

*Riaz ul Haq* Department of Biochemistry, Vanderbilt University School of Medicine, Nashville, Tennessee, USA

*Margaret H. Hardy* Department of Biomedical Sciences, University of Guelph, Guelph, Canada N1G 2W1

*Robert E. Hill* MRC Human Genetics Unit, Western General Hospital, Crewe Road, Edinburgh EH4 2XU, UK

*Birgit Hoffmann* La Jolla Cancer Research Foundation, 10901 North Torrey Pines Road, La Jolla, California 92037, USA

*Claire Horton* Anatomy and Human Biology Group, Division of Biomedical Science, Kings College, The Strand, London WC2R 2LS, UK

*Matthias Husmann* La Jolla Cancer Research Foundation, 10901 North Torrey Pines Road, La Jolla, California 92037, USA

*Juan-Carlos Izpisúa-Belmonte* EMBL, Meyerhofstrasse 1, 6900 Heidelberg, Germany

*Philippe Kastner* Laboratoire de Génétique Moléculaire des Eucaryotes du CNRS, Unité 184 de Biologie Moléculaire et de Génie Génétique de l'INSERM, Institut de Chimie Biologique, Faculté de Médecine, 11 rue Humann, 67085 Strasbourg Cédex, France

*Andrée Krust* Laboratoire de Génétique Moléculaire des Eucaryotes du CNRS, Unité 184 de Biologie Moléculaire et de Génie Génétique de l'INSERM, Institut de Chimie Biologique, Faculté de Médecine, 11 rue Humann, 67085 Strasbourg Cédex, France

*Edward J. Lammer* California Birth Defects Monitoring Program, Emeryville, California, USA

*Jürgen M. Lehmann* La Jolla Cancer Research Foundation, 10901 North Torrey Pines Road, La Jolla, California 92037, USA

*Peter LeMotte* Department of Pharmaceutical Research, F. Hoffmann-La Roche Ltd, Basel, Switzerland

*Pierre Leroy* Laboratoire de Génétique Moléculaire des Eucaryotes du CNRS, Unité 184 de Biologie Moléculaire et de Génie Génétique de l'INSERM, Institut de Chimie Biologique, Faculté de Médecine, 11 rue Humann, 67085 Strasbourg Cédex, France

*Malcolm Maden* Anatomy and Human Biology Group, Division of Biomedical Sciences, Kings College, The Strand, London WC2R 2LS, UK

*Radma Mahmood* Department of Human Anatomy, South Parks Road, Oxford OX1 2QX, UK

*Antonio Mallamaci* International Institute of Genetics and Biophysics of CNR, Via Marconi 10, 80125 Naples, Italy

*David J. Mangelsdorf* Howard Hughes Medical Institute, The Salk Institute for Biological Studies, La Jolla, California, USA

*Cathy Mendelsohn* Laboratoire de Génétique Moléculaire des Eucaryotes du CNRS, Unité 184 de Biologie Moléculaire et de Génie Génétique de l'INSERM, Institut de Chimie Biologique, Faculté de Médecine, 11 rue Humann, 67085 Strasbourg Cédex, France

*Gillian M. Morriss-Kay* Department of Human Anatomy, South Parks Road, Oxford OX1 2QX, UK

*Elisabeth E. Mufson* Department of Biochemistry, Vanderbilt University School of Medicine, Nashville, Tennessee, USA

*Paula Murphy* MRC Human Genetics Unit, Western General Hospital, Crewe Road, Edinburgh EH4 2XU, UK. Present address: Istituto di Istologia de Embriologia Generale, Universita di Roma 'La Sapienza', Via A Scarpa 14, 00161 Roma, Italy

*Harikrishna Nakshatri* Laboratoire de Génétique Moléculaire des Eucaryotes du CNRS, Unité 184 de Biologie Moléculaire et de Génie Génétique de l'INSERM, Institut de Chimie Biologique, Faculté de Médecine, 11 rue Humann, 67085 Strasbourg Cédex, France

*Maria Pannese* International Institute of Genetics and Biophysics of CNR, Via Marconi 10, 80125 Naples, Italy

*Magnus Pfahl* La Jolla Cancer Research Foundation, 10901 North Torrey Pines Road, La Jolla, California 92037, USA

*Joy M. Richman* Department of Anatomy and Developmental Biology, University College and Middlesex School of Medicine, The Windeyer Building, Cleveland Street, London W1P 6DB, UK

*Anita B. Roberts* Laboratory of Chemoprevention, National Cancer Institute, Bethesda, Maryland 20892, USA

*Annie Rowe* Medical Molecular Biology Unit, Department of Biochemistry and Molecular Biology, University College and Middlesex School of Medicine, The Windeyer Building, Cleveland Street, London W1P 6DB, UK

*Esther Ruberte* Laboratoire de Génétique Moléculaire des Eucaryotes du CNRS, Unité 184 de Biologie Moléculaire et de Génie Génétique de l'INSERM, Institut de Chimie Biologique, Faculté de Médecine, 11 rue Humann, 67085 Strasbourg Cédex, France

*Margaret G. Rush* Department of Biochemistry, Vanderbilt University School of Medicine, Nashville, Tennessee, USA

*Antonio Simeone* International Institute of Genetics and Biophysics of CNR, Via Marconi 10, 80125 Naples, Italy

*Jim C. Smith* Laboratory of Developmental Biology, National Institute for Medical Research, The Ridgeway, Mill Hill, London NW7 1AA, UK

*Michael B. Sporn* Laboratory of Chemoprevention, National Cancer Institute, Bethesda, Maryland 20892, USA

*Joseph P. Stein* The College of Medicine, State University of New York Health Science Center at Syracuse, Syracuse, New York, USA

*Anna Stornaiuolo* International Institute of Genetics and Biophysics of CNR, Via Marconi 10, 80125 Naples, Italy

*Donald G. Stump* Department of Biochemistry, Vanderbilt University School of Medicine, Nashville, Tennessee, USA

*Vilmos Thomazy* Departments of Pathology and Biochemistry, The Medical School, University of Debrecen, Debrecen, Hungary

*Cheryll Tickle* Department of Anatomy and Development Biology, University College and Middlesex School of Medicine, Cleveland Street, London W1P 6DP, UK

*Jean P. Viallet* Jeune Equipe Biologie de la Différenciation Epithéliale, Centre d'Etudes et de Recherches sur les Macromolécules Organisées, Université Joseph Fourier, 38041 Grenoble, France

*Lewis Wolpert* Department of Anatomy and Development Biology, University College and Middlesex School of Medicine, Cleveland Street, London W1P 6DP, UK

*Arthur Zelent* Laboratoire de Génétique Moléculaire des Eucaryotes du CNRS, Unité 184 de Biologie Moléculaire et de Génie Génétique de l'INSERM, Institut de Chimie Biologique, Faculté de Médecine, 11 rue Humann, 67085 Strasbourg Cédex, France

*Xiao-Kun Zhang* La Jolla Cancer Research Foundation, 10901 North Torrey Pines Road, La Jolla, California 92037, USA

# 1

# Editorial introduction

Gillian Morriss-Kay

The history of retinoid research is punctuated with events of major scientific significance. This was as true before vitamin A was first isolated in crystalline form as it is today, and is particularly notable in relation to its developmental roles. For instance, an important environment versus heredity debate was precipitated by Hale's seminal discovery in 1933 that vitamin A is essential for normal development. At that time, the ideas of Mendel, Morgan, and Bateson provided the bedrock for interpretation of any structural change from one generation to the next, so that congenital malformations and minor variation alike were interpreted as being the result of genetic mechanisms alone. The importance of specific nutritional factors in development was accepted only when Hale (1935) demonstrated that the mating of abnormal siblings, reared and maintained on an adequate diet, resulted in an entirely normal second generation.

Nearly six decades later, the environment versus heredity debate is alive again in the context of limb development, but now it is the genetic interpretation that is challenging the environmental one. The idea that a chemical 'morphogen' might control morphogenesis by creating a concentration gradient across an embryonic tissue was proposed by Turing (1952). This theory became firmly established as the model to explain the role of the 'zone of polarizing activity' (ZPA) in controlling digit pattern across the antero-posterior axis of the developing chick limb (Wolpert 1969; Summerbell 1979). Apparent identification of the putative morphogen as retinoic acid (RA) followed the discovery that locally applied RA could mimic a grafted ZPA (Tickle *et al.* 1982), and that its endogenous levels showed an anterior–posterior difference (Thaller and Eichele 1987). However, 1991 has brought new results suggesting that locally applied RA might remain local rather than alter an endogenous gradient, and that it is able to alter the pattern of gene expression in the limb bud. The morphogen theory itself is challenged by these reports. This heady area of retinoid research is specifically discussed in this book by Izpisúa-Belmonte *et al.* (Chapter 18), Ruberte *et al.* (Chapter 8), and Brockes (Chapter 9), but its influence touches many more of the articles.

In relation to development, retinoids are rather odd. First and foremost, it seems such a quirky design feature that an embryo should rely so heavily on a substance (retinoic acid) whose natural precursor (retinol, vitamin A) has to be supplied via the maternal diet. Second, the need for retinoids in development is both general (complete deficiency causes death) and specific (partial deficiency causes specific patterns of malformations). Third, retinoid excess is as devasting to normal development as is deficiency, but the pattern of malformations induced is different: some particular developmental events, or particular developing tissues, appear to require higher levels of retinoids than others.

The key to understanding some of these puzzling phenomena lies with the molecular mechanism of action of RA. In the nucleus, the number of known retinoid receptors continues to increase as more isoforms of the retinoic acid receptors (RARs) are identified. These receptors, which are closely related to the steroid hormone receptors, show specific responses to retinoic acid, as well as specific spatiotemporal patterns of distribution. In keeping with the general necessity for a minimum level of vitamin A to keep the embryo alive at all, RAR-$\alpha$1 has some of the characteristics of a housekeeping gene both in the nature of its promoter and in the fact that it is almost ubiquitously expressed. Other receptors reflect the diversity of retinoid-related effects in being highly regionalized and stage-related in the embryo. Although we know little as yet about the effects of these RA-inducible transcriptional transmodulators, it is clear that they have the potential for translating the influence of a single ligand into an enormous variety of genetic events. Further qualitative differences in the response of different cells to retinoids is possible through the specific distribution pattern of the more recently discovered family of receptors, the retinoid 'X' receptors (RXRs), whose natural ligand may be a different retinoid. These points are given substance in the chapters by Leroy *et al.* (Chapter 2), Mangelsdorf and Evans (Chapter 3), Pfahl (Chapter 4), Apfel *et al.* (Chapter 5) and elsewhere. In terms of the light thrown on questions of wider scientific significance, the discussion of evolutionary relationships of the RARs by Leroy *et al.* (Chapter 2) is particularly interesting.

In the cytoplasm, binding proteins for retinol (CRBP) and RA (CRABP) may provide a quantitative mechanism for varying the response of embryonic tissues to retinoids, by controlling the amount of retinoic acid that is available for binding to RARs. It appears likely that the presence of CRBP I indicates cells in which high levels of bound retinol (of maternal origin) are stored in order to provide a source of retinoic acid when required. In contrast, CRABP I, which is rarely coexpressed with CRBP I, may serve to bind and thereby inactivate RA in cells whose normal programme of gene expression requires that little of the available RA reaches the nucleus. The discovery that CRABP II has a different distribution from CRABP I (Ruberte *et al.*, Chapter 8), and does not correlate in an obvious way with retinoid-related congenital anomalies, means that we are now looking for a function for this binding protein. The distribution and possible functions of cytoplasmic retinoid binding proteins are discussed in the chapters by Chytil (Chapter 6), Maden (Chapter 10), Ruberte *et al.* (Chapter 8), and others.

The identification of genes whose expression is altered by RAR- or RXR-bound retinoids is a field waiting to explode, particularly where RA response elements (RARE) are concerned. In the meantime, what little we know about the effects of RA on the expression of developmentally important genes is discussed in this book in relation to homeobox genes and to the transforming growth factor beta family (TGF-$\beta$) in the chapters by Boncinelli *et al.* (Chapter 16), Murphy *et al.* (Chapter 17), Roberts *et al.* (Chapter 11), and Fitzpatrick *et al.* (Chapter 12).

The observation that retinoid excess is able to alter normal developmental processes was an accidental offshoot of studies of the effects of retinoids on bone growth (Cohlan 1953). Sadly, thirty years after this discovery, malformations were observed in human infants as an accidental side-effect following administration of isotretinoin (13-*cis*-retinoic acid) to women during the early stages of pregnancy (see Lammer and Armstrong, Chapter 21). Isotretinoin is a highly effective treatment for severe cystic acne; synthetic retinoids are making valuable contributions to the treatment of a wide variety of human diseases, reflecting the variety of biological roles of the natural molecules. One of the most important goals of applied retinoid research is to identify compounds that will have the therapeutic value of the currently available retinoid drugs without their teratogenic effects. Achievement of that goal depends on a better understanding of the roles of retinoids in both normal and abnormal development. The needs of applied research and the motivation of pure scientific curiosity are seldom so well harmonized.


REFERENCES

Cohlan, S. Q. (1953). Excessive intake of vitamin A as a cause of congenital anomalies in the rat. *Science*, **117**, 535–6.

Hale, F. (1933). Pigs born without eyeballs. *Journal of Heredity*, **24**, 105–6.

Hale, F. (1935). Relation of vitamin A to anophthalmos in pigs. *American Journal of Ophthalmology*, **18**, 1087–93.

Summerbell, D. (1979). The zone of polarizing activity: evidence for a role in normal chick limb. *Journal of Embryology and Experimental Morphology*, **50**, 217–33.

Thaller, C. and Eichele, G. (1987). Identification and spatial distribution of retinoids in the developing chick limb bud. *Nature*, **397**, 625–8.

Tickle, C., Alberts, B., Wolpert, L., and Lee, J. (1982). Local application of retinoic acid to the limb bud mimics the action of the polarizing region. *Nature*, **296**, 564–5.

Turing, A. M. (1952). The chemical basis of morphogenesis. *Philosophical Transactions of the Royal Society of London, Series B*, **237**, 37–72.

Wolpert, L. (1969). Positional information and the spatial pattern of differentiation. *Journal of Theoretical Biology*, **25**, 1–47.

# PART I

# Retinoid receptors and binding proteins: molecular aspects

# 2

# Retinoic acid receptors

Pierre Leroy, Andrée Krust, Philippe Kastner, Cathy Mendelsohn,
Arthur Zelent, and Pierre Chambon

## INTRODUCTION

The establishment of anatomical structures of the embryo throughout its
development is thought to be controlled by morphogenetic molecules, which
would instruct the cells of the three germ layers to differentiate according to their
position with respect to organizing region(s) (Slack 1987*a*; Wolpert 1989;
Summerbell and Maden 1990). Retinoic acid (RA) is generally believed to be a
natural morphogen (Maden 1982; Tickle *et al.* 1982; Slack 1987*b*; Thaller and
Eichele 1987; Smith *et al.* 1989; Eichele 1990), and plays an important role in the
development and homeostasis of vertebrates (Chytil 1984; Maden 1985;
Brockes 1989, 1990; Eichele 1989*a,b*; Durston *et al.* 1989; Sive *et al.* 1990;
Wagner *et al.* 1900; Ruiz i Altaba and Jessel 1991; and references therein).
However, recent reports raised a controversy by questioning the role of RA as a
true morphogenetic signal (Brockes 1991; Wanek *et al.* 1991; Noji *et al.* 1991).
The multiple effects exerted by RA on cultured cells or during development and
in the adult, have often been correlated with changes in the expression patterns
of specific genes (Wang and Gudas 1983; La Rosa and Gudas 1988; Murphy *et
al.* 1988; Lüscher *et al.* 1989; Vasios *et al.* 1989; Chytil and Haq 1990; Okamoto
*et al.* 1990; Simeone *et al.* 1990; Wang *et al.* 1990; Nicholson *et al.* 1990;
Dinsmore and Solomon, 1991; and references therein). These observations
suggest that RA may directly modulate gene expression through an RA-
responsive machinery, which would ultimately interpret the putative morphogenetic signal provided by RA. The fine dissection of this 'molecular machinery'
is a prerequisite to a better understanding of RA-triggered positional signalling
(see Izpisúa-Belmonte *et al.* 1991, for a discussion of this point).

Two classes of proteins that bind RA with high affinity have been identified:
the cellular RA binding proteins (CRABPs) and the nuclear RA receptors
(RARs). The CRABPs are encoded by at least two related genes (Giguère *et al.*
1990*a*; Blomhoff *et al.* 1990), and their precise function is at present poorly
understood. The RARs belong to the nuclear receptor multigene family, which
includes receptors for steroid hormones, thyroid hormones, vitamin $D_3$, and

some 'orphan' receptors (Moore 1990; O'Malley 1990). These receptors are ligand-inducible transcriptional transmodulators, which act by binding specifically to *cis*-acting elements of target genes (Green and Chambon 1988; Evans 1988; Beato 1989; Ham and Parker 1989; Diamond *et al.*1990). The identification of the RARs was a key step towards the unravelling of the molecular mechanisms involved in the RA-responsive machinery and their linkage to the control of gene expression.

The aim of this review is to summarize our current knowledge of the structure of the RARs, of their expression during embryogenesis and in the adult, and of their possible specific functions (see also Ruberte *et al.* Chapter 8).


## THREE GENES ENCODE THREE RETINOIC ACID RECEPTOR SUBTYPES

Recent work by us and others has identified human and murine cDNA clones encoding three homologous RA-binding proteins: RAR-α (Petkovich *et al.* 1987; Giguère *et al.* 1987), RAR-β (de Thé *et al.* 1987; Brand *et al.* 1988; Benbrook *et al.* 1988), and RAR-γ (Zelent *et al.* 1989; Krust *et al.* 1989; Ishikawa *et al.* 1990). The three RAR-encoding genes (α, β, γ) have been assigned to three different chromosomes in mouse (chromosomes 11, 14, 15, respectively) and in human (chromosomes 17, 3, 12, respectively) (Mattei *et al.* 1991), and no additional RAR-gene could be detected by low stringency Southern blotting of mouse and human genomic DNA with probes corresponding to a conserved part of the RA-binding domain (region E, see below) (our unpublished data). Couterparts for all three receptors have been found in amphibians (Ragsdale *et al.* 1989; Giguère *et al.* 1989; Ragsdale and Brockes 1991) and in chicken (Smith and Eichele 1991; our unpublished data). Taken together, these data indicate that the three genes encoding the RAR subfamily of nuclear receptors have been conserved throughout the evolution of higher vertebrates.

Recently, a novel subfamily of nuclear receptors has been identified, whose members (the RXRs) are RA-activated transcription enhancer factors (Mangelsdorf *et al.* 1990 and Chapter 3). However, the natural ligand for the RXRs may be a molecule related to RA, but not RA itself, as their affinity for RA seems much weaker than that of the RARs, and the synthetic retinoid TTNPB (Sporn and Roberts 1985) cannot activate RXR-α in tranfection experiments (Mangelsdorf *et al.* 1990).


## THE RETINOIC ACID RECEPTORS EXHIBIT A MODULAR STRUCTURE

Like other members of the nuclear receptor superfamily (Green and Chambon 1988), RARs have a modular structure composed of six regions designated A–F (Fig. 2.1). In the case of steroid receptors, it was shown that region C (the most highly conserved region throughout the nuclear receptor superfamily) folds into

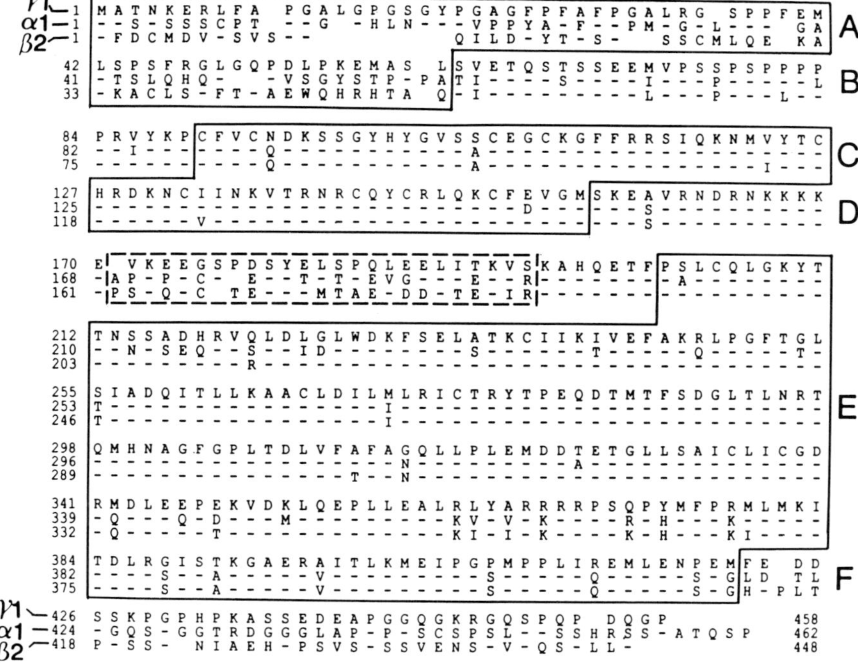

**Fig. 2.1** Alignment of murine RAR-γ1, α1 and β2 amino acid sequences. Regions A, B, C, D, E, and F are indicated, as well as the amino acid positions. Identical amino acid residues are represented by dashes. Region D2 is encased in dashed line.

a zinc-stabilized structural domain responsible for DNA binding (Evans 1988; Green and Chambon 1988; Schwabe *et al.* 1990; Härd *et al* 1990). It was also shown that the ligand-binding domain corresponds to region E, and that the A/B and E regions contain distinct trans-activation domains which are both cell type- and promoter-specific (Tora *et al.* 1988*a,b* 1989; Webster *et al.* 1988; Bocquel *et al.* 1989; Tasset *et al.* 1990; O'Donnell and Kœnig 1990). The functions of regions D and F are currently unknown. For a given RAR subtype (either α, β, or γ), the amino acid sequences of regions A–F are highly conserved from mouse to human (Krust *et al.* 1989; Zelent *et al.* 1989; and Fig. 2.2). In contrast, in a given species only the sequences of regions B, C, and E are well conserved between the three RARs, whereas those of regions A and F are not conserved (Fig. 2.3). Region D, which separates the regions C and E in the nuclear receptor superfamily and is not evolutionarily conserved in the case of steroid hormone receptors, was further subdivided into three subdomains called D1, D2, and D3 in the case of RARs (Fig. 2.1 and 2.3). Subdomains D1 and D3 appear highly conserved among the three RAR subtypes; D2 is not.

In other words, a given RAR subtype is more similar to its counterpart in

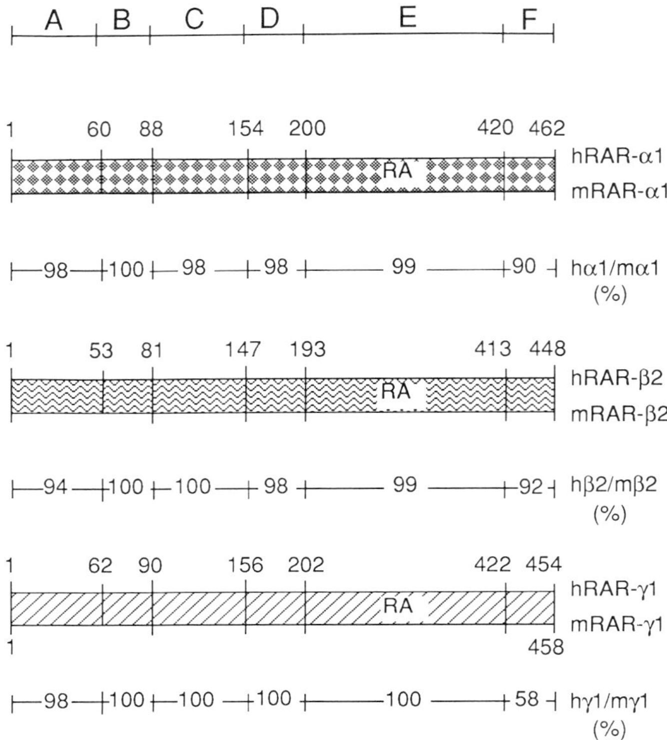



**Fig. 2.2** Homology comparison of amino acid sequences of each RAR subtype between mouse and man. Amino acid residues positions (top) corresponding to the boundaries between the different regions A to F, and percentage homology (bottom) are given for each comparison.

another species than to the two other RAR subtypes within the same species. Taken together, these observations suggest that the three RARs may be functionally distinct, and thus may regulate the expression of different sets of RA-responsive genes. This view was supported by analyses of the distribution of RAR transcripts using Northern blotting and *in situ* hybridization (see below), which showed that each mouse RAR subtype exhibits a specific pattern of expression either in adult or embryonic tissues.

THE THREE RETINOIC ACID RECEPTOR GENES GENERATE MULTIPLE ISOFORMS

An additional level in the complexity of RAR gene expression was recently revealed by the discovery of cDNA isoforms for RAR-α (seven isoforms, Leroy

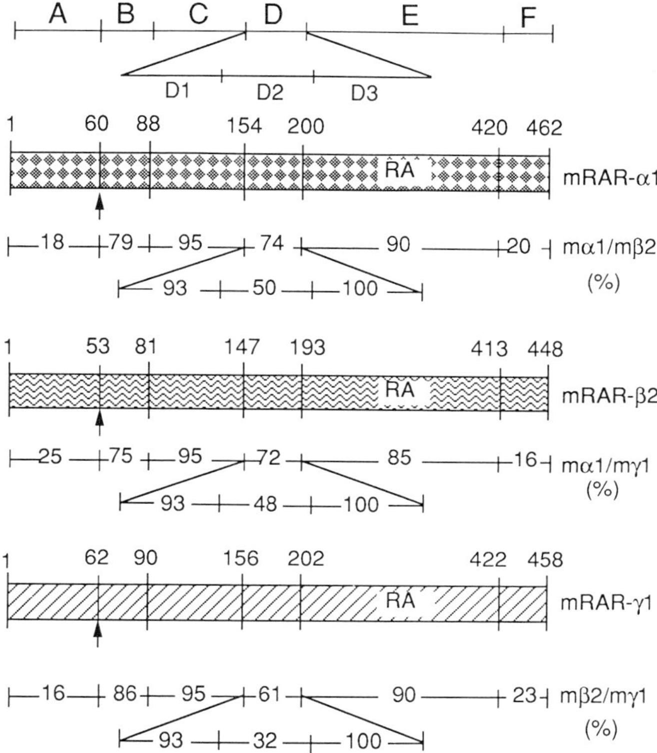



**Fig. 2.3** Homology comparison of the murine RAR subtypes with each other. The arrow corresponds to the splicing junction between the A and B regions. See also legend to Fig. 2.2.

*et al.* 1991*a*), RAR-$\beta$ (three isoforms, Zelent *et al.* 1991) and RAR-$\gamma$ (six isoforms, Kastner *et al.* 1990). For each RAR subtype (either $\alpha$, $\beta$, or $\gamma$), these isoforms share a common B–F region, but their sequences diverge N-terminal to the A/B region junction, i.e. at the exact nucleotide corresponding to the 5′ end of the B region-encoding exon (Figs 2.4–2.6). The major isoforms mRAR-$\alpha$1 and $\alpha$2, mRAR-$\beta$1/$\beta$3 and $\beta$2, and mRAR-$\gamma$1 and $\gamma$2, which account for the most frequently isolated cDNAs, encode protein isoforms with different N-terminal A regions. The significance of the less frequent isoforms, mRAR-$\alpha$3 to $\alpha$7 and mRAR-$\gamma$4 to $\gamma$7, is currently unclear (Kastner *et al.* 1990; Leroy *et al.* 1991*a*).

The expression of all three mouse RAR genes is achieved by the use of at least two promotors (Fig. 2.7) and by the processing of alternative exons (Kastner *et al.* 1990; Brand *et al.* 1990; Leroy *et al.* 1991*a*; Zelent *et al.* 1991; our unpublished results). Interestingly the mRAR-$\alpha$2 proximal promoter sequences contain an RA-response element (Leroy *et al.* 1991*b*), which closely resembles

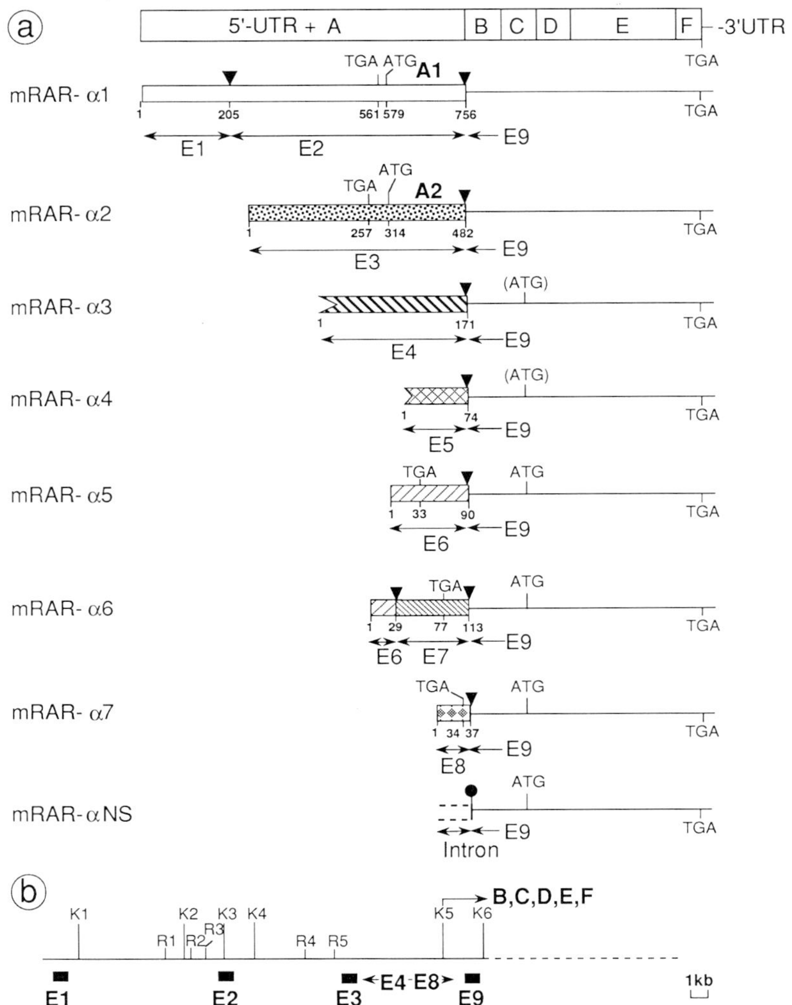

Fig. 2.4 Murine RAR-α isoforms. (*a*) Schematic representation of mRAR-α cDNA isoforms; (*b*) exonic organization in the 5′ portion of the mouse RAR-α gene (see Leroy *et al.*, 1991*a* for further details).

an RARE described by others in the human and mouse RAR-β2 promoter (de Thé *et al.* 1990; Sucov *et al.* 1990).

The DNA sequences of the major RAR isoforms are well conserved from mouse to human (Kastner *et al.* 1990; Leroy *et al.* 1991*a*; Zelent *et al.* 1991), which supports the idea that each isoform performs a specific function in mediating the RA signal. Moreover, this evolutionary conservation does not only concern region A of a given isoform, but also its 5′ untranslated region (5′-

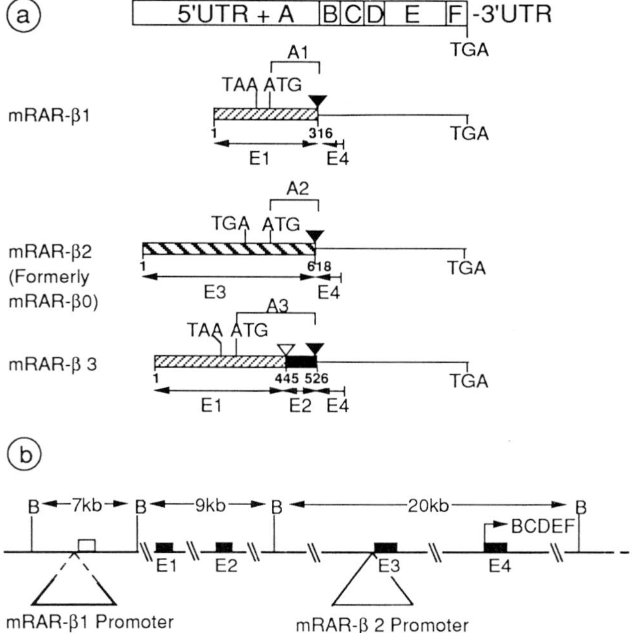



**Fig. 2.5** Murine RAR-$\beta$ isoforms (*a*) Schematic representation of the mRAR-$\beta$ cDNA isoforms; (*b*) structure of the mouse RAR-$\beta$ gene in its 5′ region (see Zelent *et al.* 1991 for further details).

UTR), thus suggesting that evolutionarily conserved mechanisms may affect the regulation of translation, and/or stabilization of RAR messenger RNAs.

THE THREE RETINOIC ACID RECEPTOR GENES SHARE A SIMILAR ORGANIZATION

For each RAR gene, the two main RAR isoforms (type 1 and type 2) are trascribed from two promoters (P1 and P2, see previous section and Fig. 2.7). In each of the three genes, the relative order and approximative location of the exons that encode the 5′-UTR and A regions of isoforms type 1 and type 2 are similar (see Fig. 2.7). Interestingly, when compared with each other, the nucleic acid sequences encoding the A regions of RAR isoforms of the same type, i.e. transcribed from either the P1 ($\alpha$1, $\beta$1/$\beta$3, $\gamma$1) or P2 ($\alpha$2, $\beta$2, $\gamma$2) promoters, are often more similar to each other than to that of RAR isoforms of the other type (Table 2.1; Kastner *et al.* 1991). Taken together these observations suggest that

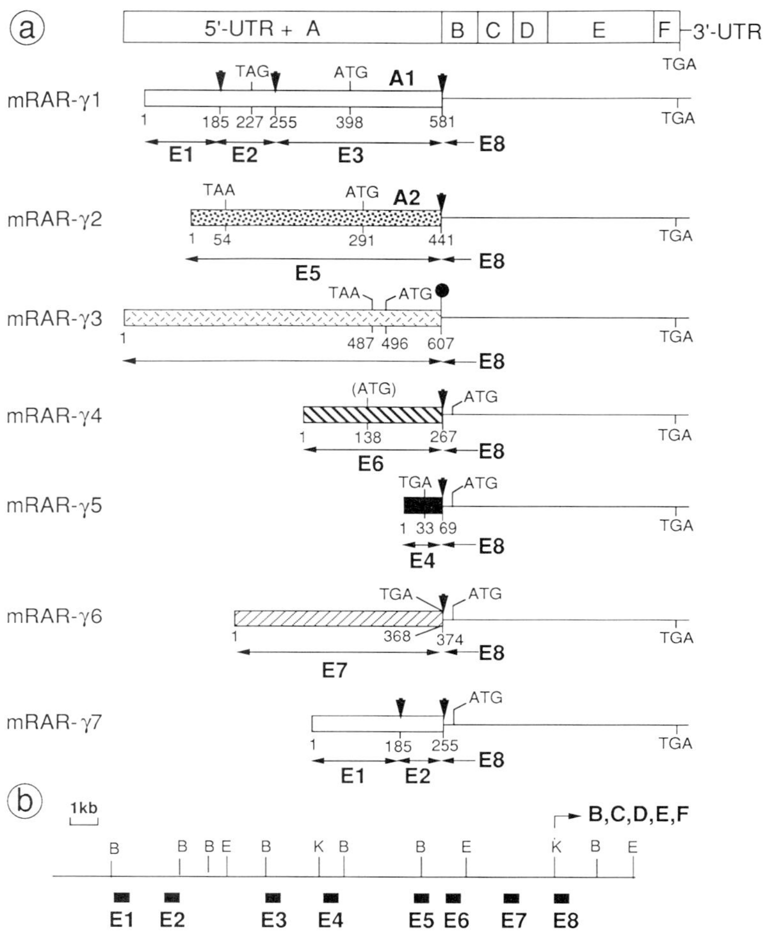



**Fig. 2.6** Murine RAR-γ isoforms. (*a*) Schematic representation of mRAR-γ cDNA isoforms; (*b*) exonic organization in the 5' portion of the mouse RAR-γ gene (see Kastner *et al.* 1990 for further details).

the three RAR genes have evolved by duplication of an ancestral gene possessing two promoters driving the transcription of the two major isoforms. However, two isoforms of different types share a degree of homology close to or higher than 40 per cent (Table 2.1). This conservation may correspond to the generation of an ancestral gene by duplication of the 5' region of an 'older' gene possessing only one promoter. The fact that three 'modern' RAR isoforms are RA-inducible (RAR-β1/β3, RAR-β2, RAR-α2, see below) suggests that the promoter sequences of this putative 'older' gene might have contained a RA-response element (RARE), which was subsequently lost in the promoters

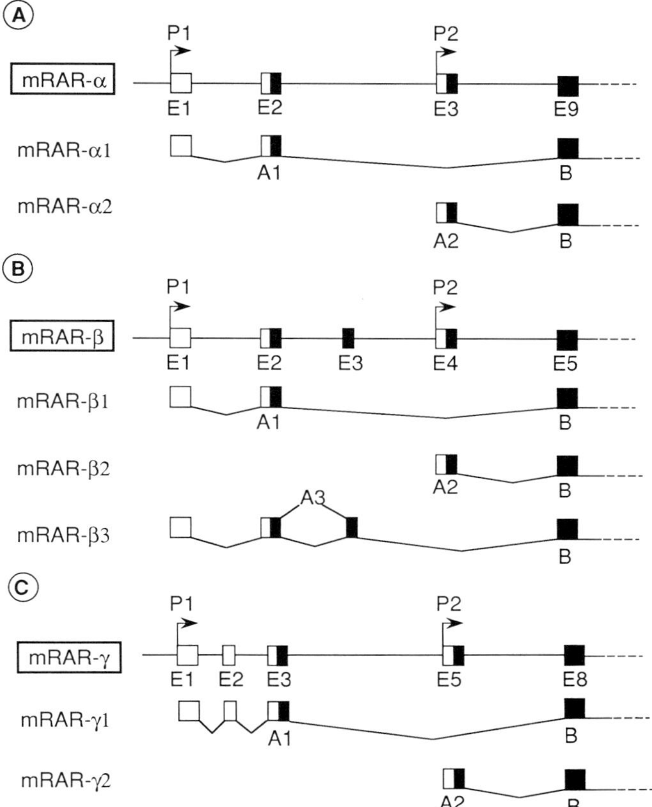



**Fig. 2.7** Schematic organization of the 5′ region of the three mouse RAR genes and of their major isoforms. (*a*), (*b*) and (*c*) correspond to RAR-α, β and γ genomic loci, respectively. For each RAR gene, the genomic organization of the 5′ region is displayed and the various exons (E) indicated by boxes. The numbering of the exons is as in Kastner *et al.* 1990; Leroy *et al.* 1991*a*; Zelent *et al.* 1991. Black boxes represent translated sequences and white boxes 5′-untranslated sequences. In each case, P1 and P2 correspond to the upstream and downstream promoters, respectively. For a given RAR subtype (α, β, or γ) A1, A2, A3, and B correspond to the A1, A2, A3, and B regions, respectively.

responsible for the transcription of non-RA-inducible RAR isoforms (RAR-α1, RAR-γ1 and γ2).

## THE MOUSE RETINOIC ACID RECEPTOR ISOFORMS ARE DIFFERENTIALLY EXPRESSED IN ADULT AND EMBRYONIC TISSUES

Northern blot analysis (Krust *et al.* 1989; Zelent *et al.* 1989) and *in situ* hybridization experiments (Dollé *et al.* 1989, 1990; Ruberte *et al.* 1990, 1991;

**Table 2.1** Nucleic acid sequence comparison of the A regions of type 1 and type 2 mouse RAR isoforms with each other. Greek letters $\alpha$, $\beta$, and $\gamma$ refer to RAR subtypes $\alpha$, $\beta$, and $\gamma$, respectively; indexes 1 and 2 refer to isoforms type 1 and type 2, respectively. Values are given as percentage homology and comparisons between two isoforms belonging to the same type are boxed

|            | $\alpha 1$ | $\alpha 2$ | $\beta 1$ | $\beta 2$ | $\gamma 1$ | $\gamma 2$ |
|------------|------|------|------|------|------|------|
| $\alpha 1$ | 100  | 43,8 | 63,0 | 40,4 | 50,0 | 39,4 |
| $\alpha 2$ | –    | 100  | 42,5 | 48,8 | 46,8 | 39,1 |
| $\beta 1$  | –    | –    | 100  | 47,0 | 43,8 | 39,8 |
| $\beta 2$  | –    | –    | –    | 100  | 39,2 | 51,6 |
| $\gamma 1$ | –    | –    | –    | –    | 100  | 42,6 |
| $\gamma 2$ | –    | –    | –    | –    | –    | 100  |

Ruberte *et al.* Chapter 8) using probes corresponding to the B–F region of each RAR subtype have revealed a wide and differential distribution of RAR transcripts both in the adult and at various stages of mouse embryogenesis. The probes used in the initial studies could not discriminate between the various RAR isoforms. To investigate the expression pattern of each RAR isoform, Northern blotting was performed by using isoform specific probes. So far, this procedure has allowed us to detect RNA transcripts corresponding to the major mouse isoforms mRAR-$\alpha$1 and $\alpha$2, mRAR-$\beta$1/$\beta$3 and $\beta$2, and mRAR-$\gamma$1 and $\gamma$2 (Figs 2.8–2.10).

Expression of mRAR-$\alpha$1 is fairly ubiquitous (Fig. 2.8) and is reminiscent of the expression of a housekeeping gene. This observation is in good agreement with the fact that the RAR-$\alpha$1 P1 promoter, which was characterized first for human RAR-$\alpha$1 and was shown to be conserved in mouse, exhibits characteristics of a housekeeping gene promoter (Brand *et al* 1990; Leroy *et al.*1991*a*). The ubiquitous presence of RAR-$\alpha$1 in a wide variety of tissues is likely to reflect the need for a RA-responsive machinery in each cell. Isoform mRAR-$\alpha$2 exhibits a more restricted pattern of expression. mRAR-$\alpha$2 transcripts are RA-inducible in embryonal carcinoma cell lines P19 and

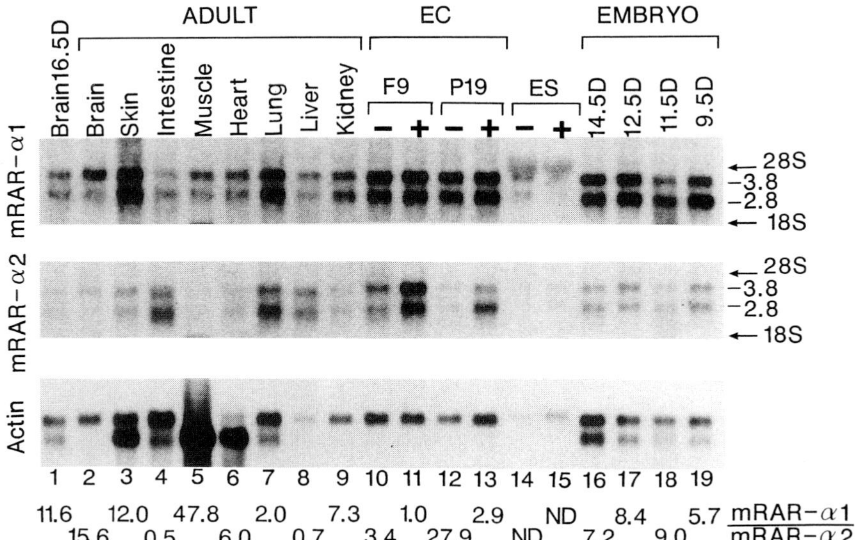

**Fig. 2.8** Northern blot analysis of mRAR-α1 and mRAR-α2 transcript distribution (see Leroy *et al.* 1991*a* for further details).

F9, which correlates with the presence of a RARE in the RAR-α2 P2 promoter (Leroy *et al.* 1991*b*).

The expression of RAR-$\beta$1 and RAR-$\beta$3 (which differs from RAR-$\beta$1 by the presence of an additional exon encoding 27 amino acids upstream from the B region, see Figs 2.5 and 2.7), has been investigated by Northern blotting as well as polymerase chain reaction (PCR)-assisted analysis (see Fig. 2.9 and Zelent *et al.* 1991). For example, isoforms mRAR-$\beta$1/$\beta$3 are abundantly expressed in adult and embryonic brain, and may therefore play some specific role in development of the central nervous system. All three mRAR-$\beta$ isoforms are RA-inducible in F9 and P19 EC cells; in the case of mRAR-$\beta$2 RA-induction this observation is consistent with the presence of a RARE in the RAR-$\beta$2 promoter (de Thé *et al.* 1990; Sucov *et al.* 1990). The presence of a RARE in RAR-$\beta$1/$\beta$3 promoter sequences remains to be investigated.

Finally the mRAR-$\gamma$1 isoform, which corresponds to the majority of mRAR-$\gamma$ transcripts in the adult, is predominantly expressed in skin, whereas mRAR-$\gamma$2 transcripts are found early in embryogenesis and in EC cells (Fig. 2.10).

Thus, it appears that all seven major RAR isoforms have a specific distribution pattern both during embryogenesis and in adult tissues, and therefore may mediate specifically some of the RA effects during development and in the adult.

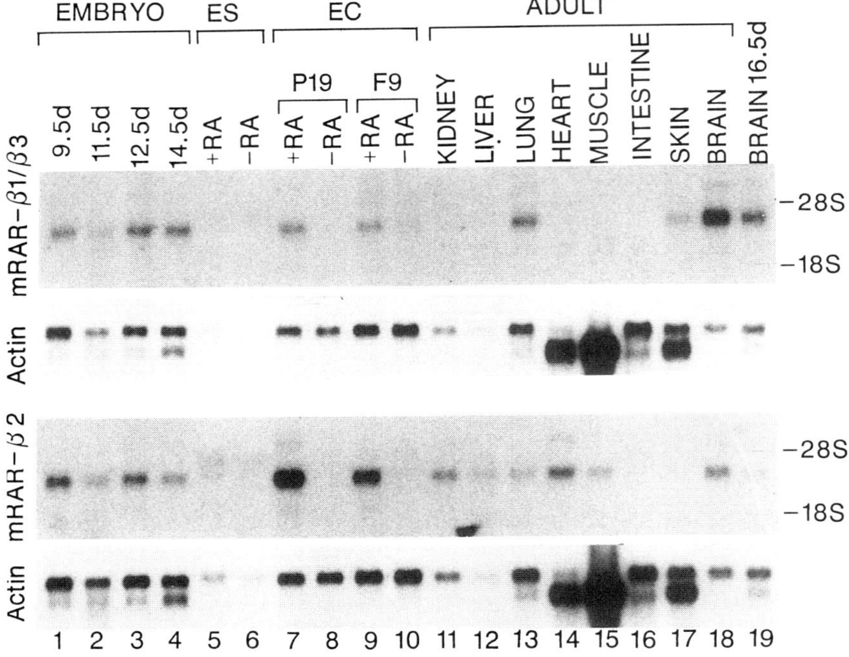

Fig. 2.9 Northern blot analysis of mRAR-$\beta1/\beta3$ and mRAR-$\beta2$ expression (see Zelent *et al.* 1991 for further details).

## THE DIVERSITY OF RETINOIC ACID RECEPTORS MAY ACCOUNT FOR THE MULTIPLE EFFECTS OF RETINOIC ACID

Their conservation through evolution and also their specific patterns of expression support the notion that the multiple RAR subtypes and their isoforms perform specific functions. In principle, these specific functions could be exerted at three levels: ligand binding, binding to RA-responsive elements of targets genes, and transcriptional activities.

The three RAR subtypes appear to be activated differentially by retinoic acid and/or synthetic retinoids (Brand *et al.* 1988; Giguère *et al.* 1990*b*; Ishikawa *et al.* 1990; Aström *et al.* 1990; Hashimoto *et al.* 1990). These observations are in keeping with the almost complete conservation (99–100 per cent) of the E region across species for a given mRAR subtype, whereas the E region of the three RARs within a given species is 85–90 per cent conserved. Thus, it is possible that the three RAR subtypes respond differentially to the presently identified natural retinoids (all-*trans*-RA, 13-*cis*-RA, 3,4,-dide-hydroretinoic acid (Thaller and Eichele, 1990) and to those that may not yet have been found. Such differential responses of the three RARs would

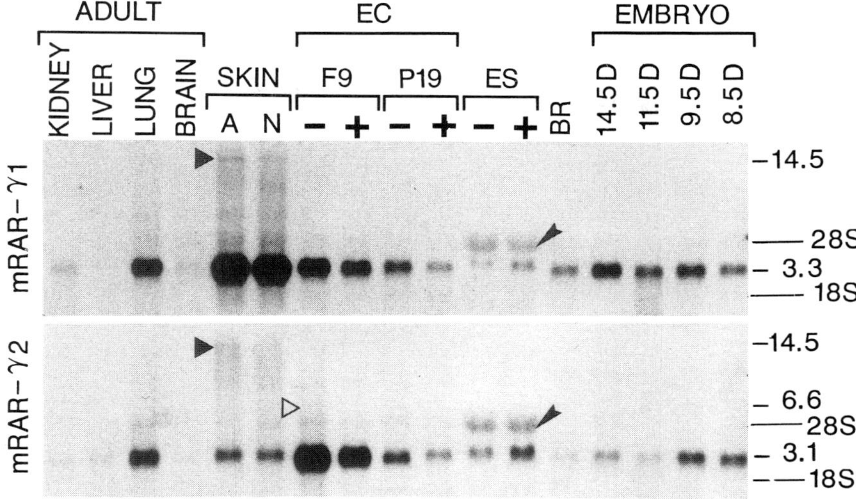


**Fig. 2.10** Northern blot analysis of mRAR-γ1 and mRAR-γ2 expression (see Kastner *et al.* 1990 for further details).

obviously increase the combinatorial possibilities for controlling transcriptional transactivation of retinoid responsive genes.

Although the C region that belongs to the DNA binding domain is highly conserved between the three RAR subtypes, each of them possesses some specific residues in this region (see Fig. 2.1), and some of these residues are conserved through evolution. For example, the serine residue 106 (mouse sequence as reference) is RAR-γ specific and conserved in mouse, man, newt (Ragsdale *et al.* 1989), and chicken (our unpublished results). These subtle conserved differences suggest that the response element recognition may differ between the three RAR subtypes. Natural RAREs have been characterized (de Thé *et al.* 1990; Sucov *et al.* 1990; Vasios *et al.* 1989, 1991; Lucas *et al.* 1991; Duester *et al.* 1991; Smith *et al.* 1991), but their possible RAR subtype-specific response remains to be established.

The amino acid sequence of the different receptors and of their isoforms diverge strikingly in their N-terminal A region. Given the evolutionary conservation of this region for a given RAR isoform, it is highly probable that this region performs an important and receptor isoform-specific function. Although no function has been found yet for these regions in the case of RARs, it has been shown that the N-terminal regions of the oestrogen, progesterone, and glucocorticoid receptors contain transcriptional activation functions (TAFs) that exhibit cell type- and promoter-specificity (see p. 9). Thus, each RAR A region may correspond to a cell type- and/or promoter-specific TAF. This possibility is particularly attractive as, together with the possible ligand-

and RA-response element specificities of the three RAR subtypes (see above), it would generate the functional diversity which is required to account at the molecular level for the pleiotropic effects of RA during development and in the adult.


ACKNOWLEDGEMENTS

We thank C. Werlé, B. Boulay, J. M. Lafontaine, and the secretarial staff for help in preparing the manuscript. We also thank C. Reibel and J. M. Garnier for technical assistance, A. Staub and F. Ruffenach for oligonucleotide synthesis.


REFERENCES

Aström, A., Pettersson, U., Krust, A., Chambon, P., and Voorhees, J. J. (1990). Retinoic acid and synthetic analogs differentially activate retinoic acid receptor dependent transcripition. *Biochemical and Biophysical Research Communications*, **173**, 339–45.

Beato, M. (1989). Gene regulation by steroid hormones, *Cell*, **56**, 335–44.

Benbrook, D., Lernhardt, E., and Pfahl, M. (1988). A new retinoic acid receptor identified from a hepatocellular carcinoma. *Nature*, **333**, 669–72.

Blomhoff, R., Green, M. H., Berg, T., and Norum, K. R. (1990). Transport and storage of vitamin A. *Science*, **250**, 399–404.

Bocquel, M. T., Kumar, V., Stricker, C., Chambon, P., and Gronemeyer, H. (1989). The contribution of the N- and C-terminal regions of steroid receptors to activation of transcription is both receptor and cell-specific. *Nucleic Acids Research*, **17**, 2581–95.

Brand, N., Petkovich, M., Krust, A., Chambon, P., de Thé, H., Marchio, A., Tiollais, P., and Dejean, A. (1988). Identification of a second human retinoic acid receptor. *Nature*, **332**, 850–3.

Brand, N., Petkovich, M., and Chambon, P. (1990). Characterization of a functional promoter for the human retinoic acid receptor-alpha (hRAR-α). *Nucleic Acids Research*, **18**, 6799–806.

Brockes, J. (1989). Retinoids, homeobox genes and limb morphogenesis. *Neuron*, **2**, 1285–94.

Brockes, J. (1990). Reading the retinoid signals. *Nature*, **345**, 766–8.

Brockes, J. (1991). We may not have a morphogen. *Nature*, **350**, 15.

Chytil, F. (1984). Retinoic acid: biochemistry, pharmacology, toxicology, and therapeutic use. *Pharmacological Review*, **36**, 93S–100S.

Chytil, F. and Haq, R. (1990). Vitamin A mediated gene expression. *Critical Reviews in Eukaryotic Gene Expression*, **1**, 61–73.

de Thé, H., Marchio, A., Tiollais, P., and Dejean, A. (1987). A novel steroid thyroid hormone receptor related gene inappropriately expressed in human hepatocellular carcinoma. *Nature*, **330**, 667–70.

de Thé, H., del Mar Vivanco-Ruiz, M., Tiollais, P., Stunnenberg, H., and Dejean, A.

(1990). Identification of a retinoic acid responsive element in the retinoic acid receptor $\beta$ gene. *Nature*, **343**, 177–80.

Diamond, M. I., Miner, J. N., Yoshinaga, S. K., and Yamamoto, K. R. (1990). Transcription factor interactions: selectors of positive or negative regulation from a single DNA element. *Science*, **249**, 1266–72.

Dinsmore, J. H. and Solomon, F. (1991). Inhibition of MAP2 expression affects both morphological and cell division phenotypes of neuronal differentiation. *Cell*, **64**, 817–26.

Dollé, P., Ruberte, E., Kastner, P., Petkovich, M., Stoner, C. M., Gudas, L. J. and Chambon, P. (1989). Differential expression of the genes encoding the retinoic acid receptors $\alpha$, $\beta$, $\gamma$ and CRABP in the developing limbs of the mouse. *Nature*, **342**, 702–5.

Dollé, P., Ruberte, E., Leroy, P., Morriss-Kay, G., and Chambon, P. (1990). Retinoic acid receptors and cellular retinoid binding proteins. I. A systematic study of their differential pattern of transcription during mouse organogenesis. *Development*, **110**, 1133–51.

Duester, G., Shean, M. L., McBride, M. S., and Stewart, M. J. (1991). Retinoic acid response element in the human alcohol dehydrogenase gene ADH3: implications for regulation of retinoic acid synthesis. *Molecular and Cellular Biology*, **11**, 1638–46.

Durston, A. J., Timmermans, J. P. M., Hage, W. J., Hendriks, H. F. J., de Vries, N. J., de Vries, Heideveld, M., and Nieuwkoop, P. D. (1989). Retinoic acid causes an anteroposterior transformation in the developing central nervous system. *Nature*, **340**, 140–4.

Eichele, G. (1989a). Retinoids and vertebrate limb pattern formation. *Trends in Genetics*, **5**, 246–51.

Eichele, G. (1989b). Retinoic acid induces a pattern of digits in anterior half wing buds that lack the zone of polarizing activity. *Development*, **107**, 863–7.

Eichele, G. (1990). Pattern formation in vertebrate limbs. *Current Opinions in Cell Biology*, **2**, 975–80.

Evans, R. (1988). The steroid and thyroid hormone receptor family. *Science*, **240**, 889–95.

Giguère, V., Ong, E. S., Segui, P., and Evans, R. M. (1987). Identification of a receptor for the morphogen retinoic acid. *Nature*, **330**, 624–9.

Giguère, V., Ong, E. S., Evans, R. M. and Tabin, C. J. (1989). Spatial and temporal expression of the retinoic acid receptor in the regenerating amphibian limb. *Nature*, **337**, 566–9.

Giguère, V., Lyn, S., Yip, P., Siu, C.-H., and Amin, S. (1990a). Molecular cloning of cDNA encoding a second cellular retinoic acid-binding protein. *Proceedings of the National Academy of Sciences of the USA*, **87**, 6233–7.

Giguère, V., Shago, M., Zirngibl, R., Tate, P., Rossant, J., and Varmuza, S. (1990b). Identification of a novel isoform of the retinoic acid receptor $\gamma$ expressed in the mouse embryo. *Molecular and Cellular Biology*, **10**, 2335–40.

Green, S., and Chambon, P. (1988). Nuclear receptors enhance our understanding of transcription regulation. *Trends in Genetics*, **4**, 309–14.

Ham, J. and Parker, M. G. (1989). Regulation of gene expression by nuclear hormone receptor. *Current Opinions in Cell Biology*, **1**, 503–11.

Härd, T., Kellenbach, E., Boelens, R., Maler, B. A., Dahlman, K., Freedman, L. P., Carlstedt-Duke, J., Yamamoto, K. R., Gustafsson, J. A. and Kaptein, R. (1990).

Solution structure of the glucocorticoid receptor DNA-binding domain. *Science*, **249**, 157–60.

Hashimoto, Y., Kagechika, H., and Shudo, K. (1990). Expression of retinoic acid receptor genes and the ligand-binding selectivity of retinoic acid receptors (RARs). *Biochemical and Biophysical Research Communications*, **166**, 1300–7.

Ishikawa, T., Umesono, K., Mangelsdorf, D. J., Aburatani, H., Stanger, B. Z., Shibasaki, Y., Imawari, M., Evans, R. M., and Takaku, F. (1990). A functional retinoic acid receptor encoded by the gene on the human chromosome 12. *Molecular Endocrinology*, **4**, 837–44.

Izpisúa-Belmonte, J.-C., Tickle, C., Dollé, P., Wolpert, L., and Duboule, D. (1991). Expression of the homeobox HOX-4 genes and the specification of position in chick wing development. *Nature*, **350**, 585–9.

Kastner, P., Krust, A., Mendelsohn, C., Garnier, J. M., Zelent, A., Leroy, P., Staub, A., and Chambon, P., (1990). Murine isoforms of retinoic acid receptor gamma with specific patterns of expression. *Proceedings of the National Academy of Science of the USA*, **87**, 2700–4.

Kastner, P., Brand, N., Krust, A., Leroy, P., Mendelsohn, C., Petkovich, M., Zelent, A., and Chambon, P. (1991). Retinoic acid nuclear receptors. In *Developmental Patterning of the Vertebrate Limb* NATO ASI series, (eds J. R. Hindchliffe, J. Hurle, and D. Summerbell), pp. 75–88. Plenum Press, London.

Krust, A., Kastner, P., Petkovich, M., Zelent, A., and Chambon, P. (1989). A third human retinoic acid receptor, hRAR-$\gamma$. *Proceedings of the National Academy of Sciences of the USA*, **86**, 5310–14.

LaRosa, G. J., and Gudas, L. J. (1988). An early effect of retinoic acid: cloning of an mRNA (Era-1) exhibiting rapid and protein synthesis-independent induction during teratocarcinoma stem cell differentiation. *Proceedings of the National Academy of Sciences of the USA*, **85**, 329–33.

Leroy, P., Krust, A., Zelent, A., Mendelsohn, C., Garnier, J.-M., Kastner, P., Dierich, A., and Chambon, P. (1991*a*). Multiple isoforms of the mouse retinoic acid receptor alpha are generated by alternative splicing and differential induction by retinoic acid. *EMBO Journal*, **10**, 59–69.

Leroy, P., Nakshatri, H., and Chambon, P. (1991*b*). The mouse retinoic acid receptor alpha-2 (mRAR-$\alpha$2) isoform is transcribed from a promoter that contains a retinoic acid response element. *Proceedings of the National Academy of Sciences USA*, **88**, 10138–42.

Lucas, P. C., O'Brien, R. M., Mitchell, J. A., Davis, C. M., Imai, E., Forman, B. M., Samuels, H. H. and Granner, D. K. (1991). A retinoic acid response element is part of a pleiotropic domain in the phosphoenolpyruvate carboxykinase gene. *Proceedings of the National Academy of Science of the USA*, **88**, 2184–8.

Lüscher, B., Mitchell, P. J., Williams, T., and Tjian, R. (1989). Regulation of transcription factor AP-2 by the morphogen retinoic acid and by second messengers. *Genes and Development*, **3**, 1507–17.

Maden, M. (1982). Vitamin A and pattern formation in the regenerating limb. *Nature*, **295**, 672–5.

Maden, M. (1985). Retinoids and the control of pattern in limb development and regeneration. *Trends in Genetics*, **1**, 103–7.

Mangelsdorf, D. J., Ong, E. S., Dyck, J. A. and Evans, R. M. (1990). Nuclear receptor that identifies a novel retinoic acid response pathway. *Nature*, **345**, 224–9.

Mattei, M.-G., Rivière, M., Krust, A., Ingvarsson, S., Vennström, B., Islam, M. Q., Levan, G., Kastner, P., Zelent, A., Chambon, P., Szpierer, J., and Szpierer, C. (1991). Chromosomal assignment of retinoic acid receptor (RAR) genes in the human, mouse and rat genomes. *Genomics*, **10**, 1061–9.

Moore, D. D. (1990). Diversity and unity in the nuclear hormone receptors: a terpenoid receptor superfamily. *The New Biologist*, **2**, 100–5.

Murphy, S. P., Garbern, J., Odenwald, W. F., Lazzarini, R. A., and Linney, E. (1988). Differential expression of the homeobox gene HOX-1.3 in F9 embryonal carcinoma cells. *Proceedings of the National Academy of Sciences of the USA*, **85**, 5587–91.

Nicholson, R. C., Mader, S., Nagpal, S., Leid, M., Rochette-Egly, C., and Chambon, P. (1990). Negative regulation of the rat stromelysin gene promoter by retinoic acid is mediated by an AP1 binding site. *EMBO Journal*, **9**, 4443–54.

Noji, S., Nohno, T., Koyama, E., Muto, K., Ohyama, K., Aoki, Y., Tamura, K., Ohsugi, K., Ide, H., Taniguchi, S., and Saito, T. (1991). Retinoic acid induces polarizing activity but is unlikely to be a morphogen in the chick limb bud. *Nature*, **350**, 83–6.

O'Donnell, A. L., Kœnig, R. J. (1990). Mutational analysis identifies a new functional domain of the thyroid hormone receptor. *Molecular Endocrinology*, **4**, 715–20.

Okamoto, K., Okazawa, H., Okuda, A., Sakai, M., Muramatsu, M., and Hamada, H. (1990). A novel octamer binding transcription factor is differentially expressed in mouse embryonic cells. *Cell*, **60**, 461–72.

O'Malley, B. (1990). The steroid receptor superfamily: more excitement predicted for the future. *Molecular Endocrinology*, **4**, 363–9.

Petkovich, M., Brand, N. J., Krust, A., and Chambon, P. (1987). A human retinoic acid receptor which belongs to the family of nuclear receptors. *Nature*, **330**, 444–50.

Ragsdale, C. W., and Brockes, J. P. (1991). Retinoic acid receptors and vertebrate limb morphogensis. In *Structure and function of hormone nuclear receptors*, (ed. M. G. Parker), pp. 269–95. Academic Press, London.

Ragsdale, C. W., Petkovich, M., Gates, P. B., Chambon, P., and Brockes, J. P. (1989). Identification of a novel retinoic acid receptor in regenerative tissues of the newt. *Nature*, **341**, 654–7.

Ruberte, E., Dollé, P., Krust, A., Zelent, A., Morriss-Kay, G., and Chambon, P. (1990). Specific spatial and temporal distribution of retinoic acid receptor gamma transcripts during mouse embryogenesis. *Development*, **108**, 213–22.

Ruberte, E., Dollé, P., Chambon, P., and Morriss-Kay, G. (1991). Retinoic acid receptors and cellular retinoid binding proteins: II. Their differential pattern of transcription during early morphogenesis in mouse embryos. *Development*, **111**, 45–60.

Ruiz i Altaba, A., and Jessel, T. (1991). Retinoic acid modifies mesodermal patterning in early Xenopus embryo. *Genes and Development*, **5**, 175–87.

Schwabe, J. W. R., Neuhaus, D., and Rhodes, D. (1990). Solution structure of the DNA-binding domain of the oestrogen receptor. *Nature*, **348**, 458–61.

Simeone, A., Acampora, D., Arcioni, L., Andrews, P. W., Boncinelli, E., and Mavilio, F. (1990). Sequential activation of HOX2 homeobox genes by retinoic acid in human embryonal carcinoma cells. *Nature*, **346**, 763–6.

Sive, H. L., Draper, B. W., Harland, R. M., and Weintraub, H. (1990). Identification of a retinoic acid-sensitive period during primary axis formation in *Xenopus laevis*. *Genes and Development*, **4**, 932–42.

Slack, J. M. W. (1987*a*). Morphogenetic gradients—past and present. *Trends in Biochemical Science*, **12**, 200–4.

Slack, J. M. W. (1987*b*). We have a morphogen! *Nature*, **327**, 553–4.

Smith, S. M., and Eichele, G. (1991). Temporal and regional differences in the expression pattern of distinct retinoic acid receptor-$\beta$ transcripts in the chick embryo. *Development*, **111**, 245–52.

Smith, S. M., Pang, K., Sundin, O., Wedden, S. E., Thaller, C., and Eichele, G. (1989). Molecular approaches to vertebrate limb morphogenesis. *Development*, (Suppl.), 121–31.

Smith, W. C., Nakshatri, H., Leroy, P., Rees, J., and Chambon, P. (1991). A retinoic acid response element is present in the mouse cellular retinol binding protein I (mCRBPI) promoter. *EMBO Journal.*, **10**, 2223–30.

Sporn, M. B., and Roberts, A. B., (1985) What is a retinoid? In *Retinoids, differentiation and disease*, Ciba Foundation Symposium 113, pp. 1–5. Pitman, London.

Sucov, H. M., Murakami, K. K., and Evans, R. M. (1990). Characterization of an autoregulated response element in the mouse retinoic acid receptor type $\beta$ gene. *Proceedings of the National Academy of Sciences of the USA*, **87**, 5392–6.

Summerbell, D. and Maden, M. (1990). Retinoic acid, a developmental signalling molecule. *Trends in Neuroscience*, **13**, 142–7.

Tasset, D., Tora, L., Fromental, C., Scheer, E., and Chambon, P. (1990). Distinct classes of transcriptional activating domains function by different mechanisms. *Cell*, **62**, 1177–87.

Thaller, C., and Eichele, G. (1987). Identification and spatial distribution of retinoids in the developing chick limb bud. *Nature*, **327**, 625–8.

Thaller, C., and Eichele, G. (1990). Isolation of 3,4-didehydroretinoic acid, a novel morphogenetic signal in the chick wing bud. *Nature*, **345**, 815–19.

Tickle, C., Alberts, B., Wolpert, L., and Lee, J. (1982). Local application of retinoic acid to the limb bud mimics the action of the polarizing region. *Nature*, **296**, 564–6.

Tora, L., Gronemeyer, H., Turcotte, B., Gaub, M.-P., and Chambon, P. (1988*a*). The N-terminal region of the chicken progesterone receptor specifies target gene activation. *Nature*, **333**, 185–8.

Tora, L., Gaub, M.-P., Mader, S., Dierich, A., Bellard, M., and Chambon, P. (1988*b*). Cell-specific activity of a GGTCA half-palindromic oestrogen-responsive element in the chicken ovalbumin gene promoter. *EMBO Journal*, **7**, 3771–8.

Tora, L., White, J., Brou, C., Tasset, D., Webster, N., Scheer, E., and Chambon, P. (1989). The human estrogen receptor has two independent nonacidic transcriptional activation functions. *Cell*, **59**, 477–87.

Vasios, G. W., Gold, J. D., Petkovich, M., Chambon, P., and Gudas, L. (1989). A retinoic acid-responsive element is present in the 5′ flanking region of the laminin B1 gene. *Proceedings of the National Academy of Sciences of the USA*, **86**, 9099–103.

Vasios, G. W., Mader, S., Gold, J. D., Leid, M., Lutz, Y., Gaub, M.-P. Chambon, P. and Gudas, L. (1991). The late retinoic acid induction of laminin B1 gene transcription involves RAR binding to the responsive element. *EMBO Journal*, **10**, 1149–58.

Wagner, M., Thaller, C., Jessel, T., and Eichele, G. (1990). Polarizing activity and retinoid synthesis in the floor plate of the neural tube. *Nature*, **345**, 819–22.

Wanek, N., Gardiner, D. M., Muneoka, K., and Bryant, S. V. (1991). Conversion by retinoic acid of anterior cells into ZPA cells in the chick wing bud. *Nature*, **350**, 81–3.

Wang, S.-Y., and Gudas, L. J. (1983). Isolation of cDNA clones specific for collagen IV and laminin from mouse teratocarcinoma cells. *Proceedings of the National Academy of Sciences of the USA*, **80**, 5880–4.

Wang, C., Kelly, J., Bowden-Pope, D. F. and Stiles, C. D. (1990). Retinoic acid promotes transcription of the PDGFα-receptor gene. *Molecular and Cellular Biology*, **10**, 6781–4.

Webster, N. J. G., Green, S., Jin, J. R. and Chambon, P. (1988). The hormone-binding domains of the estrogen and glucocorticoid receptors contain an inducible transcription activation function. *Cell*, **54**, 199–207.

Wolpert, L. (1989). Positional information revisited. *Development*, (Suppl.) 3–12.

Zelent, A., Krust, A., Petkovich, M., Kastner, P., and Chambon, P. (1989). Cloning of murine α and β retinoic acid receptors and a novel receptor γ predominantly expressed in skin. *Nature*, **339**, 714–17.

Zelent, A., Mendelsohn, C., Kastner, P., Garnier, J.-M., Ruffenach, F., Leroy, P., and Chambon, P. (1991). Differentially expressed isoforms of the mouse retinoic acid receptor beta are generated by usage of two promoters and alternative splicing. *EMBO Journal*, **10**, 71–81.

# 3

# Vitamin A receptors: new insights on retinoid control of transcription

David J. Mangelsdorf and Ronald M. Evans

## RETINOID RECEPTORS

One of the fundamental questions in retinoid biology—and in extracellular signalling processes in general—is how a group of closely related and simple compounds can mediate diverse and complex responses. Multicellular organisms have evolved two basic strategies for mediating the translation of an extracellular signal into a transcriptional response at the target cell (Fig. 3.1). In one mechanism the stimulus (e.g. a small peptide or growth factor) binds to a cell surface receptor. Transduction of this signal to the nucleus is then accomplished by one of a myriad of second messengers. The many examples of signals that operate via a second messenger are demonstrated by the high genetic and phenotypic variation of their receptors and pathways (Herschman 1989). In the second mechanism, cells process an extracellular stimulus through an intracellular receptor. In this pathway the blood-borne signals are small lipophilic molecules, such as steroids, which are either actively or passively transported through the cell membrane where they bind cytosolic or nuclear receptors. The liganded receptor complex is a functionally active transcription factor, which binds specifically to the regulatory region of target genes. In contrast to the cell surface receptors, the intracellular receptors belong to one highly conserved family of proteins (Evans 1988; Green and Chambon 1988). From the characterization of one of these receptor pathways a unified mechanism can be derived by which the whole class of intracellular receptors can be studied. To date, nuclear receptors have been discovered for steroids, thyroid hormone, and recently, the retinoids (Evans 1988; Green and Chambon 1988; Mangelsdorf *et al.* 1990).

## Retinoic acid receptors

The effects of vitamin A seen in differentiation, haematopoeisis, development, and pattern formation are due to the transcriptionally active metabolite, retinoic acid (RA) and/or closely related metabolites (Sporn *et al.* 1984). Because RA is a small lipophilic molecule that can modulate gene expression,

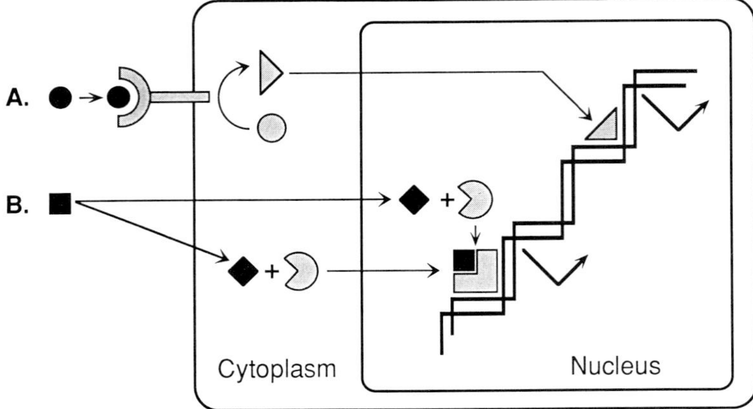

Fig. 3.1 Signal transduction in higher animals. (a) The signal, usually a soluble peptide hormone or growth factor, binds to its receptor on the target cell's surface. Ligand binding triggers a series of enzymatic events such as a change in intracellular calcium or the phosphorylation state of a protein that in turn may act as second messengers to the nucleus. (b) Lipid-soluble hormones such as steroids, vitamin D, thyroid hormone, and retinoic acid penetrate the cell membrane and bind to an intracellular receptor. In this pathway the liganded receptor itself becomes the signal to the nucleus where it directly interacts with the transcriptional machinery.

its mechanism of action has long been hypothesized to be analogous to that of steroid and thyroid hormones. This notion led to a biochemical 'hunt' for an RA receptor. In the late 1970s a cellular retinoic acid binding protein (CRABP) was discovered (Chytil and Ong 1978). The initial characterization and biochemical analysis of CRABP led many investigators to believe that these proteins were the cognate receptors for RA. However, there is no direct evidence that CRABP can regulate transcription or that it is critical for receptor function. The identification of a true RA receptor did not come about from a biochemical characterization of the protein, but rather through its evolutionary relatedness to the nuclear receptors. In 1987 Giguère et al. and Petkovitch et al. independently isolated a human cDNA encoding an RA receptor protein similar to the nuclear hormone receptors. The human RA receptor (hRAR-α) has a DNA-binding and ligand-binding domain that is structurally and functionally conserved with other members of the nuclear receptor superfamily and is able to activate transcription of target genes in an RA-dependent fashion (Evans 1988; see Fig. 3.3(a)). The discovery of this RAR was pivotal to the understanding of the mechanism of RA action because it demonstrated, for the first time, the existence of a retinoid-responsive transcription factor. Subsequently, additional RAR-related genes have been isolated and now at least three different RAR subtypes are known (see below). The ligand-binding domains of the RARs are highly conserved ( > 75 per cent amino acid identity, see below), suggesting that they all arose from a common ancestral RAR gene. These RAR isoforms are expressed in distinct patterns throughout development and in the mature organism (see below), indicating that they may mediate different functions. Thus, it appears that the diversity of effects by RA on cells can be explained in part by the diversity of RA receptors. With the discovery of multiple RAR genes one important question is whether all the actions of RA are mediated by these receptors or whether additional retinoid substrates and regulatory pathways exist.

## Retinoid X receptors

At the same time as the subfamily of RARs and their roles in retinoid action were first being investigated, several studies were being initiated to identify ligands for other receptor-like proteins. The fact that nuclear receptors share a common structure suggests the existence of a superfamily of genes whose products may be ligand-responsive transcription factors. Since 1987 at least 15 new gene products have been identified in both vertebrates and *Drosophila*, which are known members of this nuclear receptor superfamily (Milbrandt 1987; Miyajima *et al.* 1988, 1989; Giguère *et al.* 1988; Chang and Kokontis 1988; Hamada *et al.* 1989; Wang *et al.* 1989; Lazar *et al.* 1989; W. Segraves, 1991). Because initially the ligands for these receptor-like proteins are unknown they are referred to as 'orphan' receptors. The surprising discovery that there are many such orphan receptors has opened new directions in endocrine research and may lead to the identification of novel regulatory systems.

Figure 3.2 describes two methods that have been successfully employed to characterize the ligand and DNA binding specificities of orphan receptors (Giguère *et al.* 1987; Petkovitch *et al.* 1987; Mangelsdorf *et al.* 1990). Both methods are based on the observation that the molecular structure of nuclear receptors is conserved and that one receptor domain can be exchanged for another to create a functional hybrid (Green and Chambon 1987, 1988; Evans 1988; Umesono *et al.* 1988; Thompson and Evans 1989). In one strategy the DNA-binding domain of an orphan receptor is substituted with the corresponding region from the glucocorticoid receptor (Fig. 3.2(a)). The resulting chimera will now stimulate a glucocorticoid responsive promoter when exposed to the appropriate ligand. This hybrid receptor approach was used to functionally identify the hRAR-α (Giguère *et al.* 1987; Petkovitch *et al.* 1987). In the second strategy the glucocorticoid receptor DNA-binding domain is replaced with that of the orphan receptor to form a hybrid receptor responsive to the synthetic glucocorticoid, dexamethasone (Fig. 3.2(b)). This glucocorticoid-inducible chimeric receptor will now activate transcription of a reporter gene, which contains a *cis*-element responsive to the orphan receptor. Once the *cis*-responsive element for the orphan receptor is known, the need for the chimera is eliminated and the wild type receptor can be used in the ligand

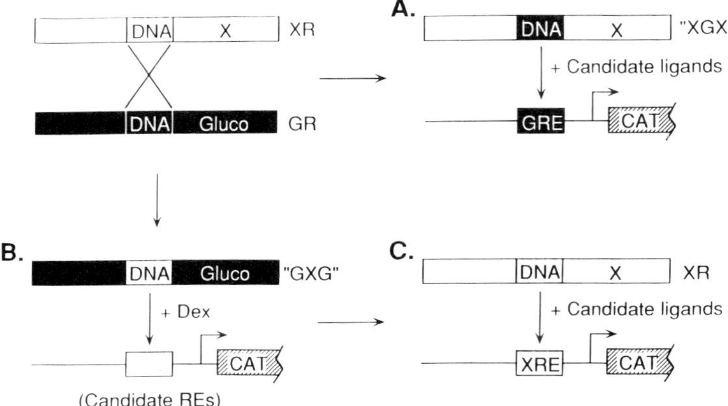


**Fig. 3.2** The *cis/trans* assay. Two strategies were developed to analyse orphan receptors. (*a*) In the first strategy the DNA-binding domain of the glucocorticoid receptor (GR) is substituted into the orphan receptor (XR) to make the chimera XGX. When coexpressed in cultured cells with a reporter plasmid containing a glucocorticoid response element (GRE), the reporter gene (e.g. CAT, chloramphenicol acetyl transferase) is expressed only in the presence of ligand X. (*b*) In the second strategy the chimera GXG is constructed and coexpressed with reporter genes containing candidate orphan receptor response elements (XREs). In the presence of the synthetic glucocorticoid, dexamethasone (Dex), the hybrid receptor GXG will only activate expression of the reporter gene when it contains the correct XR response element. (*c*) Once the XRE has been identified, the wild-type orphan receptor can be used (as in (*a*)) to screen for candidate ligands. These two assays allow the search for both the cognate ligand and response element of an orphan receptor.

screening assay (Fig. 3.2(c)). This approach allows characterization of both the DNA-binding and ligand-binding properties of an orphan receptor. Utilizing this strategy is was discovered that one of the human orphan receptors represented a second class of RA-responsive transcription factor, the retinoid X receptor (hRXR-α, Fig. 3.3(b)) (Mangelsdorf *et al.* 1990). A striking observation to come from the cloning of RXR is the apparent dissimilarity of its sequence to that of the RARs, considering they both respond to RA (Fig. 3.3). In fact, RAR is more similar to the thyroid hormone receptor than it is to RXR, suggesting that the response to RA by these two receptor systems has evolved through distinct pathways. This observation further underscores the importance and complexity of retinoid action in physiology and, coupled with the presence of a growing list of orphan receptors, leads to the speculation that other retinoid receptors may exist. Indeed, subsequent to the first reports of RAR and RXR at least three RAR subtypes (RAR-α, RAR-β, RAR-γ) and three RXR subtypes (RXR-α, RXR-β, RXR-γ) have been isolated, all of which are highly conserved among vertebrates (see Fig. 3.5).

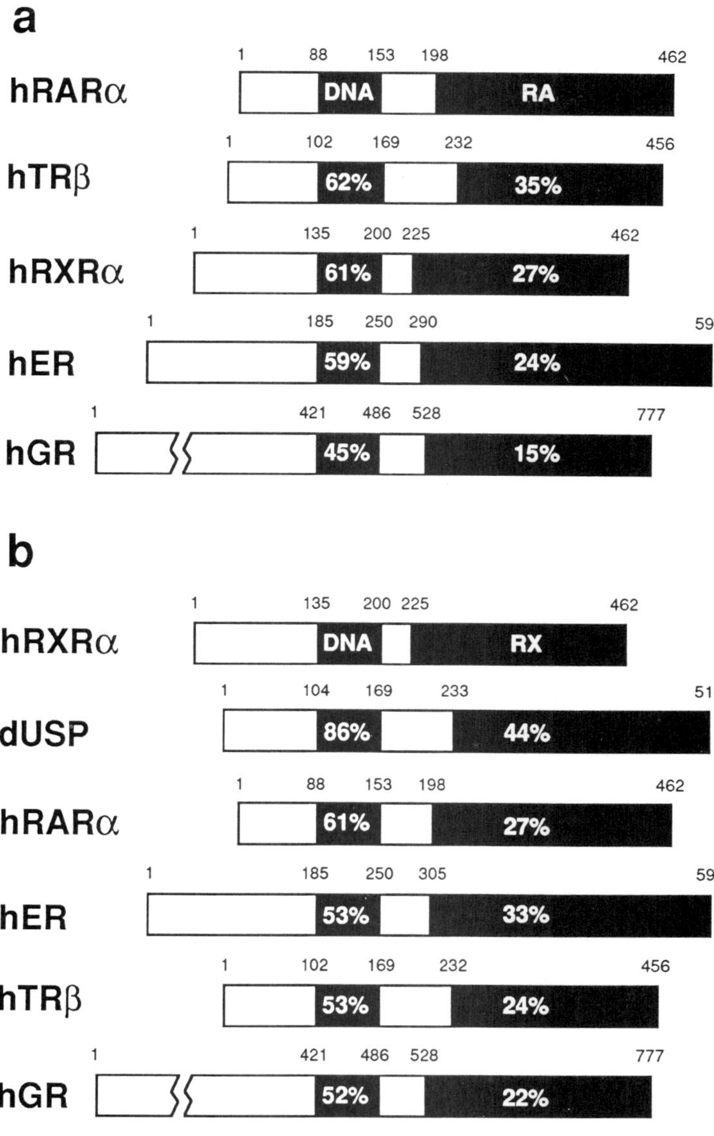


**Fig. 3.3** Schematic amino acid comparison of RAR and RXR with other members of the nuclear receptor superfamily. Primary amino acid sequences have been aligned on the basis of regions of maximum similarity, with the percentage identity indicated for the DNA-binding and ligand-binding domains in relation to (*a*) hRAR α (Petkovitch *et al.* 1987; Giguère *et al.* 1987) or (*b*) hRXR α (Mangelsdorf *et al.* 1990). hTR-*β*, human thyroid hormone receptor *β* (Weinberger *et al.* 1986); hER, human oestrogen receptor (Green *et al.* 1986); hGR, human glucocorticoid receptor (Hollenberg *et al.* 1985); dUSP, *Drosophila ultraspiracle* (Oro *et al.* 1990).

MULTIPLE RECEPTOR PATHWAYS

Because the retinoid receptors were first discovered from the molecular cloning of their cDNAs by low stringency hybridization with other known receptor sequences, very little is known about the biochemical and physiological function of these proteins. The existence of two distinct families and at least six different RA-responsive receptors further complicates the characterization process. An understanding of the differences and similarities of the RAR and RXR systems is just now becoming evident from comparative analyses based on receptor ligand specificity, target gene specificity, and the pattern of receptor expression.

## Ligand specificity

The classic model for nuclear hormone receptor action proposes that binding of the ligand to the receptor induces an allosteric change allowing the receptor–hormone complex to bind the target gene's promoter and either activate or repress transcription. As ligand activation is critical for receptor function, one potential way to dissect the roles of RAR and RXR would be to examine their ligand specificity. The initial observation that RA is the most potent inducer of both RAR and RXR suggested that it is also the natural ligand for both of these receptors. However, there is some evidence to indicate that more potent ligands may exist, and that these would discriminate between RAR and RXR. Cotransfection analyses have shown that although RAR and RXR respond with a similar rank order of potency to the naturally occurring metabolites of vitamin A (retinoic acid > retinal > retinyl acetate > retinol), there are distinct pharmacological differences with respect to the synthetic retinoids (Mangelsdorf *et al.* 1990). For example, the benzoic acid retinoid derivative, ethyl *P*-[(E)-2-(5,6,7,8-tetrahydro-5,5,8,8-tetramethyl-2-naph-thyl)-1-propenyl]-benzoic acid (TTNPB), is known to be a potent agonist of RA activity (Keeble and Maden 1989). In the cotransfection assay, however, TTNPB shows RA-like activity with hRAR-$\alpha$ but only 10 per cent of this response with hRXR-$\alpha$. Furthermore, in a comparison of dose responses, hRAR-$\alpha$ has at least a 10-fold higher sensitivity to RA than hRXR-$\alpha$, whose response has not yet saturated at levels as high as $10^{-5}$ M RA. The idea that RA may not be the highest affinity RXR ligand is further supported by binding analyses. To date we have been unable to demonstrate high-affinity binding of [$^3$H]RA to RXR extracted from cells transfected with RXR expression plasmid. In contrast, under similar conditions RA does bind with high affinity ($k_d = 4$ nM) and specificity to the RAR proteins (Ishikawa *et al.* 1990; Yang *et al.* 1991). In the case of RXR one explanation for the lack of high-affinity binding may be that RA is a lower-affinity precursor that is metabolized to a more active form. Such a metabolic pathway would be reminiscent of vitamin D action, in which it was originally postulated that 25-hydroxyvitamin $D_3$ is the active principle. It was not until the discovery of a more polar metabolite specifically bound to chromatin (Haussler *et al.* 1968) that it was realized 1, 25-dihydroxyvitamin $D_3$ was the true hormone. Biologically active RA metabolites are known to exist. Recently, Thaller and Eichele (1990) have demonstrated that the RA derivative 3,4-didehydroretinoic acid also has potent morphogenetic effects in the developing chick limb bud where it is present at a concentration six times that of RA. Moreover, didehydroretinoic acid can differentially activate RXRs and RARs in the *Xenopus* system (B. Blumberg, D. Mangelsdorf, R. Evans, and E. DeRobertis, unpublished observation). The observation that two distinct receptors respond with reciprocal specificity to similar ligands is not without precedent. An example is the oestrogen and androgen receptors. The female sex steroid, oestrogen, is produced via the aromatization of the male sex steroid, testosterone. Although these two hormones are closely related, the ligand-binding domains of their respective receptors share only marginal (25 per cent) identity (Lubahn *et al.* 1988). Based on this analogy, it is not unreasonable to presume that RARs and RXRs recognize chemically similar or identical molecules even though their ligand-binding domains are only distantly related.

Finally, it is possible that a non-retinoid ligand may also exist for RXR. In *Drosophila*, the orphan receptor *ultraspiracle* (*usp*) bears a striking resemblance to RXR in both its DNA- and ligand-binding domains (Oro *et al.* 1990; see Fig. 3.3(b)). However, *usp* does not respond to RA and, aside from effects on the visual system, *Drosophila* can be maintained for several generations on a vitamin A-deficient diet (Sang 1978). These data suggest the existence of a non-retinoid ligand that is capable of interacting with both RXR and *usp*. One of the future challenges will be to purify and characterize a natural high affinity RXR ligand. What is known is that both RAR and RXR respond to RA. An important unanswered question is whether there are other ligands (retinoid or otherwise) that distinguish the two receptor systems.


**Target gene specificity**

The primary determinant of target gene specificity and transactivation by retinoids appears to be encoded within the receptor-binding sequences of the target gene promoter. Although much of what we know of how retinoid receptors interact with DNA comes from studies of other receptor systems, there is now an intense effort to isolate and characterize the cognate *cis* elements in the regulatory region of RA-responsive genes. Several such RA response elements (RAREs) have been identified (Fig. 3.4). These RAREs are defined as the minimal DNA sequences required to bind receptor with high affinity and direct RA-specific transactivation or repression in both homologous and heterologous promoters. Given the high degree of conservation exhibited in the DNA binding domain of the nuclear receptors it is not

| RARE | Recognition sequence | Activity | | Binding | |
|---|---|---|---|---|---|
| | | RAR | RXR | RAR | RXR |
| **Common** | | | | | |
| TRE$_{pal}$ | TCA GGTCA TGACC TGA | +++ | +++ | +++ | +++ |
| TRE$_{GH}$ | GGTAA GATCA G GGACG TGACC GCA | + | + | + | N/A |
| ERE$_{vit}$ | TCA GGTCA CAG TGACC TGA | ++ | ++ | + | + |
| VDRE$_{oc}$(AP-1) | GGTGA C TCACC G GGTGA ACGGG | + | + | + | N/A |
| **Specific** | | | | | |
| βRARE | AAGG GTTCA CCGAAA GTTCA CTC | ++++ | ++ | ++++ | ++ |
| RARE$_{lam}$ | CCAGACAGGT TGACC CTTTTTCTAAGGGCT TAACC TAGC TCACC TC | + | N/A | + | N/A |

**Fig. 3.4** Retinoic acid response elements (RAREs) from RAR and RXR inducible promoters. RAREs are divided into two groups, those that share a 'common' responsiveness with other nuclear receptors and those that are 'specific' to just the retinoid receptors. The proposed half-site for receptor binding within each element that designates the GGTCA consensus is underscored with an arrow. The AP-1 site in the VDRE$_{oc}$ is underlined. The number of ( + ) signs refer to the relative trans-activity and binding affinity data for RAR and RXR and are taken from K. Umesono, D. Mangelsdorf, S. Kliewer, and R. Evans (unpublished results) and the following sources of each response element: TRE$_{pal}$, synthetic thyroid hormone response element palindrome (Glass *et al.* 1988; Umesono *et al.* 1988; Graupner *et al.* 1989); TRE$_{GH}$, rat growth hormone gene thyroid hormone response element (Glass *et al.* 1988; Umesono *et al.* 1988); ERE$_{vit}$, *Xenopus* vitellogenin A2 gene oestrogen response element (Klein-Hitpaß *et al.* 1986; de Verneuil and Metzger 1990); VDRE$_{oc}$ (AP-1), human osteocalcin gene vitamin D and AP-1 response element (Kerner *et al.* 1989; Schüle *et al.* 1990*a*); β RARE, mouse RAR β gene retinoic acid response element (de Thé *et al.* 1990*a*; Sucov *et al.* 1990); RARE$_{lam}$, mouse laminin B1 gene retinoic acid response element (Vasios *et al.* 1989; Glass *et al.* 1990).

surprising that many of these response elements are not only responsive to retinoid receptors but also to other related members of the nuclear receptor superfamily. In fact, several RAREs were first characterized based on their activation by other receptors (e.g. thyroid hormone, oestrogen, and vitamin D receptors; Fig. 3.4). One common feature present in all RAREs is the core sequence GGTCA, or a closely related derivative. It is believed that this core sequence represents a minimal half-site for DNA binding of one receptor molecule but that for efficient receptor binding and transactivation at least two such sequences should be present (Beato 1989). According to this model the receptor would bind to its full response element as a dimer (see below). This hypothesis predicts that all RAREs must contain at least two copies of the GGTCA motif and this is indeed the case (Fig. 3.4). The fact that several promoters contain more than two half-sites and these half-sites further increase the transactivation response suggests the possibility that some receptors may form more complex multimeric structures on DNA.

Thus far, most of the genes reported to be responsive to nuclear receptors are upregulated by their respective ligands. This is also true of the RA-responsive genes, although recently it has been shown that the EGF receptor gene is transcriptionally repressed by RAR (Hudson *et al.* 1990). Although the exact mechanism of receptor transactivation is not yet clear, it is known that ligand binding and DNA binding are critical and that they in some way allow the receptor to interact with the transcriptional machinery. The mechanism of transrepression by the RAR and other receptors, however, is even less well understood, but there is some evidence to suggest that it can involve pathways that are both dependent and independent of DNA binding and ligand binding (Damm *et al.* 1989; Forman *et al.* 1989; Glass *et al.* 1989; Schüle *et al.* 1990*a*). The example of the thyroid hormone receptor is particularly interesting because in the absence of hormone this receptor is capable of binding DNA and blocking basal-level transcription, whereas in the presence of hormone there is a several-fold induction of transcription above basal level (Damm *et al.* 1989). Thus, with the thyroid hormone receptor, repression and activation of the same gene are inversely dependent on the concentration of ligand.

One critical parameter that has yet to be delineated is the exact orientation with which the receptor multimer binds to DNA. Although an accurate representation will require NMR and X-ray crystallographic studies, inspection of the RAREs in Fig. 3.4 suggests that there may be more than one such orientation. There is no 'consensus' structure for an RARE, as the GGTCA half-site can occur in both a palindrome and a tandem repeat, with a variety of different spacing nucleotides between each half-element. It is clear so far that the retinoid receptors have the highest specificity for binding (K. Umesono and R. Evans, unpublished observation) with the response element isolated from the RAR-*β* gene promoter (de Thé *et al.* 1990*a*; Sucov *et al.* 1990). The *β*-RARE is further unique in that it responds exclusively to RA (Sucov *et al.* 1990) and both RAR and RXR (D. Mangelsdorf and R. Evans, unpublished observation). Based on the differences between the two retinoid receptors in the DNA-binding and ligand-binding domains we would predict that future studies will reveal hormone responsive elements that can also distinguish between RAR and RXR.


**Functions of other domains**

As stated above, in the current model of nuclear hormone activated transcription, it is the binding of a liganded receptor multimer to its DNA response element in the transcriptional control region of a target gene that leads to modulation of gene expression. Studies on both retinoid and other nuclear hormone receptors have revealed that there are now several complex processes that are known to occur in regulating this event. One of these processes is nuclear localization. In several receptors (e.g. glucocorticoid and oestrogen receptors) discrete, short stretches of basic amino acid residues have been identified in the region just C-terminal of the DNA-binding domain (Picard and Yamamoto 1987; Picard *et al.* 1990). These sequences are required for receptor migration into the nuclear compartment. Similar sequences can be found in the retinoid receptors, although their functional role has not yet been confirmed. The regulation of nuclear localization varies between receptor subfamilies. In the glucocorticoid receptor nuclear localization appears to be regulated by ligand binding, which is coincident with the release of bound HSP90 from the receptor (Bresnick *et al.* 1989). In contrast, the thyroid hormone receptor (Dalman *et al.* 1990) and RARs (S. Hollenberg, unpublished observation) do not associate with HSP90 and are resident in the nucleus in the absence of ligand (Damm *et al.* 1989; Gaub *et al.* 1989).

*Protein–protein interactions*

One function of ligand binding may be to allow dimerization of the receptor either with itself, other receptors, or other sets of transcription factors. Dimerization is critical for high affinity binding of the receptor with its cognate hormone response elements and deletion of the ligand-binding domain reduces this affinity (Kumar and Chambon 1988; Tora *et al.* 1988). A domain-like structure reminiscent of the leucine zipper motif can be found in the ligand-binding domain of both thyroid hormone receptor and RAR and it has been speculated that this region may be responsible for dimerization (Forman *et al.* 1989). That steroid hormone receptors form homodimers has been known for some time (Kumar and Chambon 1988). There is evidence that nuclear RARs also exist as dimers. In the nuclei of HL-60 cells two 95 kDa RA-binding components can be identified that migrate at only 51 kDa during denaturing gel electrophoresis (Hashimoto *et al.* 1989). One of these RA-binding components interacts specifically with RAR-$\alpha$ antibodies and the other with RAR-$\beta$ antibodies. This study suggests that RAR-$\alpha$ and RAR-$\beta$ are mutually exclusive and do not dimerize with each other; thus, the 95 kDa species may correspond to RAR homodimers formed in the presence of retinoids. Recently, however, Glass *et al.* (1989, 1990) and Yang *et al.* (1991) have demonstrated that other nuclear factors are required for RAR DNA binding. *In vitro* DNA binding studies have revealed that there may be multiple nuclear factors with which the RAR can form a heterodimeric complex (Glass *et al.* 1989). One such factor is the thyroid hormone receptor. The ability to form multiple heterodimeric structures may help explain how RARs are able to bind and transactivate through a wide variety of hormone response elements (Fig. 3.4). For example, both RAR and thyroid hormone receptor can transactivate a reporter gene that contains the response element, TRE$_{pal}$, in its promoter (Umesono *et al.* 1988). TRE$_{pal}$ is a palindrome of the consensus half-element sequence GGTCA motif; in contrast, the RARE from the RAR-$\beta$ gene is a tandem repeat of GGTCA. While RARs retain a high specificity for the $\beta$-RARE, the thyroid hormone receptor cannot transactivate or repress through this *cis* element (Sucov *et al.* 1990). These data indicate that the RAR binds to the $\beta$-RARE in a configuration different from that for the $TRE_{pal}$ and that this configuration may be due to differences in the other factors with which RAR interacts.

In addition to heterodimer function, recent studies indicate that the activity of the RAR may be modulated by the nuclear transcription factor AP-1. AP-1 is a complex of the proto-oncoproteins Jun and Fos, whose activity is stimulated by tumour promoters and growth factors. It was observed in the regulation of the human osteocalcin gene that agents stimulating AP-1 synthesis block the action of RA (Schüle *et al.* 1990*a*). Reciprocally, RA treatment is capable of inhibiting the action of AP-1. Inspection of the osteocalcin promoter revealed that within the proximity of the vitamin D and RA response element ($VDRE_{oc}$) there is a perfect AP-1 binding site (see Fig. 3.4). Although the mechanism of this mutual inhibition by RA and AP-1 is not yet clear, these two distinct regulatory systems are able to impose a type of 'check and balance' system on each other's activity. This phenomenon, whereby two different classes of regulatory factors modulate expression by interacting at a common response element, has been referred to as cross-coupling (Schüle *et al.* 1990*a*). However, a second, more recent example, which is particularly relevant because it implicates direct protein–protein interactions occur between nuclear hormone receptors and AP-1, has now been reported independently by several laboratories (Diamond *et al.* 1990; Jonat *et al.* 1990; Lucibello *et al.* 1990; Schüle *et al.* 1990*b*; Yang-Yen *et al.* 1990). In each of these studies functional antagonism of AP-1 induced expression could be demonstrated in the presence of glucocorticoid receptor. The inhibition of AP-1 activation of the collagenase promoter by glucocorticoid receptor is hormone dependent, but surprisingly, although it requires the glucocorticoid receptor DNA-binding domain, the ability of the glucocorticoid receptor to bind DNA is not required (Schüle *et al.* 1990*b*; Yang-Yen *et al.* 1990). In fact, the amino terminus and the ligand-binding domain of the glucocorticoid receptor can be exchanged with that of the RAR and the resulting hybrid receptor retains its AP-1 repression activity (Schüle *et al.* 1990*b*). Similarly, it was shown that the oncoprotein Jun could block glucocorticoid receptor-induced expression of a GRE-containing promoter, but that this inhibition did not involve the DNA-binding domain of Jun (Schüle *et al.* 1990*b*). Thus, cross-coupling appears to be a mechanism by which retinoid and other nuclear receptors can regulate the function of other transcription factors through both DNA binding and DNA binding-independent protein–protein interactions. One important future question is to determine the nature of the interaction and whether it is mediated by a conserved structural motif common to all nuclear receptors.

## Tissue distribution and expression specificity

If, as described above, RAR and RXR are both activated by RA and can transactivate similar target genes, how then is specificity generated? Alternatively, are these receptor systems functionally redundant? One way to decipher the roles of these receptors is to examine their patterns of expression and their regulation. As illustrated in Fig. 3.5, the pleiotropic effects of RA can be explained in part by the diverse pattern of RAR and RXR expression and multiplicity of target genes. Some members of both the RAR and RXR family are ubiquitously expressed, while others show specific, yet overlapping patterns. For example, in the adult, RAR-α and RXR-β are found in low levels in almost all tissues, whereas RAR-β, RAR-γ, and RXR-γ are restricted. Interestingly, the adult expression of RAR-γ is almost exclusively limited to the epidermis (Fig. 3.5). RXR-α is also abundantly expressed in epidermis and keratinocytes (D. Mangelsdorf and R. Evans, unpublished observation), indicating that these two receptors are likely to be responsible for the dermatological effects of retinoids. The overall high expression of RXRs in the adult visceral tissues, such a liver, kidney, heart, and intestine, and the relatively high dose of RA needed to activate RXRs suggests that this family of receptors may play a role in vitamin A metabolism.

Extensive work on the expression of RARs by *in situ* hybridization techniques in several laboratories (Dollé *et al.* 1989; Noji *et al.* 1989; Osumi-Yamashita *et al.* 1990; Ruberte *et al.* 1990; Smith and Eichele 1991) now reveal an emerging role for each RAR subtype during embryonic development and limb formation. In the developing mouse embryo RAR-α is almost ubiquitously expressed at low levels. RAR-β transcripts, however, are restricted to specific regions such as the interdigital mesenchyme, mesenchyme surrounding the inner ear epithelium (Dollé *et al.* 1989), and the developing nervous system (Smith and Eichele 1991). In contrast, RAR-γ gene expression can be found specifically localized in all mesenchyme giving rise to bones and facial cartilage, and differentiating squamous keratinizing epithelia (Ruberte *et al.* 1990; Osumi-Yamashita *et al.* 1990). This non-overlapping distribution of RAR-β and RAR-γ suggests that specific developmental roles exist for each. Because of its expression in the interdigital mesenchyme during the time of digit separation, it has been suggested that RAR-β could be important in regulating programmed cell death (Dollé *et al.* 1989). In the adult newt, whose limbs are capable of regenerating, RAR-β mRNA is induced to high levels following amputation (Giguère *et al.* 1989). The implication of RAR-β's involvement in neurogenesis (Simeone *et al.* 1990; Smith and Eichele 1991) is further strengthened by the finding that RA can be synthesized in the developing neural floor plate (Wagner *et al.* 1990). Likewise, the coincident expression of RAR-γ during chondrogenesis and squamous epithelial differentiation suggest that RAR-γ is crucial for transducing RA signals that transcriptionally govern these events. Although the distribution of the RXR

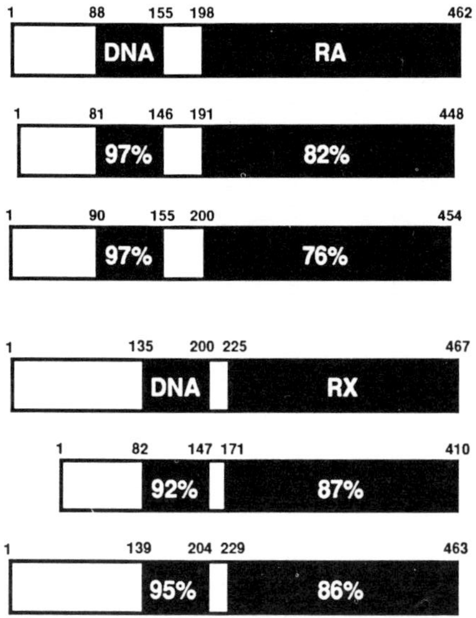

| Receptor | Species | Chromo. | Expression Pattern | |
|---|---|---|---|---|
| | | | Adult | Embryo |
| hRARα   1   88   155   198   DNA   RA   462 | h, m, n, x [1,2][3][4][18] | 17q21.1 [1] | widespread [3,11,22] | widespread [13,14] |
| hRARβ   1   81   146   191   97%   82%   448 | h, m, n [5-7][3][8] | 3p24 [6] | specific [3,11,22] (eg. muscle, prost.) | spatial and temporal [13,14] |
| hRARγ   1   90   155   200   97%   76%   454 | h, m, n, x [9,10][3][4][18] | 12 [10] | specific [3,22] (eg. skin, lung) | spatial and temporal [15,14,21] |
| mRXRα   1   135   200   225   DNA   RX   467 | h, m, x [16][19][18] | 9 [20] | specific [16,17] (eg. liver, skin, kid.) | + + [19] |
| mRXRβ   1   82   147   171   92%   87%   410 | h, m [19][17,19] | 6 [20] | widespread [17,19] | + + [19] |
| mRXRγ   1   139   204   229   95%   86%   463 | m, x [19][18] | n.d. | specific [18,19] (eg. muscle, heart) | + + [18] |

**Fig. 3.5** Comparative analysis of the RAR and RXR family of retinoid responsive transcription factors. Primary amino acid sequences have been aligned as in Fig. 3.4 in relation to either hRAR α or hRXR α. 'Species' indicates in which organisms each receptor has been identified (h, human; m, mouse; n, newt; x, *Xenopus*). 'Chromo.' indicates the human chromosome location. 'Expression Pattern' indicates the general distribution of each receptor in the adult and embryo. For a more detailed analysis see the text and the references listed below. (+ +) indicates those receptors that are relatively abundant but whose exact location is still unknown. References are as follows: 1, Petkovitch *et al.* 1987; 2, Giguère *et al.* 1987; 3, Zelent *et al.* 1989; 4, Ragsdale *et al.* 1989; 5, de Thé *et al.* 1987; 6, Brand *et al.* 1988; 7, Benbrook *et al.* 1988; 8, Giguère *et al.* 1989; 9, Krust *et al.* 1989; 10, Ishikawa *et al.* 1990; 11, de Thé *et al.* 1989; 12, Rees *et al.* 1989; 13, Dollé *et al.* 1989; 14, Osumi-Yamashita *et al.* 1990; 15, Ruberte *et al.* 1990; 16, Mangelsdorf *et al.* 1990; 17, Hamada *et al.* 1989; 18, B. Blumberg *et al.*, submitted for publication; 19, D. Mangelsdorf and R. Evans, unpublished data; 20, D. Mangelsdorf, J. Brook, D. Housman, and R. Evans, unpublished data; 21, Giguère *et al.* 1990; 22, Kastner *et al.* 1990.

receptor subfamily during embryogenesis has not yet been studied in detail, it is known that all three mouse isoforms ($\alpha$, $\beta$, and $\gamma$) are expressed abundantly in the embryo from at least day 10 post coitum to birth (D. Mangelsdorf, R. Heyman, and R. Evans, unpublished observation). The development of isoform-specific antibodies and continuing *in situ* hybridization experiments are necessary to further our understanding of the complex expression patterns of RARs and RXRs.

## RECEPTOR AUTOREGULATION

As indicated above, one mechanism to coordinate RA-responsive gene expression is through the temporal and spatial expression of retinoid receptors. It is therefore not surprising that the three RAR subtypes may all be RA inducible. The evidence is most advanced for the RAR-$\beta$ gene, which displays the most dramatic induction. Examination of its promoter has led to the identification of a RA response element ($\beta$-RARE, see Fig. 3.4), which is highly sensitive to RA (de Thé *et al.* 1990*a*; Sucov *et al.* 1990). The propagation of an RA-specific autoregulatory loop has potential biological advantages. First it allows the tissue specific pattern of RAR-$\beta$ expression to be controlled by RA synthesis. This dependence is a simple and efficient means to insure a linkage between the ligand and cellular responsiveness. Thus, although the RARE is a simple element it can be exploited to impart a complex and dynamic expression pattern. Further, the demonstration that there exists in the developing limb bud a concentration gradient of RA (Thaller and Eichele 1987) implies the existence of a parallel gradient of RAR-$\beta$. Such a gradient of receptor expression would be difficult to establish genetically. Finally, because the response of a cell will be dependent on the concentration of activated RAR, the establishment of a dual RA/RAR gradient would allow a many-fold amplified response to RA. The presence of such a gradient may explain how RA is able to induce differential gene expression and variant differentiation pathways in a concentration dependent manner. Recent evidence obtained by in situ hybridization in the developing chick (Smith and Eichele 1991) and mouse (S. Noji, personal communication) support this general hypothesis.

## PERSPECTIVES AND CONCLUSION

The discovery in 1987 by Giguère *et al.* and Petkovich *et al.* of a human RA receptor offered for the first time in a vertebrate system the hope of analysing the mechanisms of morphogenesis by identifying a set of developmentally controlled genes. This discovery was made more meaningful by the recognition that the RAR was a member of the steroid/thyroid hormone receptor superfamily. This association suggested that the mechanisms of action of retinoids are, by analogy, similar to those of steroid and thyroid hormones. Accordingly, the RAR is proposed to be a sequence-specific DNA-binding transcription factor whose activity is controlled by RA. Subsequent to the identification of the initial RAR-$\alpha$, two additional genes that comprise the RA receptor subfamily were identified ($\beta$ and $\gamma$). In 1990 Mangelsdorf *et al.* identified the retinoid X receptors (RXRs) which represent a second evolutionary pathway for mediating retinoid responsiveness. RXRs are also part of the steroid/thyroid hormone receptor superfamily but are not part of the RAR subfamily. Indeed, it appears that there are now three distinct RXR genes, also referred to as $\alpha$, $\beta$, and $\gamma$. Thus, in aggregate, six retinoid-responsive transcription factors have been identified, each of which may play a critical role in orchestrating embryonic development and adult physiology.

**Family ties**

One obvious challenge is to understand the rationale for multiple receptors in each subfamily. Do they serve interactive and interdependent functions, or are they redundant? The same sort of question can be proposed for the two retinoid subfamilies themselves. For example, is RXR a redundant pathway to the RAR, or are these two subfamilies serving fundamentally different roles? One approach to addressing these questions is to understand the temporal–spatial patterns of expression of these molecules during embryonic development and in the adult. A second is to identify the target genes of each of these molecules and indeed, simply to know whether they all recognize common sequences or have distinct specificities. Because the function of a protein must be intimately related to its structure, it can be argued that the differences between the RXRs and RARs indicate fundamentally different functions because of their inherent structural dissimilarities.

Based on homology of the RARs and RXRs, their genes must have arisen independently during evolution, possibly to cope with the increasing demands of the ever more complex evolving organisms. Although understanding the origins of these two gene families may seem tangential to understanding their significance, this may represent a limited view. By tracing these two genes back in history, we may be better able to link them with their original biological functions. Because evolution proceeds via adaptation and selection, it is likely that the original functions will be informative about their modern roles. Thus, identification of invertebrate homologues may contribute greatly in advancing this frontier.

Knowing that the receptors are sequence-specific, retinoid-dependent transcription factors does not explain how they control morphologic diversity or cell fate. For example, we do not know whether the receptors represent the tip of a hierarchical cascade that activates a limited set of regulatory genes or, alternatively, whether they directly stimulate a large fraction of downstream functions. One of the most direct ways to understand the function of an individual receptor is to identify its target genes. Ultimately, it is the genes that are being regulated that are responsible for the unique changes in cell function that are triggered by retinoids. If these genes could be identified, we will have greatly advanced our understanding of how receptors contribute to morphological diversity.

A further expansion of this idea relates to how a continuous gradient of RA can be translated into spatial patterns of gene expression associated with distinct developmental programmes. Thus, at one concentration of RA developmental programme X must be activated, while at slightly higher RA levels programme X must be suppressed while programme Y is now activated. Finally, in different cells, e.g. limb versus the nervous system, RA may activate an entirely different genetic network. Understanding how continuous gradients are translated into multiple genetic patterns is essential to understanding the entire process of morphogenesis in any complex system.

## Retinoid relatives

Although the focus of this chapter is on RA, there are many gaps in our knowledge of RA metabolism that need to be filled. For example, are there novel retinoid metabolites that selectively stimulate either the RAR or RXR? It seems likely that these two receptors, although recognizing the common molecule, RA, will show differential activity with regard to putative retinoid metabolites. Already there is evidence to suggest that RA is not the only metabolite that is responsible for all of the effects attributed to vitamin A. The recent discovery of 3,4-didehydroretinoic acid (Thaller and Eichele 1990) as an abundant molecule and its ability to effectively stimulate the activities of RXR-$\alpha$ indicates that it may play an important morphogenic role apart from RA. This discovery opens the door for other retinoid metabolites as potential signalling molecules. In this context, it will be essential to characterize the enzymes that convert retinol to RA and its various downstream metabolic products. One critical question is whether there are specific retinoid dehydrogenases that are expressed in a spatially and developmentally regulated pattern that corresponds to the known developmental sites of retinoid action. In other words, although present in high concentrations, retinol itself is essentially inert. To be active it must be converted to RA or other metabolites. Thus, the converting enzymes must be expressed in the appropriate sites and at the appropriate times throughout the body to allow local synthesis and diffusion of retinoic acid. In principle, the sites of production of such converting enzymes will provide a road map to the morphogenic organizational centres throughout the body. The cloning of such enzymes would not only be highly desirable but might provide us with the tools to begin to identify the next layer of biological organization of the embryo.

## Partner preferences

The unfortunately poor genetics of vertebrate systems implies that this route is unlikely to be revealing about RAR or RXR function. However, via the use of homologous recombination it is now possible to use cloned DNA to alter genomic sequences and to derive transgenic mice harboring the mutant allele (Capecchi 1989). This gene 'knock-out' approach in principle, allows for a systematic analysis of RAR and RXR function. It may also be useful for the introduction of mutations that modify (instead of eliminate) receptor function. This will allow the roles of individual domains to be assessed. Another approach that is now being taken is to generate functional repressors instead of activators (Espeseth *et al.* 1990; Pratt *et al.* 1990). When coexpressed in the same cell, retinoid receptors carrying the repressive or 'dominant-negative' phenotype are able to interfere with wild type receptor activity and effectively block the normal action of RA. By targeting these dominant-negative receptors to specific cell types in transgenic mice, the tissue-specific function of RARs and RXRs may be identified.

## Human disease

The cloning of the RAR gene allows an evaluation of its role in development, physiology, and human disease. Of particular interest is its potential role as a proto-oncogene, i.e. a gene that, through mutation, could contribute to oncogenic transformation. This possibility is given force by the previous discovery that the v-*erbA* oncogene, which contributes to an avian erythroleukaemia, is a mutated derivative of the thyroid hormone receptor (Sap *et al.* 1986; Weinberger *et al.* 1986). It facilitates oncogenic transformation of cells by blocking the action of the endogenous thyroid hormone receptor and, thus, prevents erythroblasts from differentiating in response to $T_3$ (Zenke *et al.* 1990). Based on these observations, it may not be surprising that an association between the RAR gene and the aetiology of acute promyelocytic leukaemia (APL) has recently been suggested.

APL is a myelocytic leukaemia that is distinguished by a reciprocal chromosome 15:17 translocation. Three recent reports demonstrate that the t(15:17) breakpoint occurs in the RAR gene itself (Borrow *et al.* 1990; de Thé *et al.* 1990*b*; Longo *et al.* 1990). This dramatic discovery suggests that a mutant form of the RAR-α gene interacts or contributes to leukaemogenesis and simultaneously provides a molecular marker for identifying leukaemic promyelocytes. This result also supports the idea that the RAR, and thus RA, may play a role in normal myeloid differentiation. Perhaps the mutant RAR blocks this normal progression. Interestingly, a clinical team in China has recently demonstrated that RA may be an effective treatment for APL (Menger *et al.* 1988). It has been known for some time that RA can inhibit the growth of myeloid leukaemia cells in culture (Douer and Koeffler 1982). Apparently,

RA triggers leukaemic cells to stop dividing, while simultaneously inducing differentiation and restoring their normal function.

The mechanism of leukaemic transformation is likely to be similar to that previously observed with *erbA*, wherein the oncogene product blocks the function of the normal receptor. Evidently RA treatment overcomes this block. In addition to the RA receptor, we know that the retinoid X receptors are also expressed in high levels in human myelocytes (D. Mangelsdorf, A. Kakizuka, and R. M. Evans, unpublished observation). Utilizing the cloned gene products should provide a better understanding of the physiological function of the RARs and RXRs in normal lymphoid differentiation, as well as of their roles in the path of physiology and treatment of APL and other human diseases. It is expected that such analysis will provide not only a better understanding of the origin of this disease, but also specific molecular markers for diagnosis and treatment which might not have been previously possible.

Thus, as can be seen from this article, the RARs and RXRs are interesting not only as molecular machines to modulate transcriptional programmes but as key molecules in the activation of developmental and physiological cascades. These studies may help to open up a new era in understanding the molecular basis not only of embryonic development but also of human disease.

ACKNOWLEDGEMENTS

The authors wish to thank Drs. Kazuhiko Umesono, Richard Heyman, and Bruce Blumberg for insightful discussion and Elaine Stevens for help in the preparation of the manuscript. Portions of this manuscript including unpublished observations, will be published subsequently in other form elsewhere. R.M.E. is an Investigator of the Howard Hughes Medical Institute at the Salk Institute for Biological Studies. This work was supported by the Howard Hughes Medical Institute, National Institutes of Health, National Cancer Institute, and the Mathers Foundation.

REFERENCES

Beato, M. (1989). Gene regulation by steroid hormones. *Cell*, **56**, 335—44.
Benbrook, D., Lernhardt, E., and Pfahl, M. (1988). A new retionic acid receptor identified from a hepatocellular carcinoma. *Nature*, **333**, 669–72.
Borrow, J., Goddard, A. D., Sheer, D., and Solomon, E. (1990). Molecular analysis of acute promyelocytic leukemia breakpoint cluster region on chromosome 17. *Science*, **249**, 1577–80.
Brand, N., Petkovich, M., Krust, A., Chambon, P., de The, H., Marchio, A., Tiollais, P., and A.Dejean, A. (1988). Identification of a second human retinoic acid receptor. *Nature*, **332**, 850–3.

Bresnick, E. H., Dalman, F. C., Sanchez, E. R., and Pratt, W. B. (1989). Evidence that the 90-kDa heat shock protein is necessary for the steroid binding conformation of the L cell glucocorticoid receptor. *Journal of Biological Chemistry*, **264**, 4992–7.

Capecchi, M. (1989). Altering the genome by homologous recombination. *Science*, **244**, 1288–92.

Chang, C. and Kokontis, J. (1988). Identification of a new member of the steroid receptor superfamily by cloning and sequence analysis. *Biochemical and Biophysical Research Communications*, **155**, 971–7.

Chytil, F. and Ong, D. E. (1978). Cellular vitamin A binding proteins. *Vitamins and Hormones*, **36**, 1–32.

Dalman, F. C., Koenig, R. J., Perdew, G. H., Massa, E., and Pratt, W. B. (1990). In contrast to the glucocorticoid receptor, the thyroid hormone receptor is translated in the DNA binding state and is not associated with hsp90. *Journal of Biological Chemistry*, **267**, 3615–18.

Damm, K., Thompson, C. C., and Evans, R. M. (1989). Protein encoded by v-*erbA* functions as a thyroid-hormone receptor antagonist. *Nature*, **339**, 593–7.

de Thé, H., Marchio, A., Tiollais, P., and Dejean, A. (1987). A novel steroid thyroid hormone receptor-related gene inappropriately expressed in human hepatocellular carcinoma. *Nature*, **330**, 667–70.

de Thé, H., Marchio, A., Tiollais, P., and Dejean, A. (1989). Differential expression and ligand regulation of the retionic acid receptor α and β genes. *EMBO Journal*, **8**, 429–33.

de Thé, H., Vivanco-Ruiz, M., Tiollais, P., Stunnenberg, H., and Dejean, A. (1990*a*). Identification of a retinoic acid responsive element in the retinoic acid receptor β receptor gene. *Nature*, **343**, 177–80.

de Thé, H., Chomienne, C., Lanotte, M., Degos, L., and Dejean, A. (1990*b*). The t(15:17) translocation of acute promyelocytic leukaemia fuses the retinoic acid receptor α gene to a novel transcribed locus. *Nature*, **347**, 558–61.

de Verneuil, H. and Metzger, D. (1990). The lack of transcriptional activation of the v-*erbA* oncogene is in part due to a mutation present in the DNA binding domain of the protein. *Nucleic Acids Research*, **18**, 4489–97.

Diamond, M. I., Miner, J. N., Yoshinaga, S. K., and Yamamoto, K. R. (1990). Transcription factor interactions: selectors of positive or negative regulation from a single DNA element. *Science*, **249**, 1266–72.

Dollé, P., Ruberte, E., Kastner, P., Petkovich, P., Stoner, C. M., Gudas, L. J., and Chambon, P. (1989). Differential expression of genes encoding α, β and γ retinoic acid receptors and CRABP in the developing limbs of the mouse. *Nature*, **342**, 702–5.

Douer, D. and Koeffler, H. P. (1982). Retinoic acid: Inhibition of the clonal growth of human myeloid leukemia cells. *Journal of Clinical Investigation*, **69**, 277–83.

Espeseth, A. S., Murphy, S. P., and Linney, E. (1989). Retinoic acid receptor expression vector inhibits differentiation of F9 embryonal carcinoma cells. *Genes and Development*, **3**, 1647–56.

Evans, R. M. (1988). The steroid and thyroid hormone receptor superfamily. *Science*, **240**, 889–95.

Forman, B. M., Yang, C., Au, M., Casanova, J., Ghysdael, J., and Samuels, H. H. (1989). A domain containing leucine-zipper-like motifs mediate novel *in vivo* interactions between the thyroid hormone and retinoic acid receptors. *Molecular Endocrinology*, **3**, 1610–26.

Gaub, M. P., Lutz, Y., Ruberte, E., Petkovich, M., Brand, N., and Chambon, P. (1989). Antibodies specific to the retionic acid human nuclear receptors α and β. *Proceedings of the National Academy of Sciences USA*, **86**, 3089–93.

Giguère, V., Ong, E. S., Segui, P., and Evans, R. M. (1987). Identification of a receptor for the morphogen retinoic acid. *Nature*, **330**, 624–9.

Giguère, V., Yang, N., Segui, P., and Evans, R. M. (1988). Identification of a new class of steroid hormone receptors. *Nature*, **331**, 91–4.

Giguère, V., Ong, E. S., Evans, R. M., and Tabin, C. J. (1989). Spatial and temporal expression of the retinoic acid receptor in the regenerating amphibian limb. *Nature*, **337**, 566–9; addendum **341**, 80.

Giguère, V., Shago, M., Zirngibl, R., Tate, P., Rossant, J., and Varmuza, S. (1990). Identification of a novel isoform of the retinoic acid receptor γ expressed in the mouse embryo. *Molecular and Cellular Biology*, **10**, 2335–40.

Glass, C. K., Holloway, J. M., Devary, O. V., and Rosenfeld, M. G. (1988). The thyroid hormone receptor binds with opposite transcriptional effects to a common sequence motif in thyroid hormone and estrogen response elements. *Cell*, **54**, 313–23.

Glass, C. K., Lipkin, S. M., Devary, O. V., and Rosenfeld, M. G. (1989). Positive and negative regulation of gene transcription by a retionic acid-thyroid hormone receptor heterodimer. *Cell*, **59**, 697–708.

Glass, C. K., Devary, O. V., and Rosenfeld, M. G. (1990). Multiple cell type-specific proteins differentially regulate target sequence recognition by the α retinoic acid receptor. *Cell*, **63**, 729–38.

Graupner, G., Wills, K. N., Tzukerman, M., Zhang, X., and Pfahl, M. (1989). Dual regulatory role for thyroid-hormone receptors allows control of retionic-acid receptor activity. *Nature*, **340**, 653–6.

Green, S. and Chambon, P. (1987). Oestradiol induction of a glucocorticoid-responsive gene by a chimaeric receptor. *Nature*, **325**, 75–8.

Green, S. and Chambon, P. (1988). Nuclear receptor enhance our understanding of transcription regulation. *Trends in Genetics*, **4**, 309–14.

Green, S., Walter, P., Kumar, V., Krust, A., Bornert, J.-M., Argos, P., and Chambon, P. (1986). Human oestrogen receptor cDNA: sequence, expression and homology to v-erb-A. *Nature*, **320**, 134–9.

Hashimoto, Y., Petkovitch, M., Gaub, P., Kagechika, H., Shudo, K., and Chambon, P. (1989). The retinoic acid receptors α and β are expressed in the human promyelocytic leukemia cell line HL-60. *Molecular Endocrinology*, **3**, 1046–52.

Hamada, K., Gleason, S. L., Levi, B.-Z., Hirschfeld, S., Apopela, E., and Ozato, K. (1989). H-2RIIBP, a member of the nuclear hormone receptor superfamily that binds to both the regulatory element of major histocompatibility class I genes and the estrogen response element. *Proceedings of the National Academy of Sciences USA*, **86**, 8289–93.

Haussler, M. R., Myrtle, J. F., and Norman, A. W. (1968). The association of a metabolite of vitamin $D_3$ with intestinal mucosa chromatin, *in vivo*. *Journal of Biological Chemistry*, **243**, 4055–64.

Herschman, H. R. (1989). Extracellular signals, transcriptional responses and cellular specificity. *Trends in Biological Sciences*, **14**, 455–8.

Hollenberg, S. M., Weinberger, C., Ong, E. S., Cerelli, G., Oro, A., Lebo, R., Thompson, E. B., Rosenfeld, M. G., and Evans, R. M. (1985). Primary structure and expression of a functional human glucocorticoid receptor cDNA. *Nature*, **318**, 635–41.

Hudson, L. G., Santon, J. B., Glass, C. K., and Gill, G. N. (1990). Ligand-activated thyroid hormone and retinoic acid receptors inhibit growth factor receptor promoter expression. *Cell*, **62**, 1165–75.

Ishikawa, T., Umesono, K., Mangelsdorf, D. J., Aburatani, H., Stanger, B. Z., Shibasaki, Y. Imawari, M., Evans, R. M., and Takaku, F. (1990). A functional retinoic acid receptor encoded by the gene on human chromosome 12. *Molecular Endocrinology*, **4**, 837–44.

Jonat, C., Rahmsdorf, H. J., Park, K.-K., Cato, A. C. B., Gebel, S., Ponta, H., and Herrlich, P. (1990). Antitumor promotion and antiinflammation: down-modulation of AP-1 (fos/jun) activity by glucocorticoid hormone. *Cell*, **62**, 1189–204.

Kastner, P., Krust, A., Mendelsohn, C., Garnier, J. M., Zelent, A., Leroy, P., Staub, A. and Chambon, P. (1990). Murine sioforms of retinoic acid receptor $\gamma$ with specific patterns of expression. *Proceedings of the National Academy of Sciences USA*, **87**, 2700–4.

Keeble, S. and Maden, M. (1989). The relationship among retinoid structure, affinity for retinoic acid-binding protein, and ability to respecify pattern in the regenerating axolotl limb. *Developmental Biology*, **132**, 26–34.

Kerner, S. A., Scott, R. A., and Pike, W. J. (1989). Sequence elements in the human osteocalcin gene confer basal activation and inducible response to hormonal vitamin $D_3$. *Proceedings of the National Academy of Sciences USA*, **86**, 4455–9.

Klein-Hitpaß, Schorpp, M., Wagner, U., and Ryffel, G. U. (1986). An estrogen-responsive element derived from the 5′ flanking region of the Xenopus vitellogenin A2 gene functions in transfected human cells. *Cell*, **46**, 1053–61.

Krust, A., Kastner, P., Petkovich, M., Zelent, A., and Chambon, P. (1989). A third human retinoic acid receptor, hRAR-$\gamma$. *Proceedings of the National Academy of Sciences USA*, **86**, 5310–14.

Kumar, V. and Chambon, P. (1988). The estrogen receptor binds tightly to its responsive element as a ligand-induced homodimer. *Cell*, **55**, 145–55.

Lazar, M. A., Hodin, R. A., Darling, D. S., and Chin, W. W. (1989). A novel member of the thyroid/steroid hormone receptor family is encoded by the opposite strand of the rat c-*erbA*α transcriptional unit. *Molecular and Cellular Biology*, **9**, 1128–36.

Longo, L., Pandolfi, P. P., Biondi, A., Rambaldi, A. Mencarelli, A., Lo Coco, F., Diverio, D., Pegoraro, L., Avanzi, G., Tabilio, A., Xangrilli, D., Alcalay, M., Donti, E., Grignani, F., and Pelicci, P. G. (1990). Rearrangements and aberrant expression of the retinoic acid receptor α gene in acute promyelocytic leukemias. *Journal of Experimental Medicine*, **172**, 1571–5.

Lubahn, D. B., Joseph, D. R., Madhabananda, S., Tan, J., Higgs, H. N., Larson, R. E. *et al.* (1988). The human androgen receptor: complementary deoxyribonucleic acid cloning, sequence analysis and gene expression in prostate. *Molecular Endocrinology*, **2**, 1265–75.

Lucibello, F. C., Slater, E. P., Jooss, K. U., Beato, M., and Müller, R. (1990). Mutual transrepression of fos and the glucocorticoid receptor: involvement of a functional domain in fos which is absent in fosB. *EMBO Journal*, **9**, 2827–34.

Mangelsdorf, D. J., Ong, E. S., Dyck, J. A., and Evans, R. M. (1990). Nuclear receptor that identifies a novel retionic acid response pathway. *Nature*, **345**, 224–9.

Meng-er, H., Yu-chen, Y., Shu-rong, C., Hin-ren, C., Jia-Xiang, L., Lin, Z., *et al.*

(1988). Use of all-*trans* retinoic acid in the treatment of actue promyelocytic leukemia. *Blood*, **72**, 567–72.

Milbrandt, J. (1987). A nerve growth factor-induced gene encodes a possible transcriptional regulatory factor. *Science*, **238**, 797–9.

Miyajima, N., Kadowaki, Y., Fukushige, S., Shimizu, S., Semba, K., Yamanashi, Y., *et al.* (1988). Identification of two novel members of *erbA* superfamily by molecular cloning: the gene products of the two are highly related to each other. *Nucleic Acids Research*, **16**, 11057–74.

Miyajima, N., Horiuchi, R., Shibuya, Y., Fukushige, S., Matsubara, K., Toyoshima, K., *et al.* (1989). Two *erbA* homologs encoding proteins with different $T_3$ binding capacities are transcribed from opposite DNA strands of the same genetic locus. *Cell*, **57**, 31–9.

Noji, S., Yamaai, T., Koyama, E., Noho, T., and Taniguchi, S. (1989). Spatial and temporal expression pattern of retinoic acid receptor genes during mouse bone development. *FEBS Letters*, **257**, 93–6.

Oro, A. E., McKeown, M., and Evans, R. M. (1990). Relationship between the product of the *Drosophila ultraspiracle* locus and the vertebrate retinoid X receptor. *Nature*, **347**, 298–301.

Osumi-Yamashita, N., Noji, S., Nohno, T., Koyama, E., Doi, H., Eto, K., *et al.* (1990). Expression of retinoic acid receptor genes in neural crest-derived cells during mouse facial development. *FEBS Letters*, **264**, 71–4.

Petkovich, M., Brand, N. J., Krust, A., and Chambon, P. (1987). A human retinoic acid receptor which belongs to the family of nuclear receptors. *Nature*, **330**, 444–50.

Picard, D. and Yamamoto, K. R. (1987). Two signals mediate hormone-dependent nuclear localization of the glucocorticoid receptor. *EMBO Journal*, **6**, 3333–40.

Picard, D., Kumar, V., Chambon, P., and Yamamoto, K. R. (1990). Signal transduction by steroid hormones: nuclear localization is differentially regulated in estrogen and glucocorticoid receptors. *Cell Regulation*, **1**, 291–9.

Pratt, M. A. C., Kralova, J., and McBurney, M. W. (1990). A dominant negative mutation of the alpha retinoic acid receptor gene in a retinoic acid-nonresponsive embryonal carcinoma cells. *Molecular and Cellular Biology*, **10**, 6445–53.

Ragsdale, C. W., Jr, Petkovich, M., Gates, P. B., Chambon, P., and Brockes, J. P. (1989). Identification of a novel retinoic acid receptor in regenerative tissues of the newt. *Nature*, **341**, 654–7.

Rees, J. L., Daly, A. K., and Redfern, C. P. F. (1989). Differential expression of the $\alpha$ and $\beta$ retinoic acid receptors in tissues of the rat. *Biochemical Journal*, **259**, 917–19.

Ruberte, E., Dolle, P., Krust, A., Zelent, A., Morriss-Kay, G., and Chambon, P. (1990). Specific spatial and temporal distribution of retinoic acid receptor gamma transcripts during mouse embryogenesis. *Development*, **108**, 213–22.

Sang, J. H. (1978). The nutritional requirements of *Drosophila*. In *The genetics and biology of Drosophila* vol. 2A, (eds M. Ashburner and T. R. F. Wright), pp. 159–92. Academic Press, London.

Sap, J., Muñoz, A., Damm, K., Goldberg, Y., Ghysdael, J., Leutz, A., Beug, H., and Vennström, B. (1986). The c-*erb-A* protein is a high- affinity receptor for thyroid hormone. *Nature*, **324**, 635–640.

Schüle, R., Umesono, K., Mangelsdorf, D. J., Bolado, J., Pike, J. W., and Evans, R. M. (1990*a*). Jun-fos and receptors for vitamins A and D recognize a common response element in the human oestocalcin gene. *Cell*, **61**, 497–504.

Schüle, R., Rangarajan, P., Kliewer, S., Ransone, L. J., Bolado, J., Yang, N. *et al.* (1990*b*). Functional antagonism between oncoprotein c-jun and the glucocorticoid receptor. *Cell*, **62**, 1217–26.

Segraves, W. A. (1991). Something old, some things new: the steroid receptor superfamily in *Drosophila*. *Cell*, **67**, 225–8.

Simeone, A., Acampora, D., Arcioni, L., Andrews, P. W., Boncinelli, E., and Mavillo, F. (1990). Sequential activation of *HOX2* homeobox genes by retinoic acid in human embryonal carcinoma cells. *Nature*, **346**, 763–6.

Smith, S. M. and Eichele, G. (1991). Temporal and regional differences in the expression pattern of distinct retinoic acid receptor-*β* transcripts in the chick embryo. *Development*, **111**, 245–52.

Sporn, M. B., Roberts, A. B., and Goodman, D. S. (eds) (1984). *The retinoids*, vols 1 and 2. Academic Press Inc., New York.

Sucov, H. M., Murakami, K. K., and Evans, R. M. (1990). Characterization of an autoregulated response element in the mouse retinoic acid receptor type *β* gene. *Proceedings of the National Academy of Sciencies USA*, **87**, 5392–6.

Thaller, C. and Eichele, G. (1987). Identification and spatial distribution of retinoids in the developing chick limb bud. *Nature*, **327**, 625–8.

Thaller, C. and Eichele, G. (1990). Isolation of 3,4-didehydroretinoic acid, a novel morphogenetic signal in the chick wing bud. *Nature*, **345**, 815–19.

Thompson, C. C. and Evans, R. M. (1989). Trans-activation by thyroid hormone receptors: functional parallels with steroid hormone receptors. *Proceedings of the National Academy of Sciences USA*, **86**, 3494–8.

Tora, L., Gronemeyer, H., Turcotte, B., Gaub, M. P., and Chambon, P. (1988). The N-terminal region of the chicken progesterone receptor specifies target gene activation. *Nature*, **33**, 185–8.

Umesono, K., Giguère, V., Glass, C. K., Rosenfeld, M. G., and Evans, R. M. (1988). Retinoic acid and thyroid hormone induce gene expression through a common responsive element. *Nature*, **336**, 262–5.

Vasios, G. W., Gold, J. D., Petkovich, M., Chambon, P., and Gudas, L. J. (1989). A retinoic acid-responsive element is present in the 5′ flanking region of the laminin B1 gene. *Proceedings of the National Academy of Sciences USA*, **86**, 9099–103.

Wagner, M., Thaller, C., Jessell, T., and Eichele, G. (1990). Polarizing activity and retinoid synthesis in the floor plate of the neural tube. *Nature*, **345**, 819–22.

Wang, L.-H., Tsai, S. Y., Cook, R. G., Beattie, W. G., Tsai, M-J., and O'Malley, B. W. (1989). COUP transcription factor is a member of the steroid receptor superfamily. *Nature*, **340**, 163–6.

Weinberger, C., Thompson, C. C., Ong, E. S., Lebo, R., Gruol, D. J., and Evans, R. M. (1986). The c-*erb*-A gene encodes a thyroid hormone receptor. *Nature*, **324**, 641–6.

Yang, N., Mangelsdorf, D. J., Schüle, R., and Evans, R. M. (1991). Characterization of DNA-binding and retinoic acid binding properties of retinoic acid receptor. *Proceedings of the National Academy of Sciences USA*, **88**, 3359–63.

Yang-Yen, H.-F., Chambard, J.-C., Sun, Y.-L., Smeal, T., Schmidt, T. J., Drouin, J., and Karin, M. 1990. Transcriptional interference between c-jun and the glucocorticoid receptor: mutual inhibition of DNA binding due to direct protein–protein interaction. *Cell*, **62**, 1205–15.

Zelent, A., Krust, A., Petkovich, M., Kastner, P., and Chambon, P. (1989). Cloning of

murine $\alpha$ and $\beta$ retinoic acid receptors $\gamma$ and a novel receptor predominantly expressed in skin. *Nature*, **339**, 714–17.
Zenke, M., Muñoz, A., Sap, J., Vennström, B., and Beug, H. (1990). V-*erbA* oncogene activation entails the loss of hormone-dependent regulator activity of c-*erbA*. *Cell*, **61**, 1035–49.

# 4

# Molecular mechanisms of retinoic acid action

Magnus Pfahl, Xiao-Kun Zhang, Jürgen M. Lehmann,
Matthias Husmann, Gerhart Graupner, and Birgit Hoffmann

The multitude of essential biological processes regulated by the vitamin A derivative retinoic acid (RA) and synthetic analogues (retinoids) (reviewed in Lotan 1980; Sporn *et al.* 1984), combined with the many beneficial and undesirable side-effects (Lippman *et al.* 1987) associated with retinoid therapies, suggest the existence of a highly complex regulatory mechanism. How can a single substance—RA—regulate a large diversity of programmes and, in addition, provoke at times opposite effects, e.g. during limb morphogenesis, where low concentrations of RA can lead to duplications of digits, while high concentrations prevent digit formation (Tickle *et al.* 1982; Summerbell 1983). The discovery of nuclear retinoic acid receptors (RAR) that belong to the steroid–thyroid hormone receptor family represented a breakthrough towards the elucidation of a molecular mechanism of RA action (Petkovich *et al.* 1987; Giguère *et al.* 1987). These receptor proteins are ligand-dependent transcription factors that contain a highly conserved 'zinc-finger' DNA binding domain and a less conserved hydrophobic ligand binding domain (Evans 1988; Green and Chambon 1988). The receptors modulate gene transcription by interacting with specific DNA sequences—the responsive elements—that are frequently located near the promoters of susceptible genes (Beato 1989).

It soon became clear that at least part of the complexity of RA action was represented at the molecular level by the existence of more than one RAR gene (Benbrook *et al.* 1988; Brand *et al.* 1988; Krust *et al.* 1989). We know now that three distinct RAR genes exist from which multiple receptor isoforms can be produced (Krust *et al.* 1989; Brand *et al.* 1990; Leroy *et al.* 1991; Zelent *et al.* 1991) by mechanisms that include alternative splicing as well as different promoter usage (Lehmann *et al.* 1991*a*; Leroy *et al.* 1991; Zelent *et al.* 1991). Thus, unlike the steroid hormone receptors, where single genes encode receptors specific for each hormone and where few isoforms appear to exist, RA interacts with three RAR subtypes and several subtype isoforms. In addition, RA derivatives appear to regulate the activity of a further subfamily of receptors called RXRs (Mangelsdorf *et al.* 1990; Mangelsdorf and Evans, Chapter 3). RA action appears to be distinct from that of certain steroid hormones, in particular the glucocorticoids where the receptor is located in the cell cytoplasm in the absence of hormone (Yamamoto 1985), whereas RARs appear to be located mostly in the cell nucleus. We have shown that RARs do not require ligand interaction to bind to their specific DNA sequences (Graupner *et al.* 1989; Hoffmann *et al.* 1990; Zhang *et al.* 1991). RARs—in addition to their ligand induced activities—may also, therefore, have ligand-independent regulatory roles, comparable to the thyroid hormone receptors (TR) that function as response element specific repressors (Graupner *et al.* 1989; Graupner *et al.* 1991*a*) and silencer proteins (Zhang *et al.* 1991) in the absence of hormone.

The three RAR subtypes, RAR-$\alpha$, $\beta$, and $\gamma$ show distinct properties and vary in their basal level activities (transcriptional activation in the absence of ligand) as well as in their response to ligands. RAR-$\gamma$ has the highest basal level activity and RAR-$\alpha$ the lowest (Lehmann *et al.* 1991*b*). The $\gamma$-receptor is also activated by the lowest concentrations of RA and RAR-$\alpha$ requires the highest concentration (Lehmann *et al.* 1991*b*; Graupner *et al.* 1991*b*). Thus it can be expected that the differential ligand responsiveness of the three RAR subtypes contributes to the largely varying RA concentrations needed for distinct biological responses. We have recently been able to show that synthetic retinoids can be characterized that show an even larger preference for individual receptor subtypes in that they are receptor subtype specific (Lehmann *et al.* 1991*b*; Graupner *et al.* 1991*b*). These compounds will represent valuable tools for the elucidation of receptor subtype specific biological roles. It will also be interesting to see whether physiological counterparts of such selectively acting compounds exist.

## WHICH GENES ARE REGULATED BY RETINOIC ACID RECEPTORS AND WHAT SEQUENCES ARE RECOGNIZED?

It was soon recognized that the synthesis of at least one of the receptors, RAR-$\beta$ was regulated by RA (de Thé *et al.* 1989). The cloning of the $\beta$-receptor promoter by us (Hoffmann *et al.* 1990) and others (de Thé *et al.* 1990; Sucov *et al.* 1990) revealed that the $\beta$-receptor is regulated by RA at the level of transcription and can induce its own synthesis through an RA-responsive element (RARE) in the $\beta$-promoter. The sequence recognized by RAR-$\beta$ and -$\gamma$ shows a direct repeat symmetry (Hoffmann *et al.* 1990) contrary to the palindromic thyroid hormone response element $T_3RE$ (Fig. 4.1(*a*)) that was found to also serve as a RARE (Umesono *et al.* 1988; Graupner *et al.* 1989; Munos-Canoves *et al.* 1990). In addition, gene activation of the $\beta$-RARE is restricted to RARs (Hoffmann *et al.* 1990) (Fig. 4.1(*b*)) while the $T_3RE$ can be activated by several receptors, including the oestrogen receptor (Graupner *et*

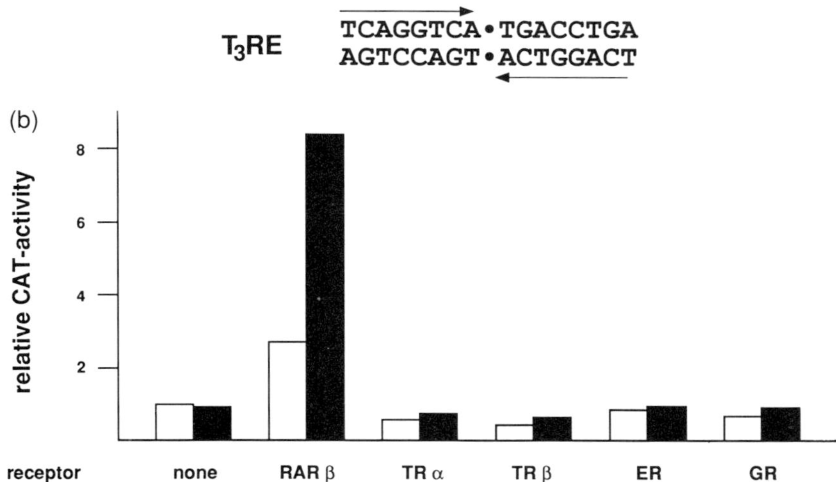


**Fig. 4.1** Characteristics of the RARE of the RAR-*β* promoter. (*a*) Comparison of the RARE sequence and symmetry of the RAR-*β* promoter with the palindromic thyroid hormone response element (T₃RE) that can also function as a RARE. (*b*) Receptor responsiveness of the *β*-RARE. CV-1 cells were cotransfected with 2 *µ*g of an RAR-*β* promoter construct containing the RARE (SmaI-BamHI-BS-CAT; position −60 to +156 described in Hoffmann *et al.* 1990) and 1 *µ*g of either one of the receptor expression vectors. Cells were grown in the presence of hormone (■): 1 *µ*M RA (RARs), 100 nM T₃ (TRs), 10 nM oestradiol (oestrogen receptor, ER), 100 nM dexamethasone (GR); or no hormone □ and 24 hours later assayed for CAT activity. A representative experiment is shown.

*al.* 1991*a*). However, RAR-mediated activation of the T₃RE is restricted in the presence of TRs (Graupner *et al.* 1989, 1991*a*; Zhang *et al.* 1991).

The analysis of the genomic organization of the RARγ gene showed that the RAR-γ1 and -γ2 isoforms are regulated and transcribed from different promoters that are located more than 15 kb apart (Lehmann *et al.* 1991*a*). Our characterization of the γ2-promoter revealed the existence of an RARE that allows an RA response of this promoter (J. M. Lehmann and M. Pfahl, unpublished results). This is consistent with the recent observation that RAR-γ expression (most likely γ2) can be rapidly induced by RA in the lung of vitamin A-starved rats (Haq *et al.* 1991). Recent data from P. Chambon's laboratory (see Leroy *et al.* Chapter 2) also indicates that an additional *β*-receptor isoform, transcribed from the *β*1 promoter, and one of the RAR-*α* isoforms are regulated by RA. Thus, isoforms from all three RAR subtypes can regulate their own synthesis and therefore belong to the class of master regulators, which is in agreement with the major role of RA in executing many important biological programmes. The ability of RARs to up-regulate synthesis of certain isoforms sets the stage for a particular sensitive mechanism where the initial response to a ligand can be amplified and/or fine-tuned, thus providing a means to interpret small differences in a ligand gradient that has been proposed to be critical in vertebrate morphogenesis for instance in limb development (Tickle *et al.* 1982; Thaller and Eichele 1990).

What can be the consequences of high RAR concentrations on gene regulation? For positive gene regulation in general one can assume that high concentrations of RAR–RA complexes allow enhanced activation of RARE-containing promoters, in particular those that contain weak responsive elements. An example for this may be the laminin B1 promoter (Vasios *et al.* 1989) that contains a weak RARE and is induced in F9 cells after RAR$\beta$ is induced (Hu and Gudas 1990). Another particularly intriguing situation may arise when RA is limited. Then the induction of an RAR promoter could raise the RAR concentration above the intracellular RA concentration. This may then lead to competition of unliganded and ligand-bound RAR complexes for their DNA binding sites and could result in a down-regulation of the RA response. Such 'peak' responses have been observed for certain homeobox genes (LaRosa and Gudas 1988; Boncinelli *et al.* Chapter 16). Mechanisms for negative gene regulation by RARs have been more difficult to unravel. Below we describe two mechanisms by which RARs repress transcription. Whereas one mode is based on the classic receptor–DNA interaction, the second mechanism of repression is independent of protein–DNA interaction. Both mechanisms of gene repression are highly sensitive to the concentrations of receptors.

**Antagonism between retinoic acid receptors**

The expression of RAR-$\beta$ and RAR-$\gamma$, in particular the isoform $\gamma1$, during mouse development occur in an almost mutually exclusive spatiotemporal pattern (Dollé *et al.* 1989, 1990; Ruberte *et al.* 1990). We therefore questioned whether RAR-$\gamma$ isoforms could also induce expression of RAR-$\beta$. Surprisingly, we observed that RAR-$\gamma1$ did not activate a reporter gene containing the $\beta$-RARE, while the $\gamma2$ isoform (that contains a different amino terminal region) was an effective activator of this response element (Fig. 4.2). In addition we could show that RAR-$\gamma1$ was a potent repressor of endogenous RAR activity in CV-1 cells and was also able to inhibit cotransfected RAR-$\beta$ and RAR-$\gamma2$ (Fig. 4.2). This repression was not due to a non-specific squelching effect, as high concentrations of cotransfected RAR-$\beta$ could overcome repression by RAR-$\gamma1$ (Husmann *et al.* 1991). Thus, the inability of RAR-$\gamma1$ to activate the RAR-$\beta$-promoter, and its repressor activity, offers at

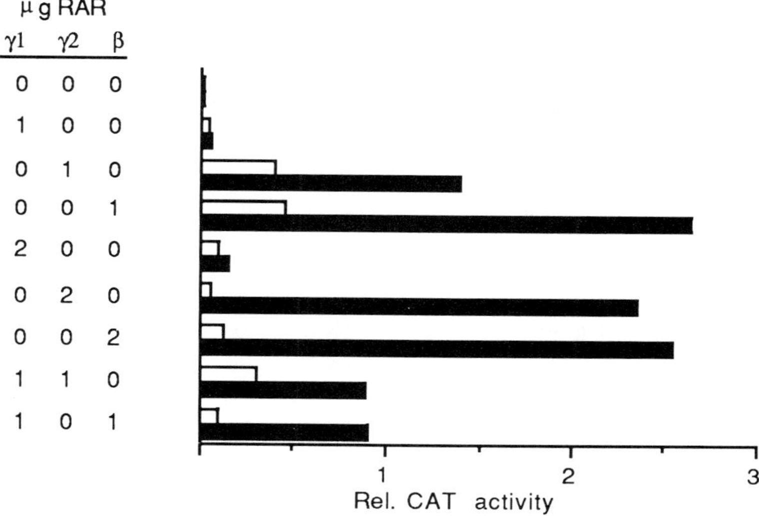



**Fig. 4.2** Transcriptional repression by RAR-$\gamma$1. CV-1 cells ($1.5 \times 10^6$ per 100 mm dish) were cotransfected by calcium phosphate precipitation with the indicated amounts of RARs plus a CAT-reporter plasmid (2 $\mu$g), driven by a promoter fragment derived from the RAR-$\beta$ gene ($-60$ to $+70$) (Hoffmann *et al.* 1990). Cultures were treated with $10^{-6}$ M RA (filled columns) or solvent (open columns) and cell lysates were obtained 48 hours after transfection. CAT activity was determined as described elsewhere (Pfahl *et al.* 1990). The mean of duplicate cultures from a typical experiment is shown. SEMs did not exceed 20 per cent.

least a partial explanation at the molecular level for the observed non-overlapping expression patterns of RAR-$\gamma$1 and RAR-$\beta$. We have shown elsewhere that RAR-$\gamma$1 and RAR-$\beta$ bind the $\beta$-RARE with similar efficiency (Hoffmann *et al.* 1990). This type of repression mechanism, in which a weak activator competes effectively with strong activators for the same response element, has been observed for a number of transcription factors and may well be the most common mechanism of transcriptional repression (Levine and Manley 1989; Karin 1990(*b*)).

## Repression of transcription by protein–protein interaction: collagenase promoter regulation by retinoic acid receptors

Retinoids are potent antiarthritic agents and their effectivity is believed to be at least in part due to their ability to down-regulate the synthesis of certain metalloproteases, including stromelysin and collagenase (Lafyatis *et al.* 1990; Nicholson *et al.* 1990). Previous studies have shown that sequences in the collagenase (Col) promoter essential for the retinoid response are identical or overlap with an AP-1 site located between $-71$ and $-61$ of the Col promoter

(Lafyatis *et al.* 1990). We have investigated the requirements of RA-dependent repression of the Col promoter in transient transfection assays.

When a Col reporter gene ($-$73Col-CAT) was transfected into HeLa cells, we observed induction of this construct by the tumour promoter 12-O-tetra-decanoyl-phorbol-13-acetate (TPA); however, repression of this induction by RA could not be observed (Fig. 4.3). This could be explained by the low

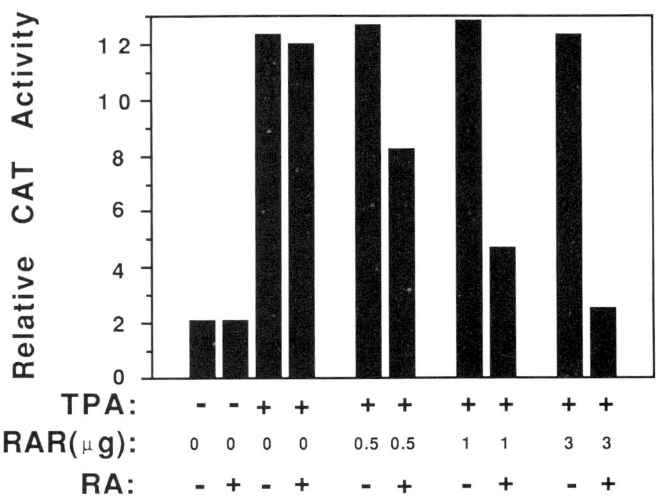


**Fig. 4.3** Repression of the collagenase promoter by RAR. The $-$73Col-CAT, a reporter construct containing the collagenase promoter sequence $-$73 to $+$60 was transfected into HeLa cells without or with the indicated amounts of RAR-$\beta$ expression vector. After transfection, the cells were grown for 24 hours in medium containing 0.5% serum followed by addition of TPAA (100 ng/ml) and/or RA ($10^{-7}$ M) as indicated.

concentrations of endogenous RAR present in HeLa cells. We therefore cotransfected an RAR-$\beta$ expression vector together with the $-$73Col-CAT reporter into these cells. In this case we observed a strong repression of $-$73Col-CAT, and the extent of the repression was directly dependent on the amount of cotransfected RAR expression vector (Fig. 4.3) and on the concentration of RA used (data not shown). In addition, repression by endogenous receptors could also be observed when the $-$73Col-CAT construct was transfected into F9 cells together with cJun and cFos expression vectors (Yang-Yen *et al.* 1991).

cJun and cFos are constituents of the transcription factor AP-1 that binds as a Jun–Jun homodimer or Jun–Fos heterodimer to the AP-1 site (also called TPA response element—TRE) (reviewed in Curran and Franza 1988; Karin 1990*a*). It has been proposed that RARs can control certain TRE-containing promoters by competing with AP-1 for overlapping response elements (Schüle *et al.* 1990*a*). We therefore investigated the possibility that RARs could bind the TRE. While cJun protein and a mixture of cJun and cFos proteins bound the TRE specifically, RAR proteins did not bind the collagenase promoter TRE (Fig. 4.4(*a*)) or a synthetic TRE (not shown). At the same time the RAR protein bound specifically to the RARE (see Fig. 4.6). When RAR was preincubated with cJun, cJun binding to the TRE was inhibited, and the degree of inhibition was dependent on the RAR concentration. This inhibition was specific, as preincubation of RAR with an anti-RAR antibody, but not with preimmune serum, prevented the inhibition of cJun by RAR (Fig. 4.4(*b*)). In further experiments we could also demonstrate that RARs could inhibit the binding of the cJun/cFos heterodimer to the TRE (Yang-Yen *et al.* 1991). These data, together with the results from the transient transfection experiments, strongly suggest that RARs can inhibit AP-1 activity by a mechanism that does not involve binding to DNA.

RETINOIC ACID RECEPTOR ACTIVITY CAN BE INHIBITED BY cJUN AND cFOS

The ability of RAR to inhibit cJun and cJun/cFos DNA binding suggested the formation of a protein complex between the two classes of proteins, inferring that an excess of cJun or cFos protein would also inhibit RAR activity. We therefore carried out a series of transfection experiments in which we could investigate the effects of cJun and cFos on gene activation by RARs. We were able to demonstrate that cJun and cFos could inhibit RAR activity in several cell types (Yang-Yen *et al.* 1991); a typical experiment is shown in Fig. 4.5. A RARE-containing reporter gene ($T_3RE_2$-CAT) was transfected into F9 cells and could be shown to be strongly activated by endogenous receptors in the presence of RA. Cotransfection of cJun and cFos expression vectors markedly reduced activation of the reporter gene.

We also investigated whether cJun or cFos protein could inhibit DNA binding of RARs. When bacterially produced cJun protein was incubated with RAR-$\beta$, a strong inhibition of RAR-$\beta$ binding to an RARE sequence was observed. cJun protein did not bind to the RARE, and preincubation of cJun with anti-Jun antibody prevented inhibition of RAR DNA binding by cJun (Fig. 4.6). Thus, the AP-1 proteins cJun and cFos inhibit gene induction by the RAR–RA complex and cJun protein can directly interfere with RAR DNA binding. Our data therefore suggest a novel mechanism by which gene transcription is inhibited by two distinct classes of transcription factors— RARs can repress genes activated by AP-1 and cJun and cFos can inhibit genes that are activated by RARs. In neither case is DNA binding of the dominant negative transcription factor required. This strongly suggests a mechanism in which direct protein–protein interaction between these two

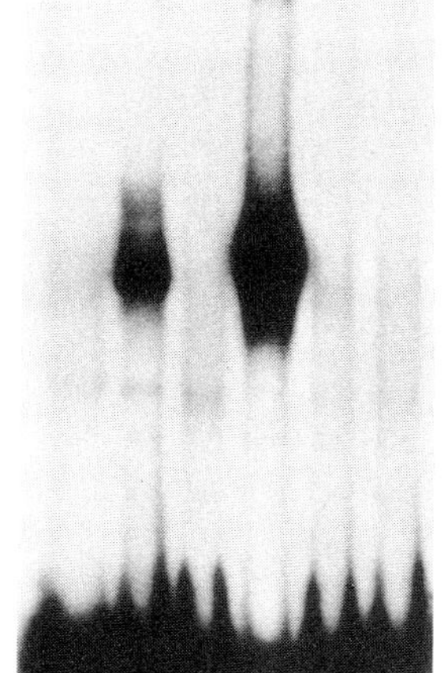


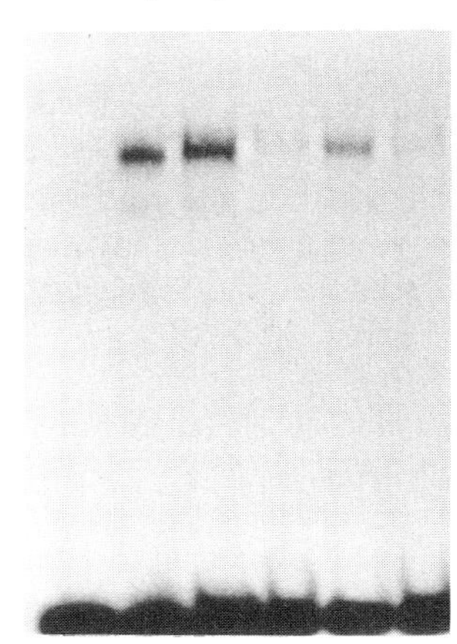


**Fig. 4.4** Inhibition of cJun DNA binding by RAR. (*a*) Analysis of DNA binding of cJun, cFos, and RAR to collagenase promoter. Equal amounts of *in vitro* synthesized cJun, cFos, and RAR proteins were incubated with [$^{32}$P]-labelled DNA fragment, which contains collagenase promoter sequences from $-73$ to $+60$. Protein–DNA complex formation was analysed by gel retardation assay. The first lane represents a control reaction containing an equal amount of unprogrammed reticulocyte lysate. The binding of RAR was analysed in the presence ($+$) or absence ($-$) of $10^{-7}$ M RA. cDNAs for RAR-$\beta$ and cJun were cloned into pBluescript (Stratagene). cFos cDNA was cloned into pGem4Z (Promega). These cDNAs were transcribed using T7, T3, or SP6 RNA polymerase, and translated in the rabbit reticulocyte lysate (Promega) as described (Pfahl *et al.* 1990). The probe for the gel retardation assay was derived from collagenase promoter ($-73$ to $+60$) and labelled with radioactivity by filling-in reaction. The gel retardation assay was carried out essentially as described (Pfahl *et al.* 1990). (*b*) Inhibition of cJun DNA binding by RAR. *In vitro* synthesized cJun protein was preincubated with 1 (1X) or 5 (5X) molar equivalents of *in vitro* synthesized RAR at 37°C for 15 minutes. RAR protein was also incubated with specific anti-RAR antibody ($\alpha$-R) or preimmune serum (NI) at room temperature for 45 minutes before incubating with cJun protein. Unprogrammed reticulocyte lysate was used to maintain equal protein concentrations in each reaction. Following this preincubation, the reaction mixtures were incubated with a [$^{32}$P]-labelled synthetic AP-1 binding site dimer probe and analyzed by gel retardation. The first lane represents a control reaction containing an equal amount of unprogrammed reticulocyte lysate.

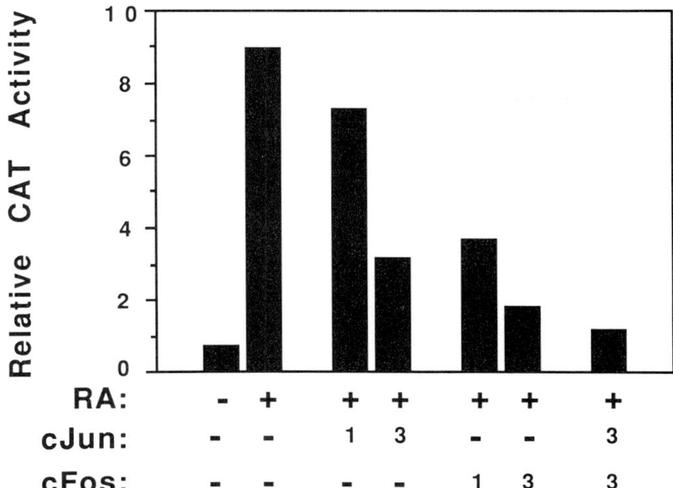


**Fig. 4.5** Inhibition of RAR activity by cJun and cFos. F9 cells were transfected with a RA responsive reporter $T_3RE_2$-CAT (Graupner *et al.* 1989) in the absence or presence of the indicated amounts of cJun and/or cFos expression vectors (in $\mu$g). After transfection, the cells were incubated in the absence ($-$) or presence ($+$) of $10^{-7}$ M RA for 24 hours before harvesting and determination of CAT activity

classes of regulatory proteins, the RARs and cJun/cFos allows mutual inhibition of activities.

cJun and cFos (AP-1) are activated by many growth factors, oncogenes and tumour promoters through the protein kinase C-mediated signal transduction pathway (reviewed in Curran and Franza 1988; Karin 1990*a*). The ability of RARs to interfere with this pathway now for the first time allows an explanation at the molecular level of the potent antineoplastic effects observed for many retinoids, and the test system described here offers a direct assay for the analysis of anti-oncogene activities of retinoids. The interaction between nuclear receptors and AP-1 is not restricted to RARs and has in fact first been shown for the glucocorticoid receptor (Diamond *et al.* 1990; Jonat *et al.* 1990; Schüle *et al.* 1990*b*; Yang-Yen *et al.* 1990). Data from our laboratory indicate that thyroid hormone receptors can substitute for RARs in repressing cJun, cFos and tumour promoter activity. In addition TRs could also be shown to antagonize the activated H-*Ras* oncogene (known to activate cFos). The mechanism described here thus offers an explanation for the well known potent anticancer effects of steroid/thyroid hormones and retinoids.

CONCLUSIONS

The RARs represent a large subfamily of nuclear receptors that are encoded by at least six genes (including three RXR genes) from which multiple receptor

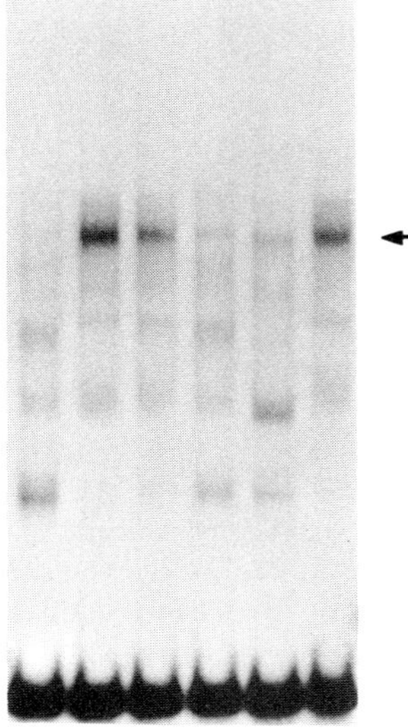


**Fig. 4.6** Inhibition of RAR DNA binding by Jun. *In vitro* synthesized RAR protein was preincubated with 1 (1X), 2 (2X), or 5 (5X) molar equivalent of bacterially produced cJun protein at 37°C for 15 minutes prior to incubation with [$^{32}$P]-labelled RARE derived from RAR-$\beta$ promoter (Hoffmann *et al.* 1990). The reaction mixtures were then analysed by gel retardation. To examine the specificity of inhibition of RAR binding by cJun protein, the cJun protein was also preincubated with anti-cJun antibody ($\alpha$-Jun) prior to mixing with labelled RARE. The first lane represents a control reaction with unprogrammed reticulocyte lysate. The arrow indicates the specific RAR–DNA complex.

isoforms can be generated by alternative splicing and alternative promoter usage. A number of molecular mechanisms have now been defined that allow the RARs to mediate RA action, including: (i) induction of transcription by binding of the RAR–RA complex to an RARE in the promoter of a responsive gene; (ii) repression of transcription by RARs through a particular isoform, as demonstrated for RAR$\gamma$-1, that may contribute to spatiotemporal restricted RAR expression patterns; (iii) RA-induced RAR synthesis that may allow magnification of small differences in RA signals and/or a largely amplified regulatory response; (iv) RAR:AP-1 interaction, a mechanism that allows interference of two distinct and usually opposing signal transduction pathways.

Whereas the first three mechanisms are all based on the classic receptor–DNA interaction model, the last mechanism is based on protein–protein interaction. The combination of mechanisms (iii) and (iv), where the RA response mediators (the RARs) are induced or amplified, offers an explanation at the molecular level of the often antagonistic responses observed at low versus high retinoid concentrations. At low RA concentrations, low to intermediate RAR levels will be induced that function mostly as RARE activators, whereas intermediate to high concentrations may induce RAR levels that are sufficient to inhibit AP-1 activity, leading to a more drastic change in the cellular signalling pathways. The molecular pathways of RA action discussed here can constitute a framework for the complex biological roles of RA. Many variations of these basic mechanisms will exist and await elucidation.

ACKNOWLEDGEMENTS

The work reported here was supported by NIH grants DK 35 085 and CA 50 676.

REFERENCES

Beato, M. (1989). Gene regulation by steroid hormones. *Cell*, **56**, 335–44.
Benbrook, D., Lernhardt, E., and Pfahl, M. (1988). A new retinoic acid receptor identified from a hepatocellular carcinoma. *Nature*, **333**, 669–72.
Brand, N. J., Petkovich, M., and Chambon, P. (1988). Characterization of a functional promoter for the human retinoic acid receptor-alpha (hRAR-α). *Nucleic Acids Research*, **18**, 6799–806.
Brand, N. J., Petkovich, M., and Chambon, P. (1990). Characterization of a functional promoter for the human retinoic acid receptor-alpha (hRAR-α). *Nucleic Acids Research*, **18**, 6799–806.
Curran, T. and Franza, B. R., Jr (1988). Fos and Jun: the AP-1 connection. *Cell*, **55**, 395–7.
de Thé, H., Marchio, A., Tiollais, P., and Dejean, A. (1989). Differential expression and ligand regulation of the retinoic acid α and β genes. *EMBO Journal*, **8**, 429–33.
de Thé, H., Vivanco-Ruiz, M. d. M., Tiollais, P., Stunnenberg, H., and Dejean, A. (1990). Identification of a retinoic acid response element in the retinoic acid receptor β gene. *Nature*, **343**, 177–80.
Dollé, P., Ruberte, E., Kastner, P., Petkovich, M., Stoner, C. M. Gudas, L. J., and Chambon, P. (1989). Differential expression of genes encoding α, β, and γ retinoic acid receptors and CRABP in the developing limbs of the mouse. *Nature*, **342**, 702–4.
Dollé, P., Ruberte, E., Leroy, P., Morriss-Kay, G., and Chambon, P. (1990). Retinoic acid receptors and cellular retinoid binding proteins I. A systematic study of their differential pattern of transcription during mouse organogenesis. *Development*, **110**, 1133–51.
Diamond, M. I., Miner, J. N., Yoshinaga, S. K., and Yamamoto, K. R. (1990).

Transcription factor interactions: selectors of positive or negative regulation from a single DNA element. *Science*, **249**, 1266–72.

Evans, R. M. (1988). The steroid and thyroid hormone receptor superfamily. *Science*, **240**, 889–95.

Giguère, V., Ong, E. S., Seigi, P., and Evans, R. M. (1987). Identification of a receptor for the morphogen retinoic acid. *Nature*, **330**, 624–9.

Graupner, G., Wills, K. N., Tzukerman, M., Zhang, X-K., and Pfahl, M. (1989). Dual regulatory role for thyroid-hormone receptors allows control of retinoic-acid receptor activity. *Nature*, **340**, 653–6.

Graupner, G., Zhang, X-K., Tzukerman, M., Wills, K. N., Hermann, T., and Pfahl, M. (1991a). Thyroid hormone receptors repress estrogen receptor activation of a TRE. *Molecular Endocrinology*, **5**, 653–6.

Graupner, G., Malle, G., Maignan, G., Lang, G., Pruniéras, M., and Pfahl, M. (1991b). 6'-substituted naphthalene-2-carboxylic acid analogs, a new class of retinoic acid receptor subtype-specific ligands: a selection by transcriptional activation assay. *Biochemical and Biophysical Research Communications*, **179**, 1554–61.

Green, S. and Chambon, P. (1988). Nuclear receptors enhance our understanding of transcription regulation. *Trends in Genetics*, **4**, 309–14.

Haq, R., Pfahl, M., and Chytil, F. (1991). Retinoic acid affects the expression of nuclear retinoic acid receptors in tissues of retinol deficient rats. *Proceedings of the National Academy of Sciences of the USA*, **88**, 8272–6.

Hoffmann, B., Lehmann, J. M., Zhang, X-K., Hermann, T., Husmann, M., Graupner, G., and Pfahl, M. (1990). A retinoic acid receptor-specific element controls the retinoic acid receptor-$\beta$ promoter. *Molecular Endocrinology*, **4**, 1727–36.

Hu, L. and Gudas, L. J. (1990). Cyclic AMP analogs and retinoic acid influence the expression of retinoic acid receptor $\alpha$, $\beta$, and $\gamma$ mRNAs in F9 teratocarcinoma cells. *Molecular and Cellular Biology*, **10**, 391–6.

Hussmann, M., Lehmann, J., Hoffman, B., Hermann, T., Tzukerman, M., and Pfahl, M. (1991). Antagonism between retinoic acid receptors. *Molecular and Cellular Biology*, **11**, 4097–4103.

Jonat, C., Rahmsdorf, H. J., Park, K-K., Cato, A. C. B., Gebel, S., Ponta, H., and Herrlich, P. (1990). Antitumor promotion and antiinflammation; down-modulation of AP-1 (Fos/Jun) activity by glucocorticoid hormone. *Cell*, **62**, 1189–204.

Karin, M. (1990a). The AP-1 complex and its role in transcriptional control by protein kinase C. *Molecular Aspects of Cellular Regulation*, **6**, 143–61.

Karin, M. (1990b). Too many transcription factors: positive and negative interactions. *The New Biologist*, **2**, 126–31.

Krust, A., Kastner, P. H., Petkovich, M., Zelent, A., and Chambon, P. (1989). A third human retinoic acid receptor, hRAR-$\gamma$. *Proceedings of the National Academy of Sciences USA*, **86**, 5310–14.

Lafyatis, R., Kim, S-J., Angel, P., Roberts, A. B., Sporn, M. B., Karin, M., and Wilder, R. L. (1990). Interleukin-1 stimulates and all-trans-retinoic acid inhibits collagenase gene expression through its 5' activator protein-1 binding site. *Molecular Endocrinology*, **4**, 973–80.

LaRosa, G. J. and Gudas, L. J. (1988). An early effect of retinoic acid: Cloning of an mRNA (Era-1) exhibiting rapid and protein synthesis-independent induction during teratocarcinoma stem cell differentiation. *Proceedings of the National Academy of Sciences USA*, **85**, 329–33.

Lehmann, J. M., Hoffmann, B., and Pfahl, M. (1991*a*). Genomic organization of the retinoic acid receptor gamma gene. *Nucleic Acids Research*, **19**, 573–8.

Lehmann, J. M., Dawson, M. I., Hobbs, P. D., Husmann, M., and Pfahl, M. (1991*b*). Identification of retinoids with nuclear receptor subtype selective activities. *Cancer Research*, **52**, 4804–4809.

Leroy, P., Krust, A., Zelent, A., Mendelsohn, C., Garnier, J-M., Kastner, P., Dierich, A., and Chambon, P. (1991). Multiple isoforms of the mouse retinoic acid receptor α are generated by alternative splicing and differential induction by retinoic acid. *EMBO Journal*, **10**, 59–69.

Levine, M. and Manley, J. L. (1989). Transcriptional repression of eukaryotic promoters. *Cell*, **59**, 405–8.

Lippmann, S. M., Kessler, J. F., and Meyskens, F. L. (1987). Retinoids as preventive and therapeutic anticancer agents. *Cancer Treatment Reports*, **71**, 391–405, 493–515.

Lotan, R. (1980). Effects of vitamin A and its analogs (retinoids) on normal and neoplastic cells. *Biochemica Biophysica Acta*, **605**, 33–91.

Mangelsdorf, D. J., Ong, E. S., Dyck, J. A., and Evans, R. M. (1990). Nuclear receptor that identifies a novel retinoic acid response pathway. *Nature*, **345**, 224–9.

Munoz-Canoves, P., Vik, D. P., and Tack, B. F. (1990). Mapping of a retinoic acid-responsive element in the promoter region of the complement factor H gene. *Journal of Biological Chemistry*, **265**, 20065–8.

Nicholson, R. C., Mader, S., Nagpal, S., Leid, M., Rochette-Egly, C., and Chambon, P. (1990). Negative regulation of the rat stromelysin gene promoter by retinoic acid is mediated by an AP-1 binding site. *EMBO Journal*, **9**, 4443–54.

Petkovich, M., Brand, N. J., Krust, A., and Chambon, P. (1987). A human retinoic acid receptor which belongs to the family of nuclear receptors. *Nature*, **330**, 444–540.

Pfahl, M., Tzukerman, M., Zhang, X-K., Lehmann, J. M., Hermann, T., Wills, K. N., and Graupner, G. (1990). Rapid procedures for nuclear retinoic acid receptor clonging and their analysis. *Methods in Enzymology*, **153**, 256–70.

Ruberte, E., Dollé, P., Krust, A., Zelent, A., Morriss-Kay, G., and Chambon, P. (1990). Specific spatial and temporal distribution of retinoic acid receptor gamma transcripts during mouse embryogenesis. *Development*, **108**, 213–22.

Schüle, R., Umesono, K., Mangelsdorf, D. J., Bolado, J., Pike, J. W., and Evans, R. M. (1990*a*). Jun-Fos and receptors for vitamins A and D recognize a common response element in the human osteocalcin gene. *Cell*, **61**, 497–504.

Schüle, R., Rangarajan, P., Kliewer, S., Ransone, L. J., Bolado, J., Yang, N., Verma, I. M., and Evans, R. M. (1990*b*). Functional antagonism between oncoprotein c-Jun and the glucocorticoid receptor. *Cell*, **62**, 1217–26.

Sporn, M. B., Roberts, A. B., and Goodman, D. S. (1984). *The retinoids* (eds M. B. Sporn, A. B. Roberts, and D. S. Goodman). Academic Press, Florida.

Sucov, H. M., Murakami, K. K., and Evans, R. M. (1990). Characterization of an autoregulated response element in the mouse retinoic acid receptor type β gene. *Proceedings of the National Academy of Sciences USA*, **87**, 5392–6.

Summerbell, D. (1983). The effect of local application of retinoic acid to the anterior margin of the developing chick limb. *Journal of Embryology and Experimental Morphology*, **78**, 269–89.

64   *M. Pfahl* et al.

Thaller, C. and Eichele, G. (1990) Retinoids and vertebrate limb pattern formation. *Nature, ( Lond.)*, **345**, 815–19.

Tickle, C., Alberts, B., Wolpert, L., and Lee, L. (1982). Local application of retinoic acid to the limb bud mimics the action of the polarizing region. *Nature, ( Lond.)*, **296**, 564–6.

Umesono, K., Giguère, V., Glass, C. K., Rosenfeld, M. G., and Evans, R. M. (1988). Retinoic acid and thyroid hormone induce gene expression through a common responsive element. *Nature*, **336**, 262–5.

Vasios, G. W., Gold, J. D., Martin, P., Chambon, P., and Gudas, L. J. (1989). A retinoic acid-responsive element is present in the 5′ flanking region of the laminin B1 gene. *Proceedings of the National Academy of Sciences USA*, **86**, 9099–103.

Yamamoto, K. R. (1985). Steroid receptor regulated transcription of specific genes and gene networks. *Annual Reviews in Genetics*, **19**, 209–52.

Yang-Yen, H-F., Chambard, J-C., Sun, Y-L., Smeal, T., Schmidt, T. J., Drouin, J., and Karin, M. (1990). Transcriptional interference between c-Jun and the glucocorticoid receptor: Mutual inhibition of DNA binding due to direct protein-protein interaction. *Cell*, **62**, 1205–15.

Yang-Yen, H-F., Zhang, X-K., Graupner, G., Tzukerman, M., Sakamoto, B., Karin, M., and Pfahl, M. (1991). Antagonism between retinoic acid receptors and AP-1: Implications for tumor promotion and inflammation. *The New Biologist* **3**, 1206–20.

Zelent, A., Mendelsohn, C., Kastner, P., Krust, A., Garnier, J-M., Ruffenach, F., Leroy, P., and Chambon, P. (1991). Differentially expressed isoforms of the mouse retinoic acid receptor *β* are generated by usage of two promoters and alternative splicing. *EMBO Journal*, **10**, 71–81.

Zhang, X-K., Wills, K., Graupner, G., Tzukerman, M., Hermann, T., and Pfahl, M. (1991). Ligand-binding domain of thyroid hormone receptors modulates DNA binding and determines their bifunctional roles. *The New Biologist*, **3**, 169–81.

5

# Differential binding and activation of synthetic retinoids to retinoic acid receptors

Christian Apfel, Marco Crettaz, and Peter LeMotte

INTRODUCTION

Retinoids, and in particular all-*trans* retinoic acid (RA), exert a wide range of biological effects. RA influences the growth and differentiation of a wide variety of normal and transformed cells. The usefulness of retinoids in dermatology in different skin disorders, such as acne and psoriasis, is well documented. Moreover, retinoids may have a potential in immunomodulation and as preventive and therapeutic anticancer agents (Bollag 1979; Bollag 1983; Lippman *et al.*, 1987).

The mechanism by which RA induces these diverse biological effects is unclear. Initially, it was suggested that they were mediated through a highly specific cellular RA-binding protein (Jetten and Jetten 1979). A crucial advance towards understanding retinoid action has been accomplished with the isolation of six specific nuclear receptors for retinoids RAR-$\alpha$ (Petkovich *et al.* 1987; Giguère *et al.* 1987), RAR-$\beta$ (Benbrook *et al.* 1988; Brand *et al.* 1988), RAR-$\gamma$ (Kastner *et al.* 1989; Krust *et al.* 1989; Giguère *et al.* 1990), RXR-$\alpha$, RXR-$\beta$, and RXR-$\gamma$ (Mangelsdorf *et al.* 1990 and Chapter 3). These receptors belong to the superfamily of steroid/thyroid hormone receptors. This family of nuclear receptors regulate gene transcription in a ligand-dependent fashion through binding to specific DNA sequences (response elements), generally located upstream of the promoter of a target gene (Umesono *et al.* 1988; Graupner *et al.* 1989; Vasios *et al.* 1989; de Thé *et al.* 1990). This results in an increased or decreased synthesis of specific proteins.

So far, it is not known whether all the different retinoid receptors bind to the same response elements or whether response elements exist that preferentially or even exclusively bind one receptor type. A second level of specificity of a response to retinoids also seems possible due to the different tissue distribution of the receptor types. The response in an organism is therefore determined by at least three parameters: (i) the tissue distribution of the receptors; (ii) the binding characteristics of the retinoid to the different receptors; and (iii) the ability of the retinoid to activate the receptors.

The first step in the action of a retinoid, namely the binding to the receptor(s), is therefore of major importance for its effects. At present we are trying to elucidate the structural requirements of retinoids for receptor binding and activation, and thus to deduce rules that will be useful in synthesis of new retinoids. For this purpose we have established a binding assay for all three retinoic acid receptors (RAR-$\alpha$, RAR-$\beta$, and RAR-$\gamma$) and a receptor transcriptional activation assay for RAR-$\alpha$ and RAR-$\beta$.

One goal has been to find selective ligands for the different receptor subtypes. We have identified one retinoid analogue that is capable of binding preferentially to RAR-$\alpha$, and one that binds preferentially to RAR-$\beta$. In general there was a good correlation betwen RAR binding and activation for both RAR-$\alpha$ and RAR-$\beta$. However, there are two exceptions. Some retinoids bind but fail to activate the receptors; nevertheless they show no antagonistic activity. Others can activate the receptors in the absence of binding activity in *in vitro* assay. Acitretin (Ro 10-1670) is such a 'proretinoid'–inactive in binding as such, but able to activate receptor function. It is hypothesized that an active metabolite of acitretin is produced by the cell which is responsible for receptor activation. The methods used in our studies are summarized below.

**Retinoid binding assay**

The binding assays were performed as previously described in detail (Crettaz *et al.* 1990). Briefly, crude extracts from COS cells (transiently transfected with the receptor cDNA) or *Escherichia coli* (containing the receptor cDNA in an expression plasmid) were incubated at 37°C with [$^3$H]-RA and various concentrations of unlabelled retinoids. The bound and free ligands were separated with charcoal dextran and the supernatants were subjected to liquid scintillation counting.

**Cell lines**

L-mouse fibroblasts were contransfected with chimeric RAR-oestrogen receptor (ER) cDNAs (RAR-$\alpha$-ER. CAS or RAR-$\beta$-ER. CAS, that contain the DNA binding region of the oestrogen receptor and therefore binds to oestrogen response elements, provided by P. Chambon (Petkovich *et al.* 1987; Brand *et al.* 1988), the *vit-tk*-CAT gene (the chloramphenicol acetyl transferase (CAT) gene under the control of the vitellogenin oestrogen response element fused to the herpes simplex thymidine kinase promoter) and the *neo* gene. Stable transfectants were selected with the antibiotic G418 (geneticin). Clones displaying a good response to RA were selected for further studies.

**Chloramphenicol acetyltransferase assay**

Experiments were done in 12- (or 24-) well dishes. At confluency, cultures were preincubated for 18–24 hours in serum-free $\alpha$-MEM medium supplemented with 0.5% Clex. The cells were then incubated with retinoids in α-MEM medium with 1 per cent bovine serum albumen (BSA) for 18–24 hours. At the end of the incubation, the cells were extracted during 10 minutes at room temperature in 0.1–0.2 ml of 0.1 M Tris–HCl, pH 7.8, 0.5 per cent Triton X-100 and 1.5 mM chloramphenicol. Cell extracts were assayed for CAT activity by incubation with 0.2 mM [³H]-acetyl-CoA (4–5 mCi/mmol), for 30 minutes at 37°C. The reaction was stopped by extraction with ethyl acetate and the upper phase counted for radioactivity.

**Construction of the expression plasmids for the full length and the ligand binding domain of retinoic acid receptor**

The mouse RAR-γ cDNA (from mouse F9 cells) was cloned with PCR and altered by site-directed mutagenesis to encode the human RAR-γ protein. The cDNA (amino acid 1 (Met) to 454 (stop)) was inserted in the *E. coli* expression vector pDS56/RBSII (lac-system, (Stüber *et al.* 1990)). For the expression of the ligand binding domain (DEF) the cDNA (amino acid 158 (Glu) to 454 (stop)) was cloned in a T7 *E. coli* expression vector (Studier and Moffat 1986)).

For details of cell culture, transfection, construction of the expression plasmids, production and purification of the ligand binding domains from RAR-α and RAR-β see Crettaz (1990).

RETINOIC ACID BINDING TO LIGAND-BINDING DOMAINS OF RAR-α, RAR-β, AND RAR-γ

We tested nuclear extracts obtained from COS cells transfected with RAR-αO (human gene) and ER-RAR-β. CAS (chimeric human gene) for specific binding of RA, but the amount of receptors obtained (determined by Scatchard analysis) was relatively low for retinoid screening purposes. Initial attempts to produce full length receptors RAR-α and RAR-β in *E. coli* gave only very low expression levels (Crettaz *et al.* 1990). Several studies on nuclear steroid receptors have demonstrated that ligand binding requires only the C-terminal portion of the protein, the ligand binding domain. We expressed, therefore, the putative ligand binding domains of RAR-α, RAR-β, and RAR-γ (DEF-domain, 262, 255, and 296 amino acids, respectively) in *E. coli*. We observed no difference in RA binding between purified and crude receptor preparations. Extracts containing the ligand-binding domains of either RAR-α, RAR-β, or RAR-γ were fully active in binding labelled RA; competition for this binding was provided by the addition of increasing concentrations of unlabelled ligand. The equilibrium binding constants ($k_d$) were 6.2 nM for RAR-α-DEF domain, 8.0 nM for RAR-β-DEF domain (details in Crettaz *et al.* 1990), and 8.4 nM for RAR-γ-DEF domain. This indicates that all three receptors—RAR-α, RAR-β, and RAR-γ—bound RA with the same affinity.

Comparing the binding constants ($k_d$) for the ligand-binding domain and for the full length receptor, we found nearly identical values for RA (data not shown). This clearly demonstrates that the ligand-binding domains of the receptors have retained the full binding properties of the native receptors, and therefore are appropriate receptor preparations for screening retinoids.

Binding to the ligand-binding domain was highly specific. Retinol and retinal did not compete with RA. In addition, other ligands for nuclear receptors, such as dexamethasone, oestradiol, and progesterone did not significantly bind to RAR-$\alpha$ or RAR-$\beta$ (Crettaz *et al.* 1990).

BINDING SPECIFICITIES OF RAR-$\alpha$, RAR-$\beta$, AND RAR-$\gamma$

For evaluation of the structural requirements of the ligand to bind to these two receptors, we measured the binding of several retinoid analogues, which differ from RA by a single molecular modification. The receptor binding activity ($IC_{50}$) is given in Table 5.1. An incorporation of the side-chain double bonds into aromatic rings (Ro 40-6976 and Ro 13-7410) results in a binding affinity comparable to RA. A saturation of side-chain double bonds at positions 7 and 8 (Ro 11-0874) has little effect, while the saturation at positions 9 and 10 (Ro 11-1813) decreases binding to both receptors. Elongation of the side chain results in a total loss of binding activity (data not shown). 13-*cis*-RA shows less binding to both receptors than RA. Any modification in the cyclohexenyl ring results in retinoid analogues with less affinity than RA (4-oxo-RA retains some activity, 4-hydroxy-RA and Ro 08-8717 were poorly active). A change in the position of the two methyl groups on the side chain (Ro 10–0191) produces a decrease in binding as well.

The reduction in binding is generally parallel for all three receptors, revealing a common physicochemical character of their binding sites. There are, however, two exceptions (Table 5.1). The arotinoid Ro 40-6055 (also known as Am 580 (Jetten *et al.* 1987) bound 22-fold better to $\alpha$ than to $\beta$ and 68-fold better to RAR-$\alpha$ than to RAR-$\gamma$. In contrast, Ro 19-0645 preferentially bound to RAR-$\beta$ (18-fold better to $\beta$ than to $\alpha$ and 8-fold better to $\beta$ than to $\gamma$).

It is important to note that both retinoids also preferentially activate the receptor to which they bind better (data not shown). This indicates that the three receptor binding sites are not completely identical. The nature of the difference between receptor binding sites is still unclear. However, it may be possible to develop even more specific ligands for RAR-$\alpha$ and RAR-$\beta$, based on Ro 40-6055 and Ro 19-0645 as lead structures and also to find a preferential RAR-$\gamma$ binder.

In addition to the receptor binding data, the *in vivo* effects of the different retinoids, as characterized by hypervitaminosis A, are given in Table 5.1. Hypervitaminosis A represents overall biological activity of a retinoid in a broad sense. It includes a variety of toxic side-effects, such as weight and hair

**Table 5.1** Binding to RAR-$\alpha$, RAR-$\beta$, and RAR-$\gamma$ and biological activity of retinoids

| Name or Roche No. | Chemical structure | Receptor binding IC$_{50}$ (nM) | | | Relative hypervitaminosis A effect* |
|---|---|---|---|---|---|
| | | RAR-$\alpha$ | RAR-$\beta$ | RAR-$\gamma$ | |
| Retinoic acid (RA) | | 14 | 14 | 14 | 1 |
| Ro 40-6976 | | 10 | 5 | 3 | 8000 |
| Ro 13-7410 arotinoid, TTNPB | | 30 | 20 | 18 | 800 |
| Ro 11-0874 | | 45 | 50 | 120 | 0.4 |
| Ro 11-1813 | | 1800 | 3000 | 5200 | 0.2 |
| Ro 04-3780 13-*cis*-RA | | 95 | 75 | 90 | 0.2 |
| Ro 12-4824 4-oxo-RA | | 260 | 410 | 410 | 0.4 |
| Ro 10-2655 | | 1000 | 1600 | 2400 | 0.8 |
| Ro 08-8717 | | 2100 | 2900 | 7000 | < 0.2 |
| Ro 10-0191 | | 2700 | 2000 | 3100 | 0.2 |
| Ro 19-0645 TTNN | | 460 | 26 | 190 | ND |
| Ro 40-6055 Am 580 | | 39 | 870 | 2650 | 0.3 |

*Relative hypervitaminosis A effect is the ratio of the IC$_{50}$ for retinoic acid to that of the retinoid analogue. ND, not determined.

loss, skin scaling, and bone fracture (Teelmann 1989, 1990). The correlation between receptor binding and *in vivo* effects is not complete. Some retinoids have no receptor binding but they are biologically active (like Ro 10-2655, see proretinoids below). Other retinoids are much more toxic than RA (around 1000-fold), but their binding affinities are approximately the same (Ro 40-6976, Ro 13-7410).

### EFFECT OF POLAR END-GROUP MODIFICATIONS ON RETINOID BINDING

Various analogues of the arotinoid Ro 13-7410 have been studied (Table 5.2). The free terminal carboxylic function was absolutely necessary for high receptor binding. Replacement of this group with a sulphinyl or phosphonyl group produced a tremendous drop in binding. Any other substitute on the arotinoid resulted in a total loss of the binding ability. Moving the carboxylic end-group from the *para* to the *meta* or *ortho* position also caused a total loss of activity. The data shows that the nature and the position of the polar end-group is very important for binding.

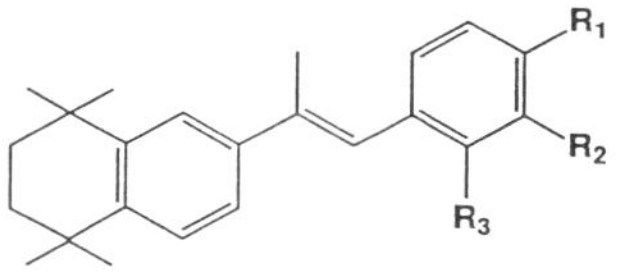

**Table 5.2** Effect of polar end-group modifications on retinoid binding to RAR-$\alpha$, RAR-$\beta$, and RAR-$\gamma$

| $R_1$ | $R_2$ | $R_3$ | Receptor binding ($IC_{50}$ in nM) | | |
|---|---|---|---|---|---|
| | | | RAR-$\alpha$ | RAR-$\beta$ | RAR-$\gamma$ |
| COOH | | | 49 | 28 | 18 |
| $SO_2^-Na^+$ | H | H | 300 | 110 | 120 |
| $PO_3H_2$ | | | 8000 | 2000 | 4200 |
| H, OH<br>F, I, Br, Cl<br>$CH_3$, $CF_3$<br>$CH(CH_3)_2$<br>$C(CH_3)_3$<br>$NO_2$, $NH_2$<br>$SO_2CH_3$<br>$SO_2C_2H_5$ | H | H | $> 10^4$ | $> 10^4$ | $> 10^4$ |
| H | COOH<br>$PO_3H_2$ | H | $> 10^4$<br>$> 10^4$ | 6300<br>$> 10^4$ | $> 10^4$<br>$> 10^4$ |
| H | H | COOH | $> 10^4$ | $> 10^4$ | $> 10^4$ |

### COMPARISON BETWEEN RECEPTOR BINDING AND RECEPTOR ACTIVATION

Retinoid receptor activation was measured as the potency of the receptor–ligand complex in stimulating the expression of the CAT reporter gene in stably transfected mouse L-cells (see p. 66). In general there was a good correlation between RAR binding and activation for both RAR-$\alpha$ and RAR-$\beta$ (data not shown). However, there are two surprising exceptions. Some retinoids bind, but fail to activate the receptors (Table 5.3), whereas others can activate the receptors in the absence of in vitro binding activity (see proretinoids below).

**Table 5.3** Retinoids that bind, but fail to activate, the RAR-$\alpha$ and RAR-$\beta$

| Product | Binding | | Activation | |
| --- | --- | --- | --- | --- |
| | RAR-$\alpha$ $IC_{50}$ (nM) | RAR-$\beta$ $IC_{50}$ (nM) | RAR-$\alpha$ | RAR-$\beta$ |
| Ro 08-8717 | 2000 | 2900 | Inactive* | Inactive |
| Ro 40-1349 | 320 | 280 | Inactive | Inactive |

*Inactive, at $10^{-6}$ M, $<20\%$ of maximal effect (measured at $10^{-6}$ M retinoic acid).

## RETINOIDS THAT BIND, BUT FAIL TO ACTIVATE, THE RECEPTOR

Table 5.3 shows two compounds that bind to the receptor but cannot activate the receptor in the gene activation assay; these compounds were therefore tested for a potential antagonistic activity. However, when incubated with RA for 24 hours, both failed to antagonize the RA effect. Surprisingly, Ro 40-1349 even intensified the maximal RA effect (data not shown). Therefore we cannot explain the lack of activation of these compounds. Possible explanations are that the penetration of these retinoids into the cell is impaired and/or that they are rapidly degraded inside the cell. For Ro 40–1349, one could speculate that this retinoid could inhibit the metabolism of RA, leading to increased cellular concentrations of RA and thus to augmented effects.

## PRORETINOIDS

To our surprise, acitretin binds very poorly to both RAR-$\alpha$ and RAR-$\beta$ ($IC_{50} > 10^{-6}$ M), whereas it is almost as active as RA in activating the receptor ($ED_{50} = 6 \times 10^{-8}$ M) (Table 5.4). Acitretin has a characteristic methoxy group in the *para* position of the phenyl group substituent. We have therefore tested a number of *para* methoxy analogues and the results are summarized in Table 5.4. None of them bind significantly to either RAR-$\alpha$ or RAR-$\beta$ but all

**Table 5.4** Binding and activation of acitretin and various 'acitretin analogues'

| Product | Binding* | | Activation | |
|---|---|---|---|---|
| | RAR-$\alpha$ | RAR-$\beta$ | RAR-$\alpha$ ED$_{50}$ (nM) | RAR-$\beta$ ED$_{50}$ (nM) |
| Ro 10-1670 *Acitretin* | Inactive | Inactive | 60 | 56 |
| Ro 10-9359 *Etretinate* | Inactive | Inactive | 110 | 380 |
| Ro 11-1906 | Inactive | Inactive | 180 | ND |
| Ro 13-7652 | S1 act | Inactive | 100 | 200 |
| Ro 19-8399 | Inactive | Inactive | 130 | 26 |
| Ro 42-1997 | Inactive | Inactive | 290 | 180 |

*S1 act, IC$_{50}$ > 2000 nM; Inactive, IC$_{50}$ > 10 000 nM; ND, not determined.

are active in the activation assays. To explain these observations, we hypothesized that acitretin and acitretin analogues are metabolized to molecules that bind to and activate RARs. We therefore attempted to demonstrate such a metabolism by incubating cells with acitretin and testing the cell extracts for binding activity. Preliminary results indicate that in various cell types acitretin is indeed transformed to metabolites with binding activity. The nature of these metabolites is as yet unknown.

## ACKNOWLEDGEMENTS

We are grateful to Antje Klem, Christian Lacoste, and Bernard Rutten for their excellent technical assistance. We thank Dr J. Eliason for fruitful discussions and helpful suggestions.

REFERENCES

Benbrook, D., Lernhardt, E., and Pfahl, M. (1988). A new retinoic acid receptor identified from a hepatocellular carcinoma. *Nature*, **333**, 669–72.

Bollag, W. (1979). Retinoids and cancer. *Cancer Chemotherapy and Pharmacology*, **3**, 207–15.

Bollag, W. (1983). Vitamin A and retinoids: from nutrition to pharmacotherapy in dermatology and oncology. *Lancet*, **1**, 860–3.

Brand, N. J., Petkovich, M., Krust, A., Chambon, P., De Thé, H., Marchio, A., Tiollais, P., and Dejean, A. (1988). Identification of a second human retinoic acid receptor. *Nature*, **332**, 850–3.

Crettaz, M., Baron, A., Siegentahler, G., and Hunziker, W. (1990). Ligand specificities of recombinant retinoic acid receptors RAR-α and RAR-β. *Biochemical Journal*, **272**, 391–7.

De Thé, H., Vivanco-Ruiz, M., Tiollais, P., Stunnenberg, H., and Dejean, A. (1990). Identification of a retinoic acid responsive element in the retinoic acid receptor gene. *Nature*, **343**, 177–80.

Giguère, V., Ong, E. S., Segui, P., and Evans, R. M. (1987). Identification of a receptor for the morphogen retinoic acid. *Nature*, **330**, 624–9.

Giguère, V., Shago, M., Zirngibl, R., Tate, P., Rossant, J., and Varmuza, S. (1990). Identification of a novel isoform of the retinoic acid receptor γ expressed in the mouse embryo. *Molecular and Cellular Biology*, **10**, 2335–40.

Graupner, G., Wills, K. N., Tzukerman, M., Zhang, X., and Pfahl, M. (1989). Dual regulatory role for the thyroid-hormone receptors allows control of retinoic acid receptor activity. *Nature*, **340**, 653–6.

Jetten, A. M., Anderson, K., Deas, M. A., Kagechika, H., Lotan, R., Rearick, J. I., and Shudo, K. (1987). New benzoic acid derivatives with retinoid activity: lack of direct correlation between biological activity and binding to cellular retinoic acid binding protein. *Cancer Research*, **47**, 3523–7.

Jetten, A. M. and Jetten, M. E. R. (1979). Possible role of retinoic acid binding protein in retinoid stimulation of embryonal carcinoma cell differentiation. *Nature*, **278**, 180–2.

Kastner, P., Krust, A., Mendelsohn, C., Garnier, J. M., Zelent, A., Leroy, P., Staub, A., and Chambon, P. (1989). Murine isoforms of retinoic acid receptor γ with specific patterns of expression. *Proceedings of the National Academy of Sciences USA*, **87**, 2700–4.

Krust, A., Kastner, P., Petkovich, A., Zelent, A., and Chambon, P. (1989). A third human retinoic acid receptor, RARγ. *Proceedings of the National Academy of Sciences USA*, **86**, 5310–14.

Lippman, S. M., Kessler, J. F., and Meyskens, F. L. (1987). Retinoids as preventive and therapeutic anticancer agents. *Cancer Treatment Reports*, **71**, 391–405.

Mangelsdorf, D. J., Ong, E. S., Dyck, J. A., and Evans, R. M. (1990). Nuclear receptor that identifies a novel retinoic acid response pathway. *Nature*, **345**, 224–9.

Petkovich, M., Brand, N. J., Krust, A., and Chambon, P. (1987). A human retinoic acid receptor which belongs to the family of nuclear receptors. *Nature*, **330**, 444–50.

Stüber, D., Matile, H., and Garotta, G. (1990). System for high level production in E. coli and rapid purification of recombinant proteins: application to epitope mapping, preparation of antibodies and structure-function analysis. In *Immunologic methods*, vol. IV (eds I. Lefkovits and B. Pernis), pp. 121–52. Academic Press, Orlando, USA.

Studier, F. W. and Moffatt, B. A. (1986). Use of bacteriophage T7 polymerase to direct selective high-level expression of cloned genes. *Journal of Molecular Biology*, **189**, 113–30.

Teelmann, K. (1989). Retinoids: Toxicology and teratogenicity to date. *Pharmacological Therapy*, **40**, 29–43.

Teelmann, K. (1990). Testing of retinoids for systemic and topic use in psoriasis and disorders of keratinization (the mouse papilloma test). In *Methods in Enzymology*, Vol. 190, (ed. L. Packer). Academic Press, London.

Umesono, K., Giguère, V., Glass, C. K., Rosenfeld, M. G., and Evans, R. M. (1988). Retinoic acid and thyroid hormone induce gene expression through a common responsive element. *Nature*, **336**, 262–5.

Vasios, G. W., Gold, J. D., Petkovich, M., Chambon, P., and Gudas, L. J. (1989). A retinoid acid-responsive element is present in the 5′ flanking region of the laminin B1 gene. *Proceedings of the National Academy of Sciences USA*, **86**, 9099–103.

# 6

# Cellular retinol binding protein: regulation of expression and putative functions

Frank Chytil, Donald G. Stump, Riaz ul Haq, Margaret G. Rush, and Elisabeth E. Mufson

Cellular retinol binding protein (CRBP) was the first protein discovered to bind retinol non-covalently in many tissues (Bashor *et al.* 1973). Subsequently, other proteins binding natural retinoids non-covalently were found in non-visual and visual tissues, including the cellular retinoic acid binding protein (CRABP) (for review see Chytil and Ong 1987). Considerable progress has been made in the purification, characterization, and cloning of these proteins, although their exact metabolic function(s) remains to be elucidated. In recent years less attention has been given to CRBP than to CRABP, most probably because the natural ligand of CRABP is retinoic acid. Retinoic acid (RA), a production of retinol oxidation, has been shown to be a potent differentiation agent (for review see Chytil 1984) used therapeutically in dermatology (Peck 1984). RA action involves activating or repressing formation of many gene products (for review see Chytil and Haq 1990). Moreover, unlike retinol, RA appears to be involved in morphogenesis (Eichele *et al.* 1985; Summerbell and Maden 1990). This article will discuss recent developments in our laboratory in further characterization of the CRBP protein and regulation of its gene.

CRBP AND CRBP II

CRBP has been studied by several laboratories for some time. More recently a new isoform of CRBP called CRBP II was first detected as a retinol-binding activity in extracts of whole neonatal rat and purified to homogeneity (Ong 1984). Both proteins—CRBP and CRBP II—have a molecular size of 15 000 Da, their endogenous ligand is all-*trans*-retinol, but this natural retinoid has higher affinity for CRBP ($k_d = 16$ nM) (Ong and Chytil 1978) than for CRBP II ($k_d = 50$ nM) (MacDonald and Ong 1987). The proteins show a distinct tissue-specific expression, CRBP being widely expressed in retinoid-sensitive tissues of the adult animal, whereas the expression of CRBP II is limited to absorptive cells of the small intestine (Crow and Ong 1985).

Recently cDNAs for CRBP from rat (Sherman *et al.* 1987), mouse (Wei *et al.* 1987) and human (Nilsson *et al.* 1988*a*) as well as for CRBP II (Demmer *et al.* 1987) and their respective genes have been characterized to a limited extent. Availability of these cDNAs has also enabled study of the regulation of the genes coding for these proteins.

## EFFECT OF SITE-DIRECTED MUTAGENESIS ON THE BINDING SPECIFICITY OF CRBP

One of the characteristics of the CRBP molecule is its binding specificity. Crude tissue extracts as well as the purified protein bind only all-*trans*-retinol and not the other natural retinoids, RA, or retinal (Bashor *et al.* 1973; Ong and Chytil 1978). To look closer into this striking feature of CRBP, experiments involving site-directed mutagenesis were performed in our laboratory (Stump *et al.* 1991).

A rat cDNA clone of CRBP (Sherman *et al.* 1987) was expressed in *Escherichia coli*. To determine amino acid residues in the CRBP molecule that may be important for the binding of all-*trans*-retinol, comparative model-building studies were performed in which strong sequence similarities were identified between CRBP and several other related proteins. Based on this analysis, specific amino acids were predicted to be important in retinol binding, and these predictions were tested using the technique of site-directed mutagenesis to subtly alter the protein's structure and binding affinity. Specifically, site-directed mutagenesis was performed to alter the glutamine 108 to arginine 108 to generate a mutant protein called Q108R. Making use of fluroescence, Q108R was found to have a 3-fold lower affinity for all-*trans*-retinol, and the fine structure of the excitation spectrum of the Q108R–all-*trans*-retinol complex was also different from that of the original wild type protein–all-*trans*-retinol complex. The mutant protein bound 13-*cis*-retinol with an excitation spectrum identical to wild type bound to 13-*cis*-retinol, but with only one-half of the fluorescence intensity. In competition binding experiments, the Q108R mutant had similar binding affinities for all-*trans*-retinol, all-*trans*-RA, 13-*cis*-RA, and retinol, while wild type CRBP was only able to bind to all-*trans*-retinol. Thus, altering a single amino acid in CRBP (glu-108 to arg-108) causes a significant change in the ligand binding specificity of the protein. Availability of considerable quantities of the wild and mutant CRBP should help in deciphering the structure of the binding site of this protein.

## CRBP GENE AND REGULATION OF ITS EXPRESSION

The gene for CRBP is present as one copy in the rat (Sherman *et al.* 1987) and human (Nilsson *et al.* 1988*a*) genomes. In mice the gene is located on chromosome 3 (Wei *et al.* 1987), in humans on chromsome 9 (Nilsson *et al.* 1988*a*), thus its location is identical with that of CRABP (Wei *et al.* 1987; Nilsson *et al.* 1988*b*). The gene for CRBP is rather large, comprising about 21 kilobases, with four exons and three introns (Nilsson *et al.* 1988*a*) considering that it codes for a protein of molecular size of 15 000 Da. The size and structure of the CRBP gene are very similar to those of the CRABP gene (Shubieta *et al.* 1987). The large size of the CRBP gene may lead to speculation that the regulation of expression of this gene is rather complex.

In recent years our laboratory has turned to the lung of the adult animal as the target of our studies. We based our decision on the classic findings of Wolbach and Howe (1925), who found that the epithelium of the trachea and bronchopulmonary tree undergo keratinizing metaplasia very early following exposure of rats to retinol deficient diet. The lung is not only one of the first organs to show signs of retinol deficiency, but is also where the repair of tissue to normal is the fastest when compared to other organs (Wolbach and Howe 1933). Moreover, our previous results indicated fast global changes in mRNA and proteins in retinol deficient rats after oral feeding of retinol or RA (Omori and Chytil 1982; Haq and Chytil 1988*a*). We therefore used short-term refeeding of retinol-deficient rats with retinol or RA as an *in vivo* model for studies on CRBP gene expression. First, we were able to establish that, 4 hours after oral administration of retinol to retinol-deficient rats, there is a large induction of CRBP mRNA (Sherman *et al.* 1987). Subsequently, we observed that administration of RA, instead of retinol, activated the CRBP gene after only 1 hour (Haq and Chytil 1988*b*). This phenomenon of a fast activation of the CRBP was also seen during F9 embryonal carcinoma cell differentiation induced by retinoic acid (Wei *et al.* 1989). More recently, we found that activation of the CRBP gene by retinoic acid does not require a state of retinol deficiency because similar induction of CRBP mRNA was observed after feeding RA to a retinol-sufficient rat (Rush *et al.* 1991). At present, we cannot distinguish whether or not this quick activation of the CRBP gene precedes, occurs simultaneously with, or follows induction of the $\beta$-RA receptor observed using the *in vivo* model (Haq *et al.* 1991). Thus, these findings on well-nourished retinol-sufficient rats indicate that the CRBP gene is not only developmentally regulated as established earlier (Ong and Chytil 1976) but that oral administration of RA used in patients with skin hyperkeratic lesions (Peck 1984) may also influence the pulmonary metabolism. Subsequent findings also show that RA is not the only agent that regulates the expression of the CRBP gene. We have observed that administration of glucocorticoids (dexamethasone) has an inhibitory effect on this gene (Rush *et al.* 1991). Dexamethasone in retinol-sufficient control rats decreased both lung and liver CRBP mRNA abundance. In animals treated with both RA and dexamethasone, the amount of CRBP message was also reduced. Thus, in the whole animal dexamethasone not only represses CRBP gene expression, but also appears to affect the action of RA.

Finally, we wondered whether the rapid induction of the CRBP mRNA by RA can be prevented by inhibitors of protein synthesis; we used a hepatoma cell line. We found that RA induces CRBP mRNA in a matter of a few hours and that retinol can cause the same effect, but more slowly. Most importantly, we also found that the activation of CRBP gene expression can be inhibited by cycloheximide. Thus, these results indicate that, at least in hepatoma cell, the early activation of the CRBP gene is preceded by event(s) that require synthesis of one or more proteins. These results are in contrast to early activation of gene expression observed by others who did not see abolition of the RA effect by an inhibitor of protein synthesis (La Rosa and Gudas 1988).

FUNCTION(S) OF CRBP IN RELATION TO THE ACTIONS OF RETINOL

The efforts to assign single or multiple functions to CRBP are, at present, a matter of conjecture; several functions could be hypothesized. The binding of retinol to CRBP renders retinol more soluble in the aqueous environment of the cell. Consequently, such binding may also, to a certain extent, prevent non-specific attachment of retinol to cellular components containing lipids, including cellular membranes. In addition, it can be envisioned that retinol may be more available to metabolic machinery catalysing esterification or hydrolysis of its storage form retinyl esters present not only in adult liver but in other organs. The hypothesis that CRBP acts as the vehicle of a shuttle mechanism bringing retinol into the nucleus (Liau *et al.* 1982) may have been weakened by experimental evidence that nuclear RA receptors prefer as ligands RA and to a lesser extent retinol (Petkovich *et al.* 1987; Crettaz *et al.* 1990).

In agreement with many reports, it is fair to conclude that expression of CRBP is cell-specific. The immunohistochemical localization of CRBP shows that this protein is at least quantitatively expressed in cells different from those expressing CRABP (Porter *et al.* 1985; Maden *et al.* 1988; Dollé *et al.* 1990). That the gene for CRBP is regulated in a different fashion from the gene for CRABP became evident when it was shown that the levels of these proteins show a different pattern in the developing rat lung (Ong and Chytil 1976). The recent suggestion that CRBP participates in the process of RA formation by storing and releasing retinol in cells where RA is required, for instance for morphogenetic purposes, is attractive (Dollé *et al.* 1990). The confirmation of this hypothesis must await better characterization, isolation, and localization of the enzyme system(s) responsible for *in vivo* formation of RA. This area of endeavour will certainly receive more attention in the near future. The clarification of CRBP function based on localization of CRBP in multicellular systems undergoing morphogenesis is even more difficult. Unlike expression of the CRABP protein, levels of CRBP do not show gradient distribution (Maden *et al.* 1988), leaving us with the question of how much retinol is actually complexed with CRBP. Variation in the extent of saturation of the protein may occur in various systems; we have reported previously that CRBP may not always be saturated with its ligand (Ong *et al.* 1976).

One very puzzling phenomenon mentioned above is the rapid induction of CRBP by RA in the lungs of retinol-deficient (Haq and Chytil 1988*b*) or retinol-sufficient (Rush *et al.* 1991) rats. It would certainly be of interest to find out whether activation of the CRBP gene by RA precedes or follows activation of expression of the $\beta$-RA receptor. Induction of the CRABP gene appears to occur later than observed for CRBP (Wei *et al.* 1989). In this respect, it should be stressed that most considerations regarding the function of CRBP are based on the assumption that the sole function of CRBP is to bind retinol. It should not be excluded *a priori* that the protein itself could be involved in an additional, thus far undefined, activity. In this respect, the amino acid sequence similarity of CRBP with some DNA binding proteins (Tsonis and Goetnick 1988) and nuclear localization of the CRBP protein by electron-microscopic immunohistochemistry is certainly intriguing (Bok *et al.* 1984). The intensive work in this area will hopefully decipher the enigma of CRBP function.

## ACKNOWLEDGEMENTS

The authors thank S. Heaver and M. Hunt for their able assistance in preparation of this manuscript.

This work was supported by the USPHS grants HD-07043, HD-9195 and HL-14214 and by the General Foods Fund.

## REFERENCES

Bashor, M. M., Toft, D. O., and Chytil, F. (1973). *In vitro* binding of retinol to rat-tissue components. *Proceedings of the National Academy of Sciences, USA*, **70**, 3483–7.

Bok, D., Ong, D. E., and Chytil, F. (1984). Immunocytochemical localization of cellular retinol binding protein (CRBP) in the retina. *Investigative Ophthalmology and Visual Sciences*, **25**, 877–83.

Chytil, F. (1984). Retinoic acid: Biochemistry, pharmacology, toxicology, and therapeutic use. *Pharmacology Review*, **36**, 93S–100S.

Chytil, F. and Haq, R. U. (1990). Vitamin A mediated gene expression. *Critical Reviews Eukaryotic Gene Expression*, **1**, 61–73.

Chytil, F. and Ong, D. E. (1987). Intracellular vitamin A-binding proteins. *Annual Review of Nutrition*, **7**, 321–5.

Crettaz, M., Baron, A., Siegenthaler, G., and Hunzikar, K. (1990). Ligand specificities of recombinant retinoic acid receptors RAR$\alpha$ and RAR$\beta$. *Biochemical Journal*, **272**, 391–7.

Crow, J. A. and Ong, D. E. (1985). Cell-specific immunohistochemical localization of a cellular retinol-binding protein (type two) in the small intestine of rat. *Proceedings of the National Academy of Sciences, USA*, **82**, 4707–11.

Demmer, L. A., Birkenmeier, E. H., Sweetser, D. A., Levin, M. S., Zollman, S., Sparkes, R. S., and Gordon, J. S. (1987). The cellular retinol binding protein II gene. Sequence analysis of the rat gene, chromosomal localization in mice and humans and documentation of its close linkage to the cellular retinol binding protein. *Journal of Biological Chemistry*, **262**, 2458–67.

Dollé, P., Ruberte, E., Leroy, P., Morriss-Kay, G., and Chambon, P. (1990). Retinoic acid receptors and cellular retinoid binding proteins. 1. A systematic study of their differential pattern of transcription during mouse organogenesis. *Development*, **110**, 1131–51.

Eichele, G., Tickle, C., and Alberts, B. M. (1985). Studies on the mechanism of retinoid induced pattern duplication in early chick limb bud. Temporal and spatial aspects. *Journal of Cellular Biology*, **101**, 1913–20.

Haq, R. U. and Chytil, F. (1988a). Early effects of retinol and retinoic acid on protein synthesis in retinol deficient rat testes. *Biochemical and Biophysical Research Communications*, **151**, 53–60.

Haq, R. U. and Chytil, F. (1988b). Retinoic acid rapidly induces lung cellular retinol-binding protein mRNA levels in retinol-deficient rats. *Biochemical and Biophysical Research Communications*, **156**, 712–14.

Haq, R. U., Pfahl, M., and Chytil, F. (1991). Retinoic acid affects the expression of nuclear retinoic acid receptors in tissues of retinol deficient rats. *Proceedings of the National Academy of Sciences, USA*, **88**, 8272–6.

LaRosa, G. J. and Gudas, L. J. (1988). An early effect of retinoic acid: cloning of an mRNA (ERA-I) exhibiting rapid and protein synthesis-independent induction during teratocarcinoma stem cell differentiation. *Proceedings of the National Academy of Sciences, USA*, **85**, 329.

Liau, G., Ong, D. E., and Chytil, F. (1982). Interaction of the retinol-cellular retinol binding protein complex with isolated nuclei and nuclear components. *Journal of Cell Biology*, **91**, 63–8.

Maden, M., Ong, D. E., Summerbell, D., and Chytil, F. (1988). Spatial distribution of cellular protein binding to retinoic acid in the chick limb bud. *Nature (Lond.)*, **335**, 733–5.

MacDonald, P. N. and Ong, D. E. (1987). Binding specificities of cellular retinol-binding protein and cellular retinol-binding protein, Type II. *Journal of Biological Chemistry*, **262**, 10550–6.

Nilsson, M. H. L., Spurr, N. K., Lundvall, J., Rask, L., and Peterson, P. A. (1988a). Human cellular retinol-binding protein gene organization and chromosomal location. *European Journal of Biochemistry*, **173**, 35–44.

Nilsson, M. H. L., Spurr, N. K., Saksema, P., Busch, C., Nordlinger, H., and Peterson, P. (1988b). Isolation and characterization of cDNA clone corresponding to bovine cellular retinoic acid-binding protein and chromosomal location of the corresponding gene. *European Journal of Biochemistry*, **173**, 45–51.

Omori, M. and Chytil, F. (1982). Mechanism of vitamin A action: Gene expression in retinol deficient rats. *Journal of Biological Chemistry*, **257**, 14370–4.

Ong, D. E. (1984). A novel retinol-binding protein from rat: Purification and partial characterization. *Journal of Biological Chemistry*, **259**, 1476–82.

Ong, D. E. and Chytil, F. (1976). Changes in levels of cellular retinol- and retinoic acid-binding proteins of liver and lung during perinatal development. *Proceedings of the National Academy of Sciences, USA*, **73**, 3976–8.

Ong, D. E. and Chytil, F. (1978). Cellular retinol-binding protein from rat liver. *Journal of Biological Chemistry*, **253**, 828–32.

Ong, D. E., Tsai, C. H., and Chytil, F. (1976). Cellular retinol-binding protein and retinoic acid-binding protein in rat testis. Effect of retinol depletion. *Journal of Nutrition*, **106**, 204.

Peck, G. L. (1984). Synthetic retinoids in dermatology. In *The retinoids*, vol. 2, (ed. M. B. Sporn, A. B. Roberts, and D. S. Goodman), pp. 391–411. Academic Press, Orlando, FL.

Petkovich, M., Brand, N. J., Krust, A., and Chambon, P. (1987). A human retinoic acid receptor which belongs to the family of nuclear receptors. *Nature*, **330**, 444–50.

Porter, S. B., Ong, D. E., Chytil, F., and Orgebin-Crist, M.-C. (1985). Localization of cellular retinol-binding protein and cellular retinoic acid-binding protein in the rat testis and epididymis. *Journal of Andrology*, **6**, 197–212.

Rush, M. G., Haq, R. U., and Chytil, F. (1991). Opposing effects of retinoic acid and dexamethasone on cellular retinol-binding protein mRNA levels in the rat. *Endocrinology*, **129**, 705–9.

Sherman, D. R., Lloyd, R. S., and Chytil, F. (1987). Rat cellular retinol-binding protein: cDNA sequence and rapid retinol-dependent accumulation of mRNA. *Proceedings of the National Academy of Sciences, USA*, **84**, 3209–13.

Shubieta, H. E., Sambrook, J. F., and McCormick, A. M. (1987). Molecular cloning and analysis of functional cDNA and genomic clones encoding bovine cellular retinoic acid-binding protein. *Proceedings of the National Academy of Sciences*, **84**, 5645–9.

Stump, D. G., Lloyd, R. S., and Chytil, F. (1991). Site-directed mutagenesis of rat cellular retinol-binding protein. *Journal of Biological Chemistry*, **266**, 4622–30.

Summerbell, D. and Maden, M. (1990). Retinoic acid, a developmental signalling molecule. *Trends in Neurosciences*, **13**, 142–7.

Tsonis, P. A. and Goetnick, P. F. (1988). Homology of cellular vitamin A-binding protein with DNA-binding proteins. *Biochemical Journal*, **249**, 933–4.

Wei, L.-N., Mertz, J. R., Goodman, D. S., and Nguyen-Huu, M. C. (1987). Cellular retinoic acid- and cellular retinol-binding proteins: Complementary deoxyribonucleic acid cloning, chromosomal assignment, and tissue specific expression. *Molecular Endocrinology*, **1**, 526–34.

Wei, L. N., Blaner, W. S., Goodman, D. S., and Nguyen-Huu, M. C. (1989). Regulation of the cellular retinoid-binding proteins and their messenger ribonucleic acid during P19 embryonal carcinoma cell differentiation induced by retinoic acid. *Molecular Endocrinology*, **3**, 454–63.

Wolbach, S. B. and Howe, P. R. (1933). Epithelial repair in recovery from vitamin A deficiency. *Journal of Experimental Medicine*, **57**, 511–26.

Wolbach, S. B. and Howe, P. R. (1925). Tissue changes following deprivation of fat-soluble A vitamin. *Journal of Experimental Medicine*, **42**, 753–77.

# PART II

# Receptors and binding proteins in embryos

# 7

# Retinoic acid receptors in the chick embryo

Annie Rowe, Nicholas S. C. Eager, Joy M. Richman, and Paul M. Brickell

## INTRODUCTION

The chicken is of particular value as an organism for studies of vertebrate embryonic development because chick embryos can be manipulated *in ovo* and the effects of such manipulations on their further development can be examined. Such studies have contributed a great deal to our view of the possible role of retinoids in vertebrate embryonic development. This is largely a result of experiments showing that the local application of all-*trans*-retinoic acid to the anterior margin of the chick wing bud results in the development of wings in which the pattern of the digits and other skeletal elements is altered. This behaviour mimics that of the polarizing region, a group of cells at the posterior margin of the limb bud that are involved in patterning the normal limb, and there is evidence that retinoids may be natural signalling substances in chick limb buds (reviewed by Tickle and Brickell 1991). For these reasons, we and others have analysed the chicken's system of retinoic acid (RA) nuclear receptors. The purpose of this article is to review what is known about this system and then to concentrate on our own recent studies of the chicken RXR and RAR-$\beta$ nuclear receptors.

## RETINOIC ACID RECEPTORS IN THE CHICKEN

The basic elements of the RA nuclear receptor system have been highly conserved between mammals and chickens. Thus, chickens are known to possess genes encoding RAR-$\alpha$ (Noji *et al.* 1991), RAR-$\beta$ (Noji *et al.* 1991; Rowe *et al.* 1991*b*; Smith and Eichele 1991), RAR-$\gamma$ (P. Chambon, personal communication) and at least one member of the RXR family of nuclear receptors, termed cRXR (Rowe *et al.* 1991*a*). The latter family may be evolutionarily more ancient than the RAR family, as an RXR-like gene has been identified in *Drosophila melanogaster*, whilst genes encoding RAR-like receptors appear to be absent from this species (Oro *et al.* 1990).

Sequence data are available for chicken RAR-$\beta$ (Noji *et al.* 1991; Rowe *et al.* 1991*b*; Smith and Eichele 1991) and cRXR (Rowe *et al.* 1991*a*) but not yet for chicken RAR-α or RAR-γ. Chicken RAR-β has a very high degree of amino acid sequence identity to human and mouse RAR-β, although less than that between human and mouse RAR-β (Table 7.1). Whilst the level of sequence identity is highest in the DNA- and ligand-binding domains, it remains high in the other domains, indicating that these are important for RAR-β function. The chicken RAR-β gene also resembles the mouse RAR-β gene (Zelent *et al.* 1991) in that its primary transcript can be alternatively spliced to yield at least three different mRNAs, encoding isoforms of the RAR-β protein, which differ in the sequence of their A domains (Smith and Eichele 1991).

The level of amino acid sequence identity between human RXR-α (Mangelsdorf *et al.* 1990), mouse RXR-β (Hamada *et al.* 1989), and cRXR is also high, and cRXR resembles human RXR-α more closely than it resembles mouse RXR-β (Table 7.2). The sequence identity between human RXR-α and cRXR is significantly less than that between the RAR-β molecules of these species, which may simply indicate that there is less need for a high degree of conservation. Alternatively, cRXR may represent a third member of the RXR gene family, rather than being chicken RXR-α or RXR-β.

## BIOLOGICAL ACTIVITY OF cRXR

When it is expressed in baby hamster kidney (BHK) cells in culture, cRXR is able to respond to retinoic acid treatment by activating transcription of a chloramphenicol acetyl transferase (CAT) reporter gene under the control of a retinoic acid responsive element (TRE$_3$-*tk*-CAT) (Fig. 7.1). Our preliminary data indicate that cRXR resembles human RXR-α (Mangelsdorf *et al.* 1990) in that it is significantly less sensitive to RA than is human RAR-β and in that it fails to respond to the synthetic retinoid TTNPB, which stimulates RAR-α, RAR-β, and RAR-γ with an efficiency approaching that of RA (Aström *et al.* 1990). The ligand-binding characteristics of cRXR and human RXR-α, which share 86 per cent amino acid identity in their ligand-binding domains, therefore appear to be broadly similar. In view of the relatively low sensitivity of these molecules to RA, it is possible that RA is not their natural ligand. If this is the case, then conservation of the ability to react with the natural ligand has also led to conservation of cross-reactivity with RA. This suggests that the natural ligand may be structurally related to RA.

## EXPRESSION PATTERN OF THE cRXR GENE

Northern blotting analysis has shown that cRXR transcripts of approximately 2.5 kb are expressed in a range of embryonic chick tissues (Rowe *et al.* 1991*a*). Transcripts are also present at low levels in a range of adult chicken tissues, including the lung, and at high levels in adult liver (Fig. 7.2). The distribution of cRXR transcripts in the adult is similar to that of the 4 kb rat RXR-α

**Table 7.1** Percentage identity between the predicted amino acid sequences of human, mouse, and chicken RAR-$\beta$ in domains A–F of the protein. Data are from Krust *et al.* (1989) and Rowe *et al.* (1991*a*)

|  | Domain | | | | | |
|---|---|---|---|---|---|---|
|  | A | B | C | D | E | F |
| Chicken/human | 90 | 96 | 98 | 89 | 97 | 86 |
| Chicken/mouse | 79 | 96 | 98 | 91 | 99 | 78 |
| Mouse/human | 94 | 100 | 100 | 98 | 99 | 92 |

**Table 7.2** Percentage identity between the predicted amino acid sequence of cRXR and those of human RXR-$\alpha$ and mouse RXR-$\beta$ in the DNA-binding (C), ligand-binding (E/F), and flanking domains of the protein. Data are from Rowe *et al.* (1991*a*)

|  | A/B | C | D | E/F |
|---|---|---|---|---|
| cRXR/human RXR-$\alpha$ | 37 | 97 | 72 | 86 |
| cRXR/mouse RXR-$\beta$ | 12 | 92 | 40 | 83 |

transcripts identified by Mangelsdorf *et al.* (1990). By *in situ* hybridization to tissue sections from stage 24 and stage 27 chick embryos, we have shown that cRXR transcripts are present at high levels in the embryonic liver and that the low level expression seen in other tissues by Northern blotting is due to high level expression in elements of the peripheral nervous system (Rowe *et al.* 1991*a*). These include dorsal root ganglia, cranial ganglia, sympathetic ganglia, enteric ganglia, and peripheral nerve tracts (Fig. 7.3). As transcripts are detectable in peripheral nerve roots, cRXR transcripts must be expressed by peripheral glial cells. However, we cannot exclude the possibility that neurons also express cRXR transcripts.

With the exception of the neurons in some cranial ganglia, the neurons and glial cells of the peripheral nervous system are the differentiated progeny of cells that have migrated from the neural crest (Le Douarin and Smith 1988). In avian embryos this is a transient structure that develops following closure of the neural tube. Neural crest cells migrate along characteristic pathways from the dorsal neural tube and localize at specific sites in the embryo, where they give rise to a broad range of cell types (Le Douarin 1982). Neural crest cells at different levels along the anterior–posterior axis of the embryo give rise to different ranges of cell types. For example, in addition to cells of the peripheral

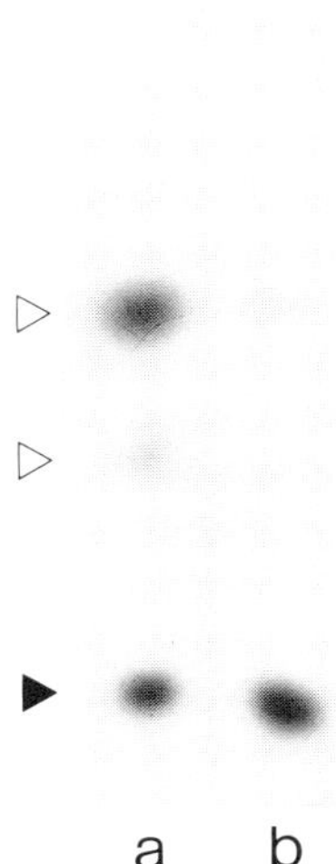



**Fig. 7.1** Retinoic-acid-dependent activation of transcription by cRXR. A plasmid containing the cRXR cDNA insert (Rowe *et al.* 1991*a*) under the control of the Moloney murine leukaemia virus long terminal repeat was cotransfected into BHK cells with the reporter plasmid TRE$_3$-*tk*-CAT. 24 hours later, cells in some dishes were treated with $10^{-6}$ M retinoic acid. After a further 48 hours, cells were harvested and assayed for CAT activity by thin-layer chromatography (Petkovich *et al.* 1987). Black arrow, unacetylated chloramphenicol; white arrows, acetylated forms. There is a much higher level of CAT activity in the retinoic-acid-treated cell extract (*a*) than in the untreated cell extract (*b*). The transfection efficiency was the same for both treated and untreated cells, as determined by cotransfection with the $\beta$-galactosidase-expressing plasmid pCH110.

nervous system, neural crest cells from the level of the trunk give rise to melanocytes and to some endocrine and paraendocrine cells, such as the chromaffin cells of the adrenal medulla and the C-cells of the thyroid gland. In contrast, cranial neural crest cells give rise to elements of the craniofacial skeleton and other connective tissue in the head, as well as to the glial cells and some of the neurons of the cranial ganglia. The cranial neural crest also gives rise to melanocytes, but has less propensity to do so than the trunk neural crest. Our preliminary data indicate that melanocytes and adrenal chromaffin cells in the chick embryo express cRXR transcripts but that these are undetectable in the cranial–neural-crest-derived mesenchymal cells of the facial primordia (E. Seleiro, A. Rowe, and P. Brickell, unpublished observations). It therefore appears that cRXR transcripts may be a general marker for cells derived from the trunk and cranial neural crest, with the exception of the cranial–neural-crest-derived cells that contribute to the skeleton and other connective tissue of the head.

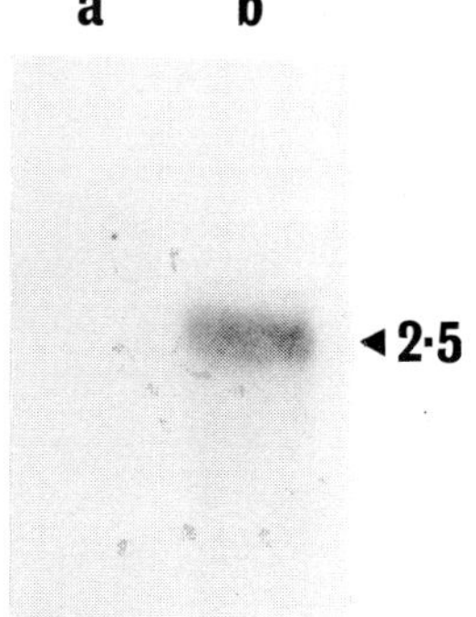


**Fig. 7.2** Northern blot of total RNA (10 μg per track) isolated from adult chicken lung (*a*) and liver (*b*) hybridized with a [$^{32}$P]-labelled RNA probe for cRXR transcripts (Rowe *et al.* 1991*a*). The size of the transcript is shown in kilobases.

The expression of cRXR transcripts by neural crest cells appears to be established early, rather than being a feature only of fully differentiated cell types. Thus, *in situ* hybridization experiments show that cRXR transcripts are present at high levels in premigratory and migrating neural crest cells at chick embryonic stages 16, 13, and 11 (Rowe *et al.* 1991*a*; our unpublished observations). Whilst we might expect all premigratory trunk neural crest cells to express cRXR transcripts, it is possible that the premigratory cranial neural crest is heterogeneous with respect to the presence of cRXR transcripts, as these are present only in a subset of cranial neural crest derivatives; this remains to be determined.

The distribution of cRXR transcripts is thus quite different from that of RAR-α, RAR-β or RAR-γ transcripts (Dollé *et al.* 1990; Ruberte *et al.* 1991), which have not been detected in the cranial nerve ganglia or in the other cranial–neural-crest-derived tissues that express cRXR transcripts. This supports the view that the RAR and RXR families of nuclear receptors have quite different biological roles.


POSSIBLE FUNCTIONS OF cRXR

The liver is the body's major site of vitamin A storage, metabolism, and mobilization (Blomhoff *et al.* 1990). The high levels of cRXR and human RXR-α transcripts in embryonic and adult liver may indicate a role for this family of receptors in regulating the expression of genes encoding enzymes involved in these processes.

It is more difficult to speculate about the role of RXR receptors in neural crest cells and their derivatives. They could be involved in mediating signals required for neural crest cell proliferation, migration and/or differentiation.

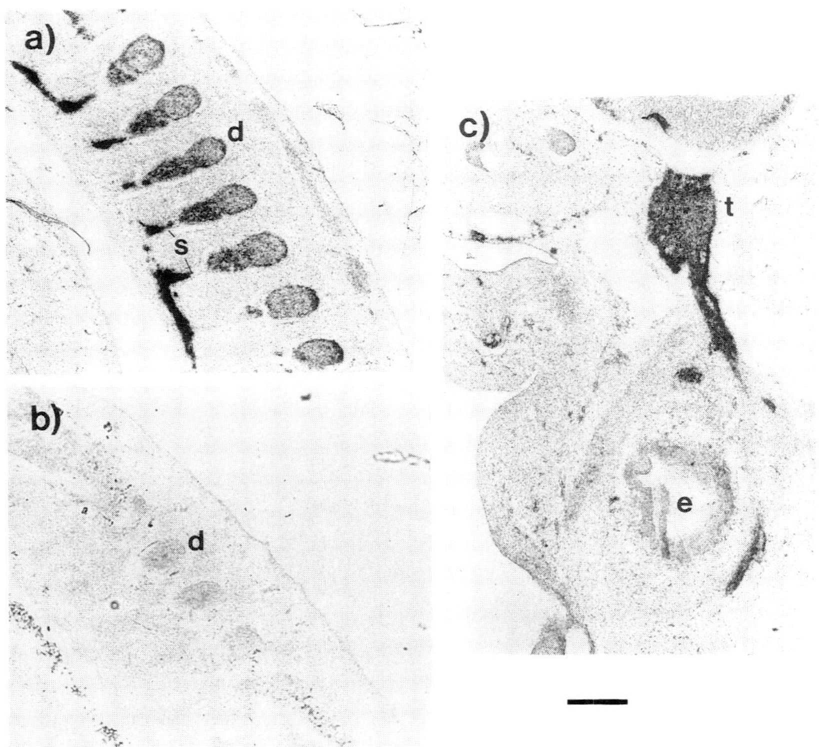

**Fig. 7.3** *In situ* hybridization of a [$^{35}$S]labelled RNA probe for cRXR transcripts (*a, c*) or a [$^{35}$S]labelled negative control RNA probe (*b*) to parasagittal sections from a stage 24 chick embryo. Probes and methods are described in Rowe *et al.* (1991*a*). d, dorsal root ganglion; e, eye; s, spinal nerves; t, trigeminal ganglion. Scale bar: 200 $\mu$m.

Alternatively, or additionally, they could mediate signals required for the maintenance and/or function of the differentiated cell types. However, there is little evidence to support the idea that retinoids are involved in these processes in the neural crest derivatives that express cRXR transcripts.

Retinoids are known to induce decreased cell-substratum adhesion in both trunk and cranial neural crest cells *in vitro* (Smith-Thomas *et al.* 1987), and it has been suggested that this could inhibit neural crest cell migration *in vivo* (Pratt *et al.* 1987). However, it remains to be determined whether these effects are mediated by retinoid-induced changes in gene expression, rather than by effects on the structure of the cell membrane. Similarly, there is no clear evidence linking retinoids with the development or functioning of the peripheral nervous system. For example, no retinoid-induced malformations of the peripheral nervous system have been described and whilst experimental vitamin A deficiency in animals results in peripheral nerve degeneration, this is secondary to a selective cessation of bone growth and does not represent a direct effect on the peripheral nervous system (Duchen and Jacobs 1984). There is evidence, however, that RA can induce nerve growth factor-dependent (NGF-dependent) survival of immature chick sympathetic neurons in culture by activating the expression of high affinity NGF receptors by these cells (H. Rohrer, personal communication). RA can also induce neurite extension in cultures of chick embryo dorsal root ganglia (Haskell *et al.* 1987), possibly by affecting NGF receptor expression on neurons or by altering the production of NGF or other factors by glial cells. It is possible that cRXR is involved in mediating these effects. However, in considering the role of cRXR in neural crest cells and their derivatives, the strong possibility that the natural ligand is not a retinoid should be borne in mind. The challenge is to identify a ligand that bears some structural relationship to RA and that is clearly required for the development and/or function of the neural crest derivatives that express cRXR transcripts.

## RAR-$\beta$ GENE EXPRESSION IN CHICK LIMB BUDS

As noted above and as discussed elsewhere in this volume (Chapter 18), local application of RA to the anterior margin of the chick limb bud results in duplication of digits, and there is evidence that RA may be a natural signalling substance in the limb bud (reviewed by Tickle and Brickell 1991). At stages when patterning can be affected by locally applied RA, RAR-$\beta$ transcripts are expressed at high levels in the proximal part of the chick limb bud, adjacent to the body, and are barely detectable above background in the distal tip of the bud (Noji *et al.* 1991; Rowe *et al.* 1991c). The changes in pattern induced by RA are thought to involve cells at the tip of the bud, and it is consequently unlikely that RAR-$\beta$ is involved in mediating these changes. In the mouse, RAR-$\alpha$ and RAR-$\gamma$ transcripts are uniformly distributed in the limb buds at equivalent embryonic stages (Dollé *et al.* 1989; Chapter 8), and so could conceivably mediate the effects of locally applied RA on patterning. It will therefore be important to determine the distribution of RAR-$\alpha$ and RAR-$\gamma$ in chick limb buds.

## RAR-$\beta$ GENE EXPRESSION IN CHICK FACIAL PRIMORDIA

In addition to causing changes in the pattern of the limb, local application of an appropriate dose of RA to the wing bud between embryonic stages 18 and 21 invariably results in the absence of the upper beak (Tamarin *et al.* 1984; Wedden and Tickle 1986; Wedden *et al.* 1988). This occurs because the RA can readily diffuse from the wing bud to the nearby face. The chick face, like that of mammals, develops from a series of primordia, each of which consists of a bud of mesenchymal cells that have migrated from the cranial neural crest, encased in an epithelial layer. The frontonasal mass, the lateral nasal processes, and the maxillary primordia give rise to the upper beak, whilst the mandibular primordia give rise to the lower beak (Fig. 7.4(a)). The basis of the defect induced by RA is a failure of outgrowth of the frontonasal mass, which then fails to make contact with the lateral nasal processes and the maxillary primordia, resulting in severe bilateral clefting of the primary palate. The development of the mandibular primordia is unaffected. Experiments in which epithelium and mesenchyme from normal and RA-treated embryos are recombined show that RA acts on the mesenchymal cells of the frontonasal mass and not on the epithelium (Wedden 1987). RAR-$\beta$ transcripts are present in the facial primordia of normal chicken embryos at stage 20, when RA is active in producing the facial defect, and also 24 and 48 hours later, at stages 24 and 28, respectively (Fig. 7.4). However, the distribution of RAR-$\beta$ transcripts at these stages is not uniform (Rowe *et al.* 1991*b*; Smith and Eichele 1991). At all three stages there are high levels of RAR-$\beta$ transcripts in the anterior part of the maxillary primordia, with much lower levels in the posterior part. There are also high levels of transcripts at all three stages in the lateral nasal processes and at the lateral edges and outermost corners of the frontonasal mass. In contrast, transcripts are relatively evenly distributed at low levels in the mandibular primordia. In all primordia, at all three stages, RAR-$\beta$ transcripts are present in the mesenchyme rather than in the epithelium. RAR-$\beta$ transcripts are also spatially restricted in the facial primordia of the mouse at equivalent stages of development, but RAR-$\alpha$ and RAR-$\gamma$ transcripts are uniformly distributed, with RAR-$\gamma$ transcripts becoming progressively restricted to chondrogenic regions as facial development proceeds (Osumi-Yamashita *et al.* 1990; Ruberte *et al.* 1990).

The presence of RAR-$\alpha$, RAR-$\beta$, and RAR-$\gamma$ transcripts in the facial primordia supports the idea that RA and/or other retinoids may be involved in normal facial development. RAR-$\beta$ could also be involved in mediating the RA-induced upper beak defect described above, as RAR-$\beta$ transcripts are present at highest levels in regions of those primordia which contribute to the upper beak.

THE REGULATION OF RAR-$\beta$ GENE EXPRESSION BY RETINOIC ACID

Experiments with a number of cultured human and mouse cell lines have shown that transcription of the RAR-$\beta$ gene can be activated by RA (de Thé *et al.* 1989; Redfern *et al.* 1990), as the gene contains an RA response element (de Thé *et al.* 1990). Local application of RA to the chick wing bud results in a large increase in the levels of RAR-$\beta$ transcripts in the vicinity of the RA soaked bead (Noji *et al.* 1991; Rowe *et al.* 1991*c*). This effect is seen as early as

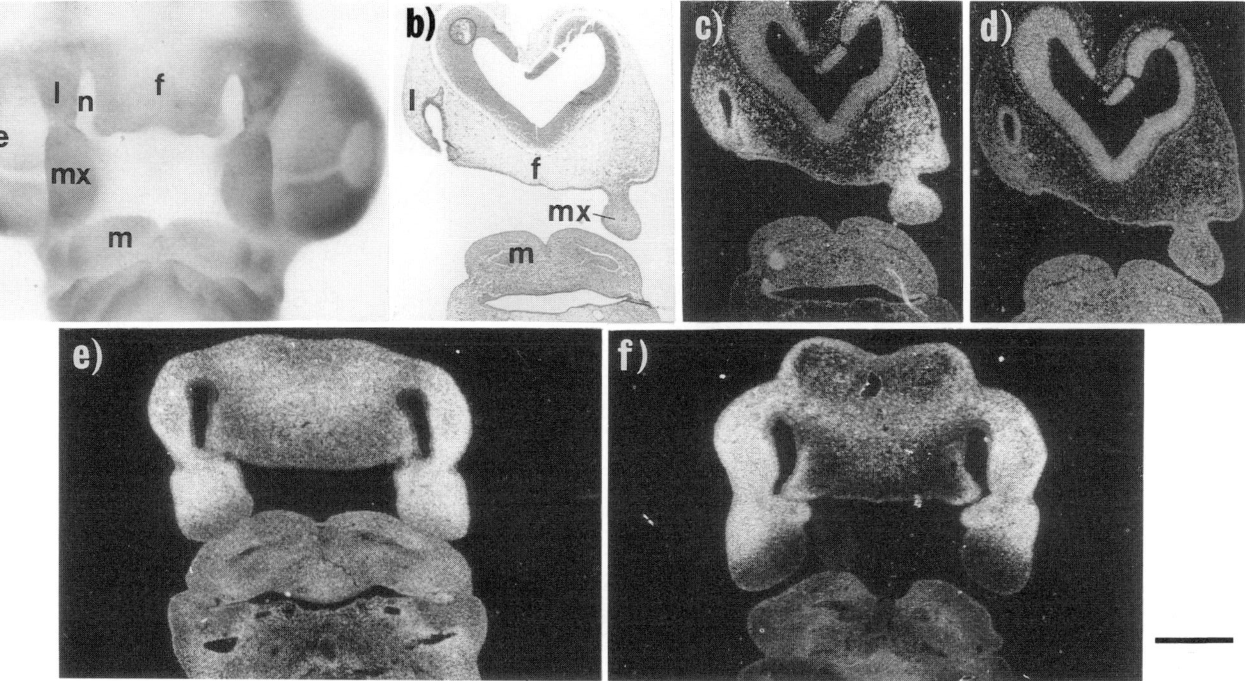



**Fig. 7.4.** (*a*) Photomicrograph of stage 24 chick embryo face showing the facial primordia. (*b–f*) *In situ* hybridization of a [$^{35}$S]labelled RNA probe for chicken RAR-$\beta$ transcripts (*b, c, e, f*) or a [$^{35}$S]labelled negative control RNA probe (*d*) to frontal sections of chick facial primordia at embryonic stage 20 (*b, c, d*), stage 24 (*e*) and stage 28 (*f*). Probes and methods are described in Rowe *et al.* (1991*b*). Sections are photographed under light-field (*b*) or dark-field (*c–f*) illumination. e, eye; f, frontonasal mass; l, lateral nasal process; m, mandibular primordium; mx, maxillary primordium; n, nasal slit. Scale bar: 500 $\mu$m.

4 hours after bead implantation, and is therefore likely to represent a direct effect of retinoic acid upon RAR-$\beta$ transcript accumulation (Noji *et al.* 1991).

Local application of RA to the wing bud, under conditions that cause the facial defect described above, also affects the distribution of RAR-$\beta$ transcripts in the maxillary primordia. By 24 hours after RA treatment, RAR-$\beta$ transcripts are no longer restricted to the anterior part of the primordia but are expressed throughout (Rowe *et al.* 1991*b*). The distribution of RAR-$\beta$ transcripts in the frontonasal mass and mandibular primordia is unaffected at this time (Rowe *et al.* 1991*b*), even though these primordia receive significant doses of locally applied retinoids (Wedden *et al.* 1987). This difference in response may be a result of differences in the levels of cellular RA-binding protein type I (CRABP) expressed by cells in the different facial primordia. Whilst CRABP is present in the maxillary primordia, higher levels are present in the frontonasal mass and mandibular primordia (Maden *et al.* 1991). CRABP is thought to act as a buffer that limits the amount of free intracellular RA available to bind to RARs, so that RAR-mediated activation of RAR-$\beta$ gene transcription might occur in response to lower levels of RA in the maxillary primordia than in the frontonasal mass and mandibular primordia.

It is not clear whether these changes in RAR-$\beta$ transcript distribution are related to the changes in limb and face development that follow RA treatment. However, these results do indicate that temporal or spatial variations in the levels of endogenous retinoids and of CRABP in normal embryos could modulate the levels of RAR-$\beta$ in some cells. This could be important in determining the nature of the cellular response to retinoids, in terms of the target genes that are activated.


RETINOIC ACID RECEPTORS AND THE NEURAL CREST


As discussed above, cRXR transcripts may be a general marker for neural-crest-derived cells, with the exception of the cranial–neural-crest derived cells that give rise to the craniofacial skeleton and other connective tissue in the head. In contrast, RAR-$\alpha$, RAR-$\beta$, and RAR-$\gamma$ transcripts are present in the cranial–neural-crest-derived mesenchymal cells of the facial primordia but have not been detected in other neural-crest-derived tissues such as the peripheral nervous system. Transcripts encoding cRXR and the RARs are also expressed by cells that are not derived from the neural crest; cRXR transcripts in the liver and RAR transcripts in a broad range of other tissues (Dollé *et al.* 1990). RAR-$\beta$ transcripts are not distributed uniformly in the mesenchymal cells of the facial primordia and it is possible that this reflects the fact that mesenchymal cells at different anatomical sites are derived from different regions of the cranial neural crest. Cells in the anterior and posterior parts of the maxillary primordia, which express very different levels of RAR-$\beta$

transcripts (Rowe *et al.* 1991*b*), are certainly derived from different regions of the cranial neural crest (A. Lumsden, personal communication). We have so far been unable to detect RAR-$\beta$ transcripts in the premigratory cranial neural crest, but they are certainly present in migrating cranial neural crest cells in the chick (our unpublished observations). The role of RAR-$\beta$ in these migrating cells, like that of cRXR, remains to be determined.

Interestingly, CRABP is also expressed in cells derived from the neural crest in chick embryos (Maden *et al.* 1989, 1991; Vaessen *et al.* 1990) and mouse embryos (Perez-Castro *et al.* 1989; Dencker *et al.* 1990; Dollé *et al.* 1990; Maden *et al.* 1990). In the chick, CRABP is expressed by migrating trunk neural crest cells, although expression in cranial neural crest cells is only apparent after they have arrived at their destination (Maden *et al.* 1991). The timing of the onset of CRABP expression in these cells may not be critical, however, as mouse cranial neural crest cells appear to express CRABP during migration (Maden *et al.* 1990; Ruberte *et al.* 1991). High levels of CRABP transcripts and of CRABP are present in the mesenchyme of the facial primordia and in the cranial, dorsal root, sympathetic, and enteric ganglia of the chick and mouse (Maden *et al.* 1989, 1990; Dencker *et al.* 1990; Vaessen *et al.* 1990). This pattern of expression has been taken to indicate that CRABP may be involved in the differentiation of neural-crest-derived cells rather than in their migration (Maden *et al.* 1991).

The general impression is that expression of cRXR transcripts by neural-crest-derived cells precedes expression of CRABP, except in the craniofacial mesenchyme, which does not express cRXR transcripts. An assessment of the relationship between cRXR and CRABP in the cell types that appear to express both molecules must await the unequivocal identification of the natural ligand of cRXR. If this is a retinoid, then cRXR and CRABP may be part of the same signalling system in these cell types.


CONCLUSIONS


The basic elements of the RA receptor system are highly conserved between mammalian and avian species, in terms of the number of genes present, the primary structure of their products and the patterns of their expression. However, this conclusion is based on limited studies of the chicken RAR-$\beta$ gene and it will be important to extend the comparison to RAR-$\alpha$ and RAR-$\gamma$. Studies of the distribution of transcripts encoding RAR-$\beta$ and cRXR in chick embryos suggest a role for these molecules in a range of normal developmental processes, including those involving neural crest cells. Finally, studies of chick embryos have shown that expression of the RAR-$\beta$ gene is inducible by RA in some cells *in vivo*, and it will be important to determine the value of this phenomenon in the developing embryo.

ACKNOWLEDGEMENTS

We are grateful to Cheryll Tickle for stimulating discussions and for her critical reading of this manuscript, and to Sarah Wedden for the photograph shown in Fig. 7.4(a). A.R. is supported by a grant from The Wellcome Trust and N.S.C.E. is the recipient of a Medical Research Council studentship.

REFERENCES

Aström, A., Pettersson, U., Krust, A., Chambon, P., and Voorhees, J. J. (1990). Retinoic acid and synthetic analogs differentially activate retinoic acid receptor dependent transcription. *Biochemical and Biophysical Research Communications*, **173**, 339–45.

Blomhoff, R., Green, M. H., Berg, T., and Norum, K. R. (1990). Transport and storage of vitamin A. *Science*, **250**, 399–404.

Dencker, L., Annerwall, E., Busch, C., and Eriksson, U. (1990). Localization of specific retinoid-binding sites and expression of cellular retinoic-acid-binding protein (CRABP) in the early mouse embryo. *Development*, **110**, 343–52.

de Thé, H., Marchio, A., Tiollais, P., and DeJean, A. (1989). Differential expression and ligand regulation of the retinoic acid receptor $\alpha$ and $\beta$ genes. *EMBO Journal*, **8**, 429–33.

de Thé, H., del Mar Vivanco-Ruiz, M., Tiollais, P., Stunnenberg, H., and DeJean, A. (1990). Identification of a retinoic acid responsive element in the retinoic acid $\beta$ gene. *Nature*, **343**, 177–80.

Dollé, P., Ruberte, E., Kastner, P., Petkovich, M., Stoner, C. M., Gudas, L., and Chambon, P. (1989). Differential expression of genes encoding $\alpha$, $\beta$, and $\gamma$ retinoic acid receptors and CRABP in the developing limbs of the mouse. *Nature*, **342**, 702–5.

Dollé, P., Ruberte, E., Leroy, P., Morriss-Kay, G., and Chambon, P. (1990). Retinoic acid receptors and cellular binding proteins I. A systematic study of their differential pattern of transcription during mouse organogenesis. *Development*, **110**, 1133–51.

Duchen, L. W. and Jacobs, J. M. (1984). Nutritional deficiencies and metabolic disorders. In *Neuropathology*, (ed. J. H. Adams, J. A. N. Corsellis, and L. W. Duchen), pp. 573–626. Edward Arnold, London.

Hamada, K., Gleason, S. L., Levi, B-Z., Hirschfeld, S., Appella, E., and Ozato, K. (1989). H-2RIIBP, a member of the nuclear hormone receptor superfamily that binds to both the regulatory element of major histocompatibility class I genes and the oestrogen response element. *Proceedings of the National Academy of Sciences of the USA*, **86**, 8289–93.

Haskell, B. F., Stach, R. W., Werrbach-Perez, K., and Perez-Polo, J. R. (1987). Effect of retinoic acid on nerve growth factor receptors. *Cell and Tissue Research*, **247**, 67–73.

Krust, A., Kastner, P., Petkovich, M., Zelent, A., and Chambon, P. (1989). A third human retinoic acid receptor hRAR-$\gamma$. *Proceedings of the National Academy of Sciences of the USA*, **86**, 5310–4.

Le Douarin, N. M. (1982). *The neural crest*. Cambridge University Press.

Le Douarin, N. M. and Smith, J. (1988). Development of the peripheral nervous system from the neural crest. *Annual Review of Cell Biology*, **4**, 375–404.

Maden, M., Ong, D. E., Summerbell, D., Chytil, F., and Hirst, E. A. (1989). Cellular retinoic acid-binding protein and the role of retinoic acid in the development of the chick embryo. *Developmental Biology*, **135**, 124–32.

Maden, M., Ong, D. E., and Chytil, F. (1990). Retinoid-binding protein distribution in the developing mammalian nervous system. *Development*, **109**, 75–80.

Maden, M., Hunt, P., Eriksson, U., Kuroiwa, A., Krumlauf, R., and Summerbell, D. (1991). Retinoic acid-binding protein, rhombomeres and the neural crest. *Development*, **111**, 35–44.

Mangelsdorf, D. J., Ong, E. S., Dyck, J. A., and Evans, R. M. (1990). Nuclear receptor that identifies a novel retinoic acid response pathway. *Nature*, **345**, 224–9.

Noji, S., Nohno, T., Koyama, E., Muto, K., Ohyama, K., Aoki, Y., Tamura, K., Ohsugi, K., Ide, H., Taniguchi, S., and Saito, T. (1991). Retinoic acid induces polarizing activity but is unlikely to be a morphogen in the chick limb bud. *Nature*, **350**, 83–6.

Oro, A. E., McKeown, M., and Evans, R. M. (1990). Relationship between the product of the *Drosophila ultraspiracle* locus and the vertebrate retinoid X receptor. *Nature*, **347**, 298–301.

Osumi-Yamashita, N., Noji, S., Nohno, T., Koyama, E., Doi, H., Eto, K., and Taniguchi, S. (1990). Expression of retinoic acid receptor genes in neural crest-derived cells during mouse facial development. *FEBS Letters*, **264**, 71–4.

Perez-Castro, A. V., Toth-Rogler, L. E., Wei, L., and Nguyen-Huu, M. C. (1989). Spatial and temporal pattern of expression of the cellular retinoic acid-binding protein and the cellular retinol-binding protein during mouse embryogenesis. *Proceedings of the National Academy of Sciences of the USA*, **86**, 8813–7.

Petkovich, M., Brand, N. J., Krust, A., and Chambon, P. (1987). A human retinoic acid receptor which belongs to the family of nuclear receptors. *Nature*, **330**, 444–50.

Pratt, R. M., Goulding, E.H., and Abbott, B. D. (1987). Retinoic acid inhibits migration of cranial neural crest cells *in vitro*. *Journal of Craniofacial Genetics and Developmental Biology*, **7**, 205–17.

Redfern, C. P. F., Daly, A. K., Latham, J. A. E., and Todd, C. (1990). The biological activity of retinoids in melanoma cells: induction of expression of retinoic acid receptor-$\beta$ by retinoic acid in S91 melanoma cells. *FEBS Letters*, **273**, 19–22.

Rowe, A., Eager, N. S. C., and Brickell, P. M. (1991*a*). A member of the RXR nuclear receptor family is expressed in neural-crest-derived cells of the developing chick peripheral nervous system. *Development*, **111**, 771–8.

Rowe, A., Richman, J. M., and Brickell, P. M. (1991*b*). Retinoic acid treatment alters the distribution of retinoic acid receptor-$\beta$ transcripts in the embryonic chick face. *Development*, **111**, 1007–16.

Rowe, A., Richman, J. M., Ochanda, J. O., and Brickell, P. M. (1991*c*). Retinoic acid treatment alters the pattern of retinoic acid receptor beta expression in the embryonic chick limb. In *Developmental patterning of the vertebrate limb*, (ed. J. R. Hinchliffe, J. Hurle, and D. Summerbell). Plenum, New York.

Ruberte, E., Dollé, P., Krust, A., Zelent, A., Morriss-Kay, G., and Chambon, P. (1990). Specific spatial and temporal distribution of retinoic acid receptor gamma transcripts during mouse embryogenesis. *Development*, **108**, 213–22.

Ruberte, E., Dollé, P., Chambon, P., and Morriss-Kay, G. (1991). Retinoic acid receptors and cellular retinoid binding proteins II. Their differential pattern of transcription during early morphogenesis in mouse embryos. *Development*, **111**, 45–60.

Smith, S. M. and Eichele, G. (1991). Temporal and regional differences in the expression pattern of distinct retinoic acid receptor-*β* transcripts in the chick embryo. *Development*, **111**, 245–52.

Smith-Thomas, L., Lott, I., and Bronner-Fraser, M. (1987). Effects of isotretinoin on the behavior of neural crest cells *in vitro*. *Developmental Biology*, **123**, 276–8.

Tamarin, A., Crawley, A., Lee, J., and Tickle, C. (1984). Analysis of upper beak defects in chicken embryos following treatment with retinoic acid. *Journal of Embryology and Experimental Morphology*, **84**, 105–23.

Tickle, C. and Brickell, P. M. (1991). Retinoic acid in limb development. *Seminars in Developmental Biology*, **2**, 189–97.

Vaessen, M-J., Meijers, J. H. C., Bootsma, D., and Guerts van Kessel, A. (1990). The cellular retinoic-acid-binding protein is expressed in tissues associated with retinoic-acid-induced malformations. *Development*, **110**, 371–8.

Wedden, S. E. (1987). Epithelial-mesenchyme interactions in the development of chick facial primordia and the target of retinoid action. *Development*, **99**, 341–51.

Wedden, S. E., Lewin-Smith, M. R., and Tickle, C. (1987). Analysis of the effects of retinoids on cartilage differentiation in micromass cultures of chick facial primordia and the relationship to a specific facial defect. *Developmental Biology*, **122**, 78–89.

Wedden, S. E., Ralphs, J. R., and Tickle, C. (1988). Pattern formation in the facial primordia. *Development*, **103**, Supplement, 31–40.

Wedden, S. E. and Tickle, C. (1986). Quantitative analysis of the effect of retinoids on facial morphogenesis. *Journal of Craniofacial Genetics and Developmental Biology*, **2**, 169–78.

Zelent, A., Mendelsohn, C., Kastner, P., Krust, A., Garnier, J-M., Ruffenach, F., Leroy, P., and Chambon, P. (1991). Differentially expressed isoforms of the mouse retinoic acid receptor *β* are generated by usage of two promoters and alternative splicing. *EMBO Journal*, **10**, 71–81.

8

# Retinoic acid receptors and binding proteins in mouse limb development

Esther Ruberte, Harikrishna Nakshatri, Philippe Kastner, and
Pierre Chambon

## INTRODUCTION

The vertebrate limb appears as an outgrowth in the flank of the embryonic body. Initially it consists of a mesenchymal core covered by a layer of ectoderm. During the elongation of the limbs the apparently homogeneous population of the mesenchymal cells gives rise to different cell types: cartilage, bone, muscle, and other tissues, that develop in appropriate spatial relationships to constitute the developed limb. From grafting experiments that have been performed mainly in the chick, it appears that two mechanisms are involved in the formation of the proximodistal (from the base to the tip) and the anteroposterior (from the thumb to the little finger) pattern of the limbs.

Development of the proximodistal axis involves the mesenchymal cells in the distal tip of the limbs, which through a mesenchymal–ectodermal interaction, induce the overlying ectoderm to thicken. This results in the formation of a pseudostratified columnar epithelium known as the apical ectodermal ridge (AER). This ridge appears to be essential for the outgrowth of the limbs and grafting experiments have shown that its removal results in a truncated limb (Zwilling 1956; Saunders and Gasseling 1968). Once the AER is formed it exerts an inductive influence on the underlying mesenchyme, which grows rapidly and differentiates. Positional values along the proximodistal axis are defined by the time spent in an area of undifferentiated mesenchyme under the tip of the bud, known as the progress zone (Summerbell *et al.* 1973). Cells leaving the progress zone at early times differentiate into proximal structures, while the cells that leave the progress zone later give rise to distal structures.

The anteroposterior positional values are provided by a region of mesenchyme located at the posterior margin, known as the zone of polarizing activity (ZPA). Grafting of this region into an anterior position induces a duplication in the anteroposterior axis, which results in a mirror symmetrical duplication of the digits; thus instead of the normal set of three digits (2, 3, and 4) a new pattern with six digits arises (4, 3, 2, 2, 3, and 4) (Saunders and Gasseling 1968; Tickle *et al.* 1975). It has been suggested that the ZPA releases a morphogen that diffuses across the limb bud to form a gradient with the highest concentration in the posterior margin (Tickle *et al.* 1975).

EFFECTS OF RETINOIDS ON LIMB PATTERN FORMATION

In mammals, vitamin A (retinol) and its metabolites are essential for vision, reproduction, normal embryo morphogenesis, and differentiation of a number of cell types. Retinol is taken from the plasma and stored in the liver, from which it is subsequently delivered to the target tissues (for a review see Blomhoff *et al.* 1990). Retinol can be converted to retinoic acid (RA) via retinal; the reaction is irreversible. In most instances, RA can substitute for vitamin A under hypovitaminosis conditions, but some physiological processes, such as vision and spermatogenesis, specifically require retinol.

It has been known for a long time that vitamin A and its derivatives, collectively known as retinoids, play an important role in vertebrate limb development and in amphibian limb regeneration (for reviews see Brockes 1989; Eichele 1989). There is some experimental evidence that supports the idea that RA could be the morphogen released by the ZPA. When a bead soaked in RA is implanted in the anterior margin of the chick limb, it induces duplications of the anteroposterior axis similar to those obtained by ZPA grafting (Tickle *et al.* 1982; Summerbell 1983). The effects obtained by such RA treatment are dose-dependent, as soaking a bead in low concentrations of RA leads to additional digits two and/or three (2, 2, 3, and 4 or 3, 2, 2, 3, and 4), whereas soaking a bead in high concentrations of RA results in additional digits two, three, and four (4, 3, 2, 2, 3, and 4 or 4, 3, 2, 3, and 4) (Tickle *et al.* 1982; Summerbell 1983; Tickle *et al.* 1985). This suggests that the formation of a given digit is dependent on the concentration of RA to which the mesenchymal cells are exposed, and that in order to form a pattern of digits two, three, and four, the cells of the mesenchymal limb bud are exposed to a graded concentration of RA. It has indeed been shown that the application of RA to the anterior margin of the limb bud results in an RA concentration gradient (Tickle *et al.* 1985; Eichele and Thaller 1987). In addition there is some evidence of a gradient of RA through the chick limb bud, with the highest concentrations in the posterior margin (Eichele and Thaller 1987; Thaller and Eichele 1987).

A recent publication, however, suggests a different mechanism by which exogenously added RA could act on the chick limb. The application of a bead soaked in RA to the anterior margin of the limb bud would induce the anterior cells to form a polarizing region. This newly formed ZPA would then respecify the anteroposterior axis of the limb (Wanek *et al.* 1991).

RETINOIC ACID RECEPTORS

Three genes encoding the retinoic acid receptors RAR-$\alpha$ (Giguère *et al.* 1987; Petkovich *et al.* 1987), RAR-$\beta$ (Benbrook *et al.* 1988; Brand *et al.* 1988) and

RAR-$\gamma$ (Krust *et al.* 1989; Zelent *et al.* 1989) have been characterized. The three receptor genes show a high degree of homology and several RAR isoforms can be generated from each of them (Kastner *et al.* 1990; Leroy *et al.* 1991; Zelent *et al.* 1991; for a review, see Leroy *et al.* Chapter 2). Sequence comparisons have demonstrated that the RARs belong to the superfamily of steroid/thyroid hormone nuclear receptors, which act as inducible transcriptional transregulators by binding to cognate *cis*-acting DNA response elements (enhancers in the case of positive regulation) (for a review see Green and Chambon 1988). The three RARs bind RA and synthetic retinoids with different affinities (Aström *et al.* 1990), and it is assumed that they could differentially activate or repress the transcription of subsets of RA-responsive genes. Thus, the existence of multiple receptors and isoforms may account for the highly pleiotropic effects of a simple molecule such as RA.

## CELLULAR RETINOID BINDING PROTEINS

In addition to the nuclear receptors there are several cellular cytoplasmic retinol- and RA-binding proteins (CRBP and CRABP, respectively) (Ong and Chytil 1978*a,b*; for additional references see Dollé *et al.* 1990). They belong to a superfamily of low molecular weight proteins, which all bind hydrophobic ligands. Two CRBPs (I and II) have been described. Their function is unknown, but it has been proposed that they are involved in the metabolism of retinol and its conversion into retinoic acid (for a review see Blomhoff *et al.* 1990).

Similarly, there are at least two CRABPs, I and II, which bind RA with high affinity (for review and references see Blomhoff *et al.* 1990; Giguère *et al.* 1990). It has been proposed that CRABPs could transport RA from the cytoplasm to the nucleus, where it would be transferred to the nuclear receptors (Takase *et al.* 1979, 1986). Alternatively, it has been suggested that the role of CRABPs could be controlling the actual concentration of free RA in a given cell (Robertson 1987; Maden *et al.* 1988; Smith *et al.* 1989; Boylan and Gudas 1991; Ruberte *et al.* 1991). Maden *et al.* (1988) have described a graded distribution of CRABP I in the chick limb bud, with the highest levels in the anterior margin of the limb, thus opposite to the putative concentration gradient of total RA (see above), in such a way that the presence of CRABP may result in a steeper gradient of free RA available for binding to RARs.

## DISTRIBUTION OF RAR, CRABP, AND CRBP TRANSCRIPTS IN THE DEVELOPING MOUSE LIMB

The patterns of distribution of RNA transcripts of the three nuclear receptors, of CRABP I and II, and of CRBP I have been studied by *in situ* hybridization at various stages of mouse limb development (Dollé *et al.* 1989*a*; Ruberte *et al.*

1990; Dollé *et al.* 1990; Ruberte *et al.* 1991; our unpublished results). The RAR probes used to date do not discriminate between the RNA transcripts corresponding to the different isoforms of each RAR.

At day 10.5 of gestation both fore- and hindlimb buds are present in the mouse embryo. They are composed of undifferentiated mesenchyme, surrounded by an ectodermal layer, and the most distal cells are proliferating actively (progress zone) under the influence of the newly formed AER. The transcripts of two nuclear receptors, RAR-α and RAR-γ, appear to be evenly distributed throughout the mesenchymal cells of the limb buds. In contrast RAR-β transcripts are restricted to the very proximal region of the limbs, with CRABP I showing a reciprocal distribution (Fig 8.1(a)). Only RAR-α could be detected in the ectodermal layer. At this stage no CRBP I transcripts can be detected in the limb buds (Fig. 8.1(a)).

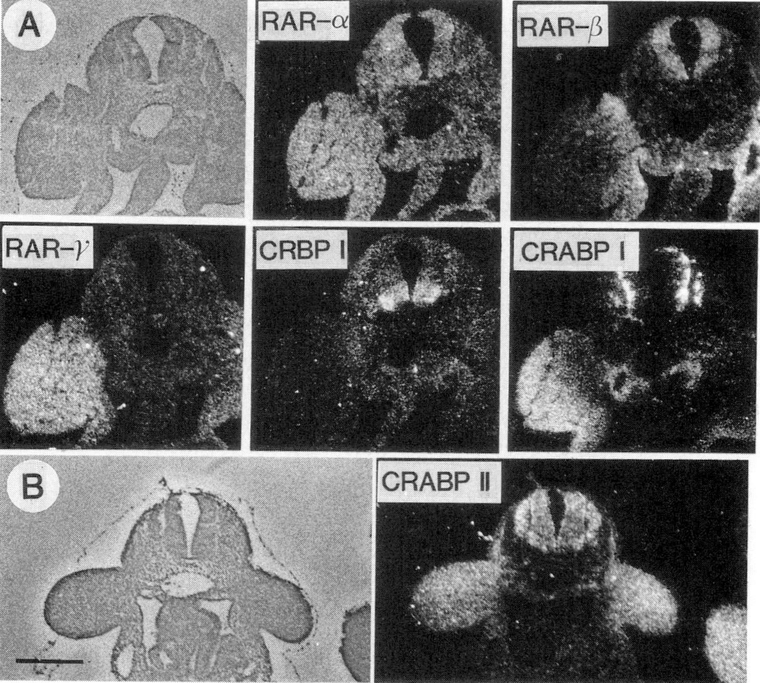


**Fig. 8.1** Comparison of the expression domains of the three RARs and retinoid binding proteins in the mouse limb buds at the stage of undifferentiated mesenchyme. (*a*) Frontal sections through the hindlimb of a 11.5 day embryo; (*b*) frontal section through the forelimbs of a 10.5 day embryo. Scale bar: 250 μm.

Serial sections almost perpendicular to the anteroposterior axis of the forelimb at day 10.5 of development are shown in Fig. 8.2. Section A is the most anterior; however, since the planes of the sections are slightly oblique, the left limb (L) is always cut more anteriorly than the right (R). The

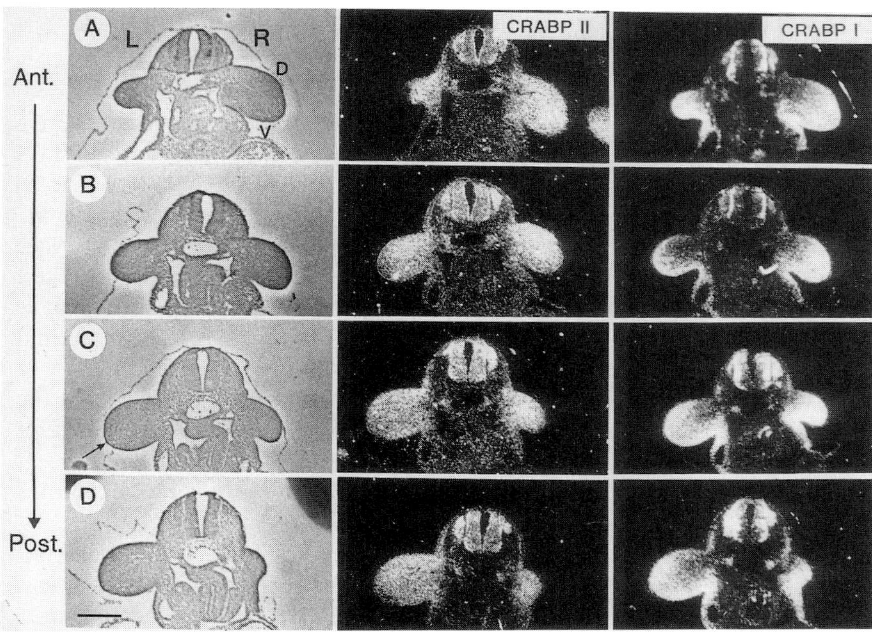


**Fig. 8.2** Serial sections through the anteroposterior axis of the forelimb buds of a 10.5 day mouse embryo. Consecutive sections have been hybridized with the CRABP I and II probes. As the plane of the sections is slightly oblique, the left limb (L) represents a more anterior plane of section than the right limb (R). Ant., anterior; Post., posterior, D, dorsal; V, ventral; arrow, AER. Scale bar: 250 $\mu$m.

consecutive sections have been hybridized with CRABP I and II probes. There is a proximal to distal graded increase in CRABP I transcripts in the anterior region (Fig. 8.2 (b–d), left limb), whereas they are more homogeneously distributed in the posterior region (Fig. 8.2 (a–c), right limb), in a such a way that the anterior portion of the limb core appears devoid of CRABP I transcripts. CRABP I transcripts are also present in the dorsal portion of the AER (not illustrated, see Dollé *et al.* 1989*a*). The immunohistochemistry study of Maden *et al.* (1988) has similarly revealed a higher concentration of CRABP I in the distal region of the chick limb. Furthermore, these authors have reported an anteroposterior gradient of CRABP I with opposite polarity to that of RA. We do not see an equivalent gradient of CRABP I transcripts in the mouse. In 10.5 day limbs, CRABP II transcripts appear to be more evenly distributed than those of CRABP I, although they are more abundant in the dorsal than in the ventral portion of the limbs (Figs 8.1(b) and 8.2). In the distal portion of the limbs, the rapidly proliferating mesenchymal cells are particularly rich in CRABP I transcripts, but are poorer in CRABP II transcripts than the rest of the mesenchymal cells (Figs 8.1 and 8.2).

At approximately day 12.5 p.c. the limbs are paddle-shaped and the first signs of cytodifferentiation appear. RAR-α transcripts are almost ubiquitously distributed, and will remain so at later developmental stages. In contrast, RAR-γ transcripts become progressively restricted to the presumptive precartilaginous blastema (Fig. 8.3). RAR-β transcripts are restricted to a group of cells located close to the precartilaginous blastema (opaque patch) and in the interdigital mesenchyme. Interestingly, both these regions include cells that will undergo programmed cell death (Sulik and Dehart 1988; Alles and Sulik 1989). At the same developmental stage CRBP I transcripts are found in areas surrounding the precartilaginous blastema, and a signal is detected in the interdigital mesenchyme as well as in the marginal ectodermal ridge (Fig. 8.3 and data not shown).

At the same stage both CRABP I and II transcripts are excluded from the precartilaginous condensations that express RAR-γ; the CRABP II transcript distribution represents an almost negative image of the limb bone models (Fig. 8.3(b)), whereas the CRABP I transcript distribution is more complex (Fig. 8.3(a)).

At 14.5 days p.c., RAR-β transcripts are found in the interdigital mesenchyme, as well as in cells adjacent to the cartilage of the phalanges. RAR-γ transcripts are detected in the cartilage cells of every bone model (Fig. 8.4(a)). CRBP I transcripts are still found in the periphery of the cartilaginous models as well as in the interdigital mesenchyme (Fig. 8.4(a)). Transcripts for both CRABP I and II are detected in the periphery of the cartilaginous bone models, particularly in the developing tendons (Fig. 8.4(a) and (b), and data not shown).


POSSIBLE FUNCTION OF MULTIPLE RETINOIC ACID RECEPTORS AND RETINOID BINDING PROTEINS

*In situ* hybridization analyses reveal that the transcript distribution of the three nuclear RA receptors and of the cellular RA and retinol binding proteins exhibit specific spatiotemporal patterns during limb development, which strongly suggests that each of the receptors and binding proteins performs a different function. Note, however, that the presence of a transcript does not necessarily mean that the corresponding protein is synthesized, as regulatory post-transcriptional mechanisms may exist.

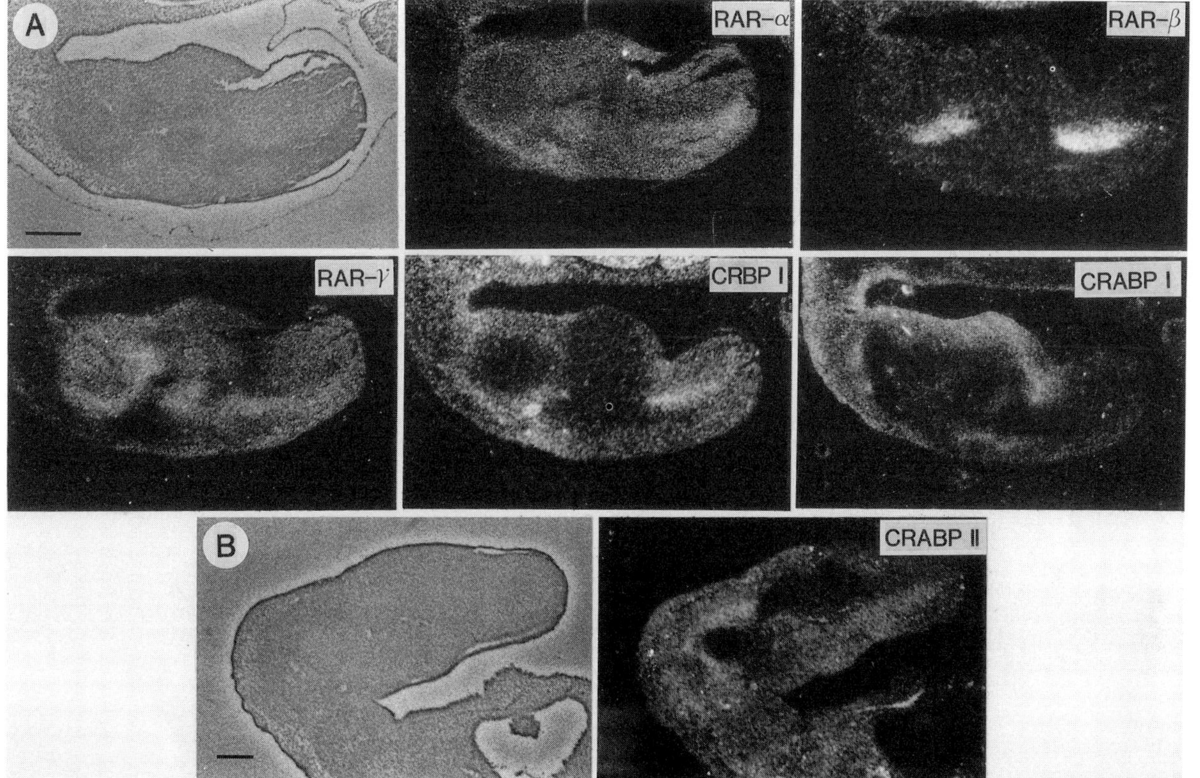


**Fig. 8.3**  Comparison of the expression domains of the RARs, CRABPs and CRBP I genes in the limb of a 12.5 day embryo. At this stage the first signs of cytodifferentiation appear. (*a*) Longitudinal section of a hindfoot plate; (*b*) longitudinal section of a forelimb. Scale bar: 250 $\mu$m.

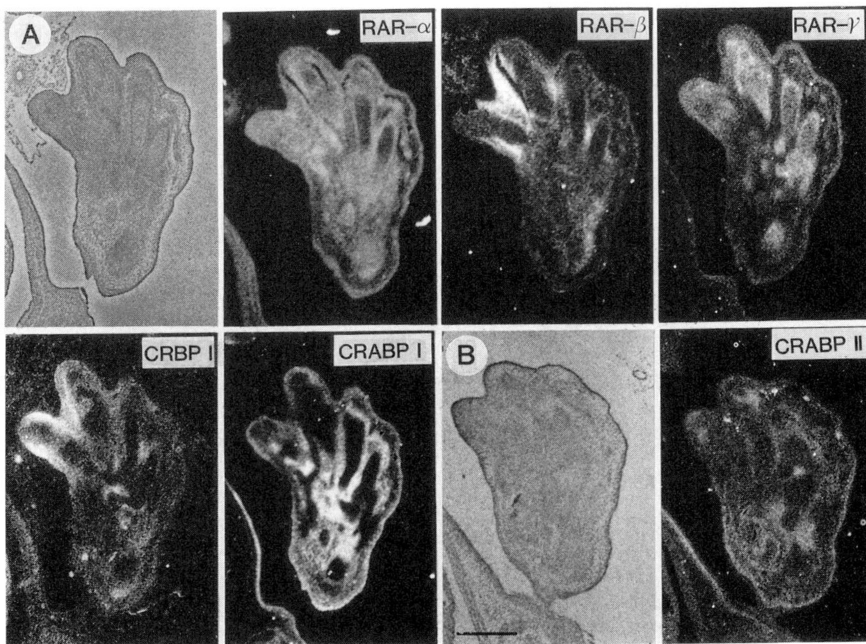



**Fig. 8.4** Comparison of the transcript domains of the three RARs, CRABP I and II, and CRBP I in hindfoot of a 14.5 day embryo. (*a*) Consecutive sections from the same embryo; (*b*) equivalent section from an embryo of the same stage. Scale, bar: 500 $\mu$m.

Transcripts coding for RAR-α are ubiquitously distributed in the limb at all stages of development, indicating that all cells are likely to contain at least one of the RARs. In this respect, although the present *in situ* hybridization analyses do not allow us to to differentiate between the transcripts coding for the different isoforms, we note that the promoter from which the RAR-α1 isoform is transcribed has a structure typical of a housekeeping gene (Brand *et al.* 1990; see Leroy *et al.* Chapter 2). The distribution of RAR-α, RAR-β and RAR-γ transcripts in the limbs are not always mutually exclusive, and thus some cells contain transcripts coding for two or all three RARs. In the early limb bud, for example, RAR-α and RAR-γ transcripts are coexpressed throughout the mesenchymal cells, whereas all three RARs are coexpressed only in the most proximal region of the limb.

To date, the expression of a number of genes has been reported to respond to RA, e.g. some of the homeogenes (Mavilio *et al.* 1988; Simenone *et al.* 1990; for additional references see Ruberte *et al.* 1991) and genes encoding some components of the extracellular matrix (for references see Nicholson *et al.* 1990; Vasios *et al.* 1991). The different patterns of distribution of RAR transcripts in the developing limb support the proposal that the three RARs differentially control the transcription of different subsets of RA-responsive genes. Indeed, as RAR-α transcripts are apparently present in all cells, there would be no need for coexpression of RAR-β and/or RAR-γ, if all three RARs were to act similarly on all RA-responsive genes. Future studies will show whether the three RAR subtypes (and possibly their isoforms) differentially recognize the various response elements of RA-responsive genes, and/or modulate differentially the transcription of these genes depending on their promoter context, i.e. the nature of the other regulatory factors that control their transcription. Obviously, such differential recognition and promoter context-specific activity would provide at least part of the combinatorial mechanism required to account for the pleiotropic effects of RA during embryogenesis. Note that further combinatorial possibilities may result from differential affinity of the three RARs for retinoids (Aström *et al.* 1990), and from the possible formation of heterodimers (Forman and Samuels 1990) between the various RAR subtypes and isoforms, which may also exhibit different target gene specificity. Such combinatorial regulations may explain how a given RAR subtype can participate in the control of different programmes of gene expression in cells with different developmental fates. For example, in a 14.5 day limb, RAR-β transcripts are present in the cells of the interdigital mesenchyme, which will probably contribute to the tendons and other connective tissues. The final programme of gene expression controlled by RAR-β in these cells could be dictated by the specific regulatory factors that are coexpressed in the interdigital region and in the periphery of the digits.

The observation that CRABP I and II transcripts exhibit different patterns of distribution suggests that they each perform specific functions in the developing limb. However, in agreement with the results of previous studies on cells in culture (Chytil and Ong 1985), CRABP I and II do not appear to be indispensable for the transduction of the RA signal, as they are not present in the limb cartilage cells that express RAR-γ and are known to be targets for retinoids (Kochhar *et al.* 1984; Langille *et al.* 1989; for further references see Ruberte *et al.* 1990). it has been proposed that CRABPs play an important role in controlling the levels of free RA available for binding to the nuclear RA receptors (Robertson 1987; Maden *et al.* 1988; Smith *et al.* 1989; Ruberte *et al.* 1991). In this respect, it is interesting to note that both CRABP I and II transcripts are not evenly distributed in the mesenchyme of the limb bud, which suggests that the mesenchymal cells could be exposed to a complex pattern of RA concentrations, which may specify positional cues.

Homeogenes are likely candidates for the interpretation of RA signals, as at least some of the homeogenes have been shown to respond to RA both *in vitro* (Mavilio *et al.* 1988; Simeone *et al.* 1990; for additional references see Ruberte *et al.* 1990) and *in vivo* (Izpisúa-Belmonte 1991; Nohno *et al.* 1991). With the exception of *Hox-4.1*, each gene of the *Hox-4* complex has a unique restricted expression domain in the posterior part of the developing mouse (Dollé *et al.*

1989*b*) and chick (Nohno *et al.* 1991) limb according to its position in the complex. Moreover, there is a sequential triggering of expression of the *Hox-4* genes from the posterodorsal region of the limb, which apparently coincides with the ZPA. There is also a graded expression of each gene along the anteroposterior axis, which is highest in the posterior region and decreases anteriorly. Strikingly, grafting a polarizing region or a local application of RA (Izpisúa-Belmonte *et al.* 1991; Nohno *et al.* 1991) to the anterior margin of chick wing buds, results in mirror-image patterns of *Hox-4* gene expression, which correlate with the subsequent appearance of mirror-image duplications of digits. These results are consistent with the idea that RA is involved in the specification of the expression of the *Hox-4* complex. As discussed above, the pattern of expression of CRABP I and II is complex. Both of them are maximally expressed in the posterior region of the limb, but CRABP I expression is the highest in the distal region where CRABP II expression is the lowest. Interestingly, the region of maximal overlap of expression of the CRABP I and II genes corresponds to the region of appearance of *Hox-4* gene expression. Whether and how the RA binding proteins I and II are involved in RA-induced expression of the *Hox-4* complex genes during limb development will require the generation of mice in which the expression of CRABP I and II has been altered.

ACKNOWLEDGEMENTS

We thank Gillian Morriss-Kay for fruitful discussions and Cathy Mendelsohn and Mark Leid for critical reading of the manuscript. This work was supported by grants from the CNRS, the INSERM, the Association pour la Recherche sur le Cancer and the Fondation pour la Recherche Médicale Française.

REFERENCES

Alles, A. J. and Sulik, K. K. (1989). Retinoic-acid-induced limb-reduction defects: perturbation of zones of programmed cell death as a pathogenetic mechanism. *Teratology*, **40**, 163–71.

Aström, A., Pettersson, U., Krust, A., Chambon, P., and Voorhees, J. J. (1990). Retinoic acid and synthetic analogs differentially activate retinoic acid receptor dependent transcription. *Biochemical and Biophysical Research Communications*, **173**, 339–45.

Benbrook, D., Lernhardt, E., and Pfahl, M. (1988). A new retinoic acid receptor identified from a hepatocellular carcinoma. *Nature*, **330**, 669–72.

Blomhoff, R., Gree, M. H., Berg, T., and Norum, K. R. (1990). Transport and storage of vitamin A. *Science*, **250**, 339–404.

Boylan, J. F. and Gudas, L. J. (1991). Overexpression of the cellular retinoic acid binding protein-I (CRABP-I) results in a reduction in differentiation-specific gene expression in F9 teratocarcinoma cells. *Journal of Cell Biology*, **112**, 965–79.

Brand, N., Petkovich, M., Krust, A., Chambon, P., de Thé, H., Marchio, A., Tiollais, P., and Dejean, A. (1988). Identification of a second human retinoic acid receptor *Nature*, **332**, 850–3.

Brand, N. J., Petkovich, M., and Chambon, P. (1990) Characterization of a functional promoter for the human retinoic acid receptor-alpha (hRAR-α). *Nucleic Acids Research*, **18**, 6799–806.

Brockes, J. (1989). Retinoids, homeobox genes, and limb morphogenesis. *Neuron*, **2**, 1285–94.

Chytil, F. and Ong, D. E. (1985). In *The retinoids*, vol. 2, (eds M. B. Sporn, A. B. Roberts, and D. S. Goodman), pp. 90–123. Academic Press, Orlando, FL.

Dollé, P., Ruberte, E., Kastner, Ph., Petkovich, M., Stoner, C. M., Gudas, L., and Chambon, P. (1989*a*). Differential expression of genes encoding α, β and γ retinoic acid receptors and CRABP in the developing limbs of the mouse. *Nature*, **342**, 702–5.

Dollé, P., Izpisúa-Belmonte, J. C., Falkenstein, H., Renucci, A., and Duboule, D. (1989*b*). Coordinate expression of the murine Hox-5 complex homeo-box-containing genes during limb pattern formation. *Nature*, **342**, 767–72.

Dollé, P., Ruberte, E., Leroy, P., Morriss-Kay, G., and Chambon, P. (1990). Retinoic acid receptors and cellular binding proteins I. A systematic study of their differential pattern of transcription during the mouse organogenesis. *Development*, **110**, 1133–51.

Eichele, G. (1989). Retinoids and vertebrate limb pattern formation. *Trends in Genetics*, **5**, 246–51.

Eichele, G. and Thaller, C. (1987). Characterisation of concentration gradients of morphogenetically active retinoids in the chick limb bud. *Journal of Cell Biology*, **105**, 1917–23.

Forman, B. M. and Samuels, H. H. (1990). Interactions among a subfamily of nuclear hormone receptors: the regulatory zipper model. *Molecular Endocrinology*, **4**, 1293–301.

Giguère, V., Ong, E. S., Segui, P., and Evans, R. M. (1987). Identification of a receptor for the morphogen retinoic acid. *Nature*, **330**, 624–9.

Giguère, V., Lyn, S., Yip, P., Siu, C. H., and Amin, S. (1990). Molecular cloning of cDNA encoding a second cellular retinoic acid-binding protein. *Proceedings of the National Academy of Science USA*, **87**, 6233–7.

Green, S. and Chambon, P. (1988). Nuclear receptors enhance our understanding of transcription regulation. *Trends in Genetics*, **4**, 309–14.

Izpisúa-Belmonte, C., Tickle, C., Dollé, P., Wolpert, L., and Duboule, D. (1991). Expression of the homeobox Hox-4 genes and the specification of position in chick wing development. *Nature*, **350**, 585–9.

Kastner, Ph., Krust, A., Mendelsohn, C., Garnier, J. M., Zelent, A., Leroy, P., Staub, A., and Chambon, P. (1990). Murine isoforms of retinoic acid receptor γ with specific patterns of expression. *Proceedings of the National Academy of Science USA*, **87**, 2700–4.

Kochhar, D. M., Penner, J. D., and Tellone, C. (1984). Comparative teratogenic activities of two retinoids: Effects on palate and limb development. *Teratogenesis, Carcinogenesis and Mutagenesis*, **4**, 377–87.

Krust, A., Kastner, P., Petkovich, M., Zelent, A., and Chambon, P. (1989). A third human retinoic acid receptor, hRAR-γ. *Proceedings of the National Academy of Science USA*, **86**, 5310–14.

Langille, R. M., Paulsen, D. F., and Solursh, M. (1989). Differential effects of physiological concentrations of retinoic acid in vitro on chondrogenesis and myogenesis in chick craniofacial mesenchyme. *Differentiation*, **40**, 84–92.

Leroy, P., Krust, A., Zelent, A., Mendelsohn, C., Garnier, J. M., Kastner, P., Dierich, A., and Chambon, P. (1991). Multiple isoforms of the mouse retinoic acid receptor α are generated by alternative splicing and differential induction by retinoic acid. *EMBO Journal*, **10**, 59–69.

Maden, M., Ong, D. E., Summerbell, D., and Chytil, F. (1988). Spatial distribution of cellular protein binding to retinoic acid in the chick limb bud. *Nature*, **335**, 733–5.

Mavilio, F., Simeone, A., Boncinelli, E., and Andrews, P. W. (1988). Activation of four homeobox gene clusters in human embryonal carcinoma cells induced to differentiate by retinoic acid. *Differentiation*, **37**, 73–9.

Nicholson, R. C., Mader, S., Nagpal, S., Leid, M., Rochette-Egly, C., and Chambon, P. (1990). Negative regulation of the rat stromelysin gene promoter by retinoic acid is mediated by an AP1 binding site. *EMBO Journal*, **13**, 4443–54.

Nohno, T., Noji, S., Koyama, E., Ohyama, K., Myokai, F., Kuroiwa, A., Saito, T., and Taniguchi, S. (1991). Involvement of the Chox-4 chicken homeobox genes in determination of anteroposterior axial polarity during limb development. *Cell*, **64**, 1197–205.

Ong, D. E. and Chytil, F. (1978a). Cellular retinol binding protein from rat liver. *Journal of Biological Chemistry*, **253**, 828–32.

Ong, D. E. and Chytil, F. (1978b). Cellular retinoic acid binding protein from rat testis. *Journal of Biological Chemistry*, **253**, 4551–4.

Petkovich, M., Brand, N. J., Krust, A., and Chambon, P. (1987). A human retinoic acid receptor which belongs to the family of nuclear receptors. *Nature*, **330**, 444–50.

Robertson, M. (1987). Towards a biochemistry of morphogenesis. *Nature*, **330**, 420–1.

Ruberte, E., Dollé, P., Krust, A., Zelent, A., Morriss-Kay, G., and Chambon, P. (1990). Specific spatial and temporal distribution of retinoic acid receptor gamma transcripts during mouse embryogenesis. *Development*, **108**, 213–22.

Ruberte, E., Dollé, P., Chambon, P., and Morriss-Kay, G. (1991). Retinoic acid receptors and cellular retinoid binding proteins II. Their differential pattern of transcription during early morphogenesis in mouse embryos. *Development*, **111**, 45–60.

Saunders, J. W. and Gasseling, M. T. (1968). Ectodermal–mesenchymal interactions in the origin of limb symmetry. In *Epithelial–mesenchymal interactions*, (eds R. Fleischmajer and R. E. Billingham), pp. 78–97. Williams and Wilkins, Baltimore.

Simeone, A., Arampora, D., Arcioni, L., Andrews, P. W., Boncinelli, E., and Mavilio, F. (1990). Sequential activation of human HOX-2 homeobox genes by retinoic acid in embryonal carcinoma cells. *Nature*, **346**, 763–6.

Smith, S. M., Pang, K., Sunding, O., Wedden, S. E., Thaller, C., and Eichele, G. (1989). Molecular approaches to vertebrate limb morphogenesis. *Development*, **107** (Suppl.), 121–31.

Sulik, K. K. and Dehart, D. B. (1988). Retinoic-acid-induced limb malformations resulting from apical ectodermal ridge cell death. *Teratology*, **37**, 527–37.

Summerbell, D. (1983). The effect of local application of retinoic acid to the anterior margin of the developing chick limb. *Journal of Embryology and Experimental Morphology*, **78**, 269–89.

Summerbell, D., Lewis, J. H., and Wolpert, L. (1973). Positional information in chick limb bud morphogenesis. *Nature*, **244**, 492–6.

Takase, S., Ong, D. E., and Chytil, F. (1979). Cellular retinol-binding protein allows specific interaction of retinol with the nucleus in vitro. *Proceedings of the National Academy of Science USA*, **76**, 2204–20.

Takase, S., Ong, D. E., and Chytill, F. (1986). Transfer of RA from complex with cellular retinoic acid binding protein to the nucleus. *Archives of Biochemistry and Biophysics*, **247**, 328.

Thaller, C. and Eichele, G. (1987). Identification and spatial distribution of retinoids in the developing chick limb bud. *Nature*, **327**, 625–8.

Tickle, C., Summerbell, D., and Wolpert, L. (1975). Positional signalling and specification of digits in chick limb morphogenesis. *Nature*, **254**, 199–202.

Tickle, C., Alberts, B., Wolpert, L., and Lee, J. (1982). Local application of retinoic acid to the limb bud mimics the action of the polarizing region. *Nature*, **296**, 564–5.

Tickle, C., Lee, J., and Eichele, G. (1985). A quantitative analysis of the effect of all-*trans*-retinoic acid on the pattern of chick wing development. *Developmental Biology*, **109**, 82–95.

Vasios, G., Mader, S., Gold, J. D., Leid, M., Lutz, Y., Gaub, M.-P., Chambon, P., and Gudas, L. (1991). The late retinoic acid induction of laminin B1 gene transcription involves RAR binding to the responsive element. *EMBO Journal*, **5**, 1149–58.

Wanek, N., Gardiner, D. M., Muneoka, K., and Bryant, S. V. (1991). Conversion by retinoic acid of anterior cells into ZPA cells in the chick wing bud. *Nature*, **350**, 81–3.

Zelent, A., Krust, A., Petkovich, M., Kastner, Ph., and Chambon, P. (1989). Cloning of murine α and β retinoic acid receptors and a novel receptor γ predominantly expressed in the skin. *Nature*, **339**, 714–17.

Zelent, A., Mendelsohn, C., Kastner, Ph., Krust, A., Garnier, J. M., Ruffenach, F., Leroy, P., and Chambon, P. (1991). Differentially expressed isoforms of the mouse retinoic acid receptor β are generated by usage of two promoters and alternative splicing. *EMBO Journal*, **10**, 71–81.

Zwilling, E. (1956). Interaction between limb bud ectoderm and mesoderm in the chick embryo, I. Axis estabishment. *Journal of Experimental Zoology*, **132**, 157–72.

# 9

# Effect of retinoic acid on regenerating urodele limbs

Jeremy Brockes

## INTRODUCTION

Limb regeneration is an important system for evaluating the morphogenetic effects of retinoids. In turn, the activity of retinoids on epimorphic regeneration promises to shed much light on basic aspects of this process. In this brief review I shall outline the main events of limb regeneration and the effects of retinoic acid (RA) on axial specification, as well as discussing hypotheses for the mechanism of these effects and the prospects offered by the identification of nuclear RA receptors in the limb and its regenerate. I have not considered the effects on avian limb development in this account because these have been considered elsewhere in this volume.

## LIMB REGENERATION IN URODELE AMPHIBIANS

The urodele (tailed) amphibians are vertebrates that can regenerate their limbs as adults, e.g. the newt and the axolotl. In anurans such as the frog, the ability to regenerate a limb is lost progressively at around the stage of metamorphosis, but tadpole limb regeneration has been studied and was the system in which the effects of retinoids on limb morphogenesis were discovered (Niazi and Saxena 1978). After amputation of the urodele limb at any level on the proximodistal axis (PD) the epidermal cells at the circumference migrate rapidly over the cut surface to seal the limb. They form a structure called the wound epidermis, which is thought to play an important role in outgrowth of the regenerate (Wallace 1981). The wound epidermis is formed, at least to the extent of monolayer formation, during the first 12–24 hours. During the subsequent period, the mesenchymal tissue underlying the wound epidermis in the vicinity of the amputation plane undergoes an important progressive change in identity. This process, loosely referred to as dedifferentiation (Ferretti and Brockes 1991), results in the local formation of blastemal cells, the progenitors of the mesenchymal tissue of the regenerate. The blastemal cells divide, initially under control of the nerve supply, and subsequently undergo differentiation and morphogenesis to form the regenerate.

Blastemal cells are mesenchymal progenitors. They give rise to the skeletal elements, the connective tissue, and the muscle, i.e. to those cell types that are the substrate for pattern formation in the limb. It is noteworthy from this overall description that limb regeneration depends on two fundamental and somewhat interlinked processes—the derivation of blastemal cells from tissue in the immediate vicinity of the amputation plane, followed by events of division, differentiation, and morphogenesis that are reminiscent of development. One way in which cell origin and subsequent fate are linked is particularly important for understanding the action of RA. Blastemal cells that arise after amputation at a particular location on the PD axis only give rise to the missing structures, i.e. to the structures distal to this location. It is sometimes said that such cells possess a positional memory that prevents them from giving rise to proximal structures (Stocum 1984). This property behaves as a continuously graded variable running along the PD axis, but its molecular basis is unknown. Proximal and distal blastemas display differences in the expression of homeobox genes (Brown and Brockes 1991) and a member of the cytokeratin family (Ferretti *et al.* 1991), as well as behaving differently in cellular assays (Stocum and Crawford 1987), but the relevance of these for positional memory is unclear. It is also not understood whether this property is inherited directly from the cells of origin in the limb, or if it is induced by the action of a position dependent signal operating in the early regenerate.


EFFECTS OF RETINOIDS ON LIMB REGENERATION


When RA or precursor retinoids are delivered to the early regenerate, either by application to the water surrounding the animal or by injection into the peritoneal cavity, the blastema is initially inhibited from further division. After the retinoid level decreases, either by transferring the animals to fresh water or following metabolic breakdown, the blastema resumes division and gives rise to structures proximal to its point of origin (Maden 1982). Thus, a wrist blastema, which normally forms a hand, may give rise to an entire arm. These effects, which are graded and dependent on the concentration of RA, are not produced by any other molecules to date. A variety of studies have clarified several important aspects of the effect. RA is able to proximalize the blastema only when applied during the phase of dedifferentiation when the blastemal cells are generated (Maden *et al.* 1985). The effects do not depend on the proximity of the stump for their appearance and an RA-treated blastema can be transplanted to a neutral location and still give rise to proximal structures.

An important aspect of the action of RA in urodeles is the distinction between limb development and regeneration. The use of monoclonal antibodies as cell markers has revealed several differences between the mesenchymal cells of the developing bud and the regenerating blastema (Ferretti and Brockes 1991). The latter, for example, contain a particular pair of cytokeratin filaments that are absent in the limb bud (Ferretti *et al.* 1989). The morphogenetic effects of RA in proximalizing the blastema are not shown by the bud. The doses of RA that evoke morphogenetic duplications in regenerating limbs produce deletions or truncations in developing limbs (Scadding and Maden 1986*a*). This difference in the response of developing and regenerating limbs raises the question of the relationship between the morphogenetic effects on axial specification and the long observed teratogenic effects. A completely satisfying account of the action of retinoids in the limb must in the end encompass both aspects of their effects.

Another issue in understanding retinoid effects on regeneration is respecification of the transverse axes of the limb. In the regenerating anuran tadpole limb, RA produces mirror-image duplications in the anteroposterior (AP) axis (Niazi and Saxena 1978; Scadding and Maden 1986*b*) as well as serial duplications in the PD axis, but AP effects are not generally observed in the urodele. It has been shown that RA is able to posteriorize limbs with surgically constructed double anterior sections (Kim and Stocum 1986), and more recently, to ventralize double dorsal regenerates (Ludolph *et al.* 1990). Both of these latter effects are unidirectional in the sense that they are not exerted on double posterior or double ventral constructions, and in consequence these are unable to regenerate. Thus, after 'normal' amputation the missing distal values are replaced, and RA treatment is able to proximalize them in a dose-dependent fashion. The underlying cellular and molecular mechanisms remain unclear but it is possible to make some general comments.


POSSIBLE MECHANISMS OF MORPHOGENETIC EFFECTS OF RETINOIC ACID


One possibility that is tacitly assumed by some accounts is that RA can act directly on blastemal cells to change their positional value. The response would thus be a cell autonomous property. A related but distinct possibility is that RA promotes some change in an interacting population, for example the epidermis, that respecifies the blastemal cells. This is somewhat similar to the proposal in the chick limb bud that a local implant of RA induces neighbouring mesenchymal cells to become a polarizing region that signals the formation of a new posterior boundary (Wanek *et al.* 1991). A somewhat different model is that RA does not act on pattern formation directly, but rather promotes local dedifferentiation to form a larger blastema. The positional effects would follow from interactions between the blastemal cells that establish a more extensive complement of positional values. This would be broadly consistent with the timing of action of RA on the blastema, in that the effects are exerted during the phase of dedifferentiation.

In general terms, it is now thought that the effects of RA are mediated by controlling gene expression. This conviction stems from the identification of steroid/thyroid hormone-like nuclear receptors that interact directly with RA and mediate changes in transcription of genes that have an appropriate response element. One approach to clarifying the mechanistic questions discussed above is to identify the receptor or receptors that control the morphogenetic effects (Ragsdale and Brockes 1990).

RETINOIC ACID RECEPTORS IN THE URODELE LIMB AND BLASTEMA

There are three members of the RAR family identified to date in mouse and human. They differ in all regions of the molecule but most markedly in the A and F regions. As the A region is subject to a variety of splicing reactions, the most diagnostic region for a particular isotype is the F region. When newt limb blastemal cDNA libraries were analysed for the presence of RAR clones, a clear homologue of mouse and human RAR-$\alpha$ was isolated and shown to be expressed in the newt limb (Ragsdale *et al.* 1989). A partial clone that appears to be a homologue of human RAR-$\beta$ has also been obtained (Giguère *et al.* 1989). The third isotype in the newt limb appears to be distinct and is called RAR-$\delta$ (Ragsdale *et al.* 1989). Delta appears to be a relative of $\gamma$ but its identity to the F region of human RAR-$\gamma$ is only 26 per cent at the amino acid level.

Delta is the most abundant RAR in the newt limb and blastema and it increases on amputation at both RNA and protein levels (Ragsdale and Brockes 1990). Two different isoforms that result from alternative A regions have been identified. The major isoform, $\delta$-1, is unrelated to all of the other A region sequences found in human and mouse. The second isoform, $\delta$-2, is related to the $\gamma$-2 isoform and in view of the failure to detect a clear newt gamma clone, it seems that $\delta$ may be the newt equivalent of $\gamma$. Because of its sequence identity and high level of expression in the limb compared to other tissues, the $\delta$-1 isoform seems to be the most promising candidate for mediating some of the effects of RA on the urodele limb. It is, however, a significant challenge for the future to obtain experimental evidence about the functional roles of the different receptors in the limb.

The modular structure of the RARs suggests several approaches to interfering with their function in limb cells. For example, it is possible to construct mutant receptors that lack the EF regions and hence do not interact with RA (Espeseth *et al.* 1989). If such molecules are transfected into tissue culture cells they block the function of resident RARs, possibly by binding to response elements in the DNA and acting as repressors. Another possibility is the construction of chimeric receptors in which the EF region is replaced with the hormone-binding domain from another nuclear receptor, for example the thyroid hormone receptor. Under these circumstances the biological effects of the RAR come under the control of the new ligand.

While such approaches are rapidly applicable to the study of retinoid effects in culture, it is more challenging to bring them to bear on the problems of limb regeneration in the animal. To do this we have established limb blastemal cells in cell culture (Ferretti and Brockes 1988). The cells continue to express characteristic markers for blastemal cells and they can be passaged in culture without obvious crisis or senescence. It is possible to introduce plasmids into such cells by microinjection of either the nucleus or cytoplasm. The cultured cells can be appropriately marked and implanted as a pellet underneath the wound epidermis of an early blastema, from where they are efficiently recruited into the regenerate. Our current efforts are directed at pursuing such approaches with a variety of mutant RAR constructions. Such experiments may help to clarify some of the uncertainties in our understanding of retinoid effects on the regenerating limb.

REFERENCES

Brown, R. and Brockes, J. P. (1991). Identification and expression of a regeneration-specific homeobox gene in the newt limb blastema. *Development*, **111**, 489–96.
Espeseth, A. S., Murphy, S. P., and Linney, E. (1989). Retinoic acid receptor expression vector inhibits differentiation of F9 embryonal carcinoma cells. *Genes and Development*, **3**, 1647–56.
Ferretti, P. and Brockes, J. P. (1988). Culture of newt cells from different tissues and their expression of a regeneration-associated antigen. *Journal of Experimental Zoology*, **247**, 77–91.
Ferretti, P. and Brockes, J. P. (1991). Cell origin and identity in limb development and regeneration. *Glia*, **4**, 214–24.
Ferretti, P., Brockes, J. P., and Brown, R. (1991). A newt type II keratin restricted to normal and regenerating limbs and tails is responsive to retinoic acid. *Development*, **111**, 497–507.
Ferretti, P., Fekete, D. M., Patterson, M., and Lane, E. B. (1989). Transient expression of simple epithelial keratins by mesenchymal cells of regenerating newt limb. *Developmental Biology*, **133**, 415–24.
Giguère, V., Ong, E. S., Evans, R. M., and Tabin, C. J. (1989). Spatial and temporal expression of the retinoic acid receptor in the regenerating amphibian limb. *Nature*, **337**, 566–9.
Kim, W.-S. and Stocum, D. L. (1986). Retinoic acid modifies positional memory in the anteroposterior axis of regenerating axolotl limbs. *Developmental Biology*, **114**, 170–9.
Ludolph, D. C., Cameron, J. A., and Stocum, D. L. (1990). The effect of retinoic acid on positional memory in the dorsoventral axis of regenerating axolotl limbs. *Developmental Biology*, **140**, 41–52.
Maden, M. (1982). Vitamin A and pattern formation in the regenerating limb. *Nature*, **295**, 672–5.
Maden, M., Keeble, S., and Cox, R. A. (1985). The characteristics of local application of retinoic acid to the regenerating axolotl limb. Roux's *Archives of Developmental Biology*, **194**, 228–35.

Niazi, I. A. and Saxena, S. (1978). Abnormal hind limb regeneration in tadpoles of the toad, Bufo andersoni, exposed to excess vitamin A. *Folia Biologica (Krakow)*, **26**, 3–11.

Ragsdale, C. W. and Brockes, J. P. (1990). Retinoic acid receptors and vertebrate limb morphogenesis. In *Structure and function of hormone nuclear receptors*, (ed. M. G. Parker). Academic Press, London.

Ragsdale, C. W., Petkovich, M., Gates, P. B., Chambon, P., and Brockes, J. P. (1989). Identification of a novel retinoic acid receptor in regenerative tissues of the newt. *Nature*, **341**, 654–7.

Scadding, S. R. and Maden, M. (1986a). Comparison of the effects of vitamin A on limb development and regeneration in the axolotl, Ambystoma mexicanum. *Journal of Embryology and Experimental Morphology*, **91**, 19–34.

Scadding, S. R. and Maden, M. (1986b). Comparison of the effects of vitamin A on limb development and regeneration in Xenopus laevis tadpoles. *Journal of Embryology and Experimental Morphology*, **91**, 35–53.

Stocum, D. L. (1984). The urodele limb regeneration blastema. Determination and organization of the morphogenetic field. *Differentiation*, **27**, 13–28.

Stocum, D. L. and Crawford, K. (1987). Use of retinoids to analyse the cellular basis of positional memory in regenerating amphibian limbs. *Biochemie et Cell Biologie*, **65**, 750–61.

Wallace, H. (1981). *Vertebrate limb regeneration*. Wiley, Chichester.

Wanek, N., Gardiner, D. M., Muneoka, K., and Bryant, S. V. (1991). Conversion by retinoic acid of anterior cells into ZPA cells in the chick wing bud. *Nature*, **350**, 81–3.

# 10

# Retinoid binding proteins in the developing vertebrate nervous system

Malcolm Maden, Emily Gale, Claire Horton, and Jim C. Smith

## RETINOIC ACID AND CENTRAL NERVOUS SYSTEM DEVELOPMENT

It has been known for more than 40 years that the development of the central nervous system (CNS) depends crucially on the correct balance of vitamin A in the maternal diet. Nutritional and teratological studies on the effects of hyper- and hypovitaminosis A have revealed devastating consequences for the embryo (Kalter and Warkany 1959). The organ system primarily affected is the nervous system and its derivative, the neural crest. It is quite likely, therefore, that the CNS requires specific levels of vitamin A, perhaps in the form of retinoic acid (RA), for its normal development.

Two recent pieces of evidence support this suggestion. First, more precise teratological studies have identified specific CNS regions whose development can be suppressed by RA while the remaining parts of the CNS develop normally. In the zebrafish embryo, RA administration results in the loss of an area of CNS between the caudal midbrain and rostral hindbrain, which would normally form the cerebellum (Holder and Hill 1991). In the *Xenopus* embryo, increasing RA doses cause the loss of increasing amounts of CNS commencing with the forebrain (Durston *et al.* 1989), and a disruption of rostral hindbrain segmentation (Papalopulu *et al.* 1991). In the rat embryo the same area of rostral hindbrain is deleted (Morriss and Thorogood 1978), perhaps in humans too (Lammer *et al.* 1985 and Chapter 21) and neural crest migration is inhibited both *in vitro* (Thorogood *et al.* 1982) and *in vivo* (Pratt *et al.* 1987).

The second piece of evidence is that we have recently identified endogenous RA in the CNS using high pressure liquid chromatography (HPLC) (Hunter *et al.* 1991). To overcome the problems of small amounts of tissue we used the spinal cord of the larval axolotl (*Ambystoma mexicanum*), an animal that retains various embryonic characteristics into adult life, including the ability to produce new neurons. The level of RA in the spinal cord detected by HPLC was comparable to that found in the chick limb bud. Furthermore, we also demonstrated that RA stimulates neurite outgrowth from explants of the spinal cord with a particular threshold response.

Ideally, of course, one would like to measure the levels of endogenous RA in, for example, the embryonic midbrain or each of the hindbrain rhombomeres, but this is not yet possible. What is possible, however, is to look at the distribution of the retinoic acid receptors (RARs) and the retinoid-binding proteins in specific domains of the CNS. As RA or retinol are their endogenous ligands this method may give a good indication of the tissues that require retinoids for their development.

RETINOIC ACID RECEPTORS AND RETINOID-BINDING PROTEINS IN THE CENTRAL NERVOUS SYSTEM

The distribution of RAR-$\alpha$, $\beta$, and $\gamma$ has been described in the developing mouse CNS (Dollé *et al.* 1990; Ruberte *et al.* 1991) and of RAR-$\beta$ and RXR-$\alpha$ in the embryonic chick CNS (Smith and Eichele 1991; Rowe *et al.* 1991). Each is present in specific domains, suggesting that RA, their ligand, must play important roles in relation to the control of the pattern of gene activity during CNS development. The same can be said of the two retinoid-binding proteins—cellular retinoic acid-binding protein (CRABP) and cellular retinol-binding protein (CRBP)—which also have specific distributions in the developing CNS. These are the subject of this chapter.

Before describing their distributions it is important to note that there are two different forms of each binding protein: CRBP I, CRBP II, CRABP I, and CRABP II. CRBP I (15.7 KDa) (Bashor *et al.* 1973) is widespread in embryos and adults. CRBP II (16 KDa) is present almost exclusively in the mucosal absorptive cells of the adult rat small intestine (Crow and Ong 1985). As its function is the uptake of retinol from the gut it is not relevant to the CNS. CRABP I (15.5 KDa) (Ong and Chytil 1975) and CRABP II (16.2 KDa) (Kitamoto *et al.* 1988; Bailey and Siu 1988; Giguère *et al.* 1990) are closely related proteins that differ in the N-terminal 25 amino acids by three amino acids in the chick and by seven amino acids in the rat and mouse. With regard to the distributions described below, the *in situ* hybridizations published to date have all been specific to CRABP I (Vaessen *et al.* 1989, 1990; Dollé *et al.* 1990; Ruberte *et al.* 1991). CRABP II *in situ*s are currently being performed (Ruberte *et al.* Chapter 8). On the basis of our immunocytochemical studies we have suggested (Maden *et al.* 1990) that CRABP I is specific to embryonic neural tissues (including the neural crest) and CRABP II is present in other mesenchymal tissues, e.g. the limb bud. This accords with immunoblot studies in the chick (Kitamoto *et al.*, 1989), the identical *in situ* patterns of Vaessen *et al.* (1989, 1990) in the CNS and branchial arches and with our most recent studies in the chick with an antibody specific to CRABP I (Maden *et al.* 1991). But it does not accord with the *in situ* descriptions of CRABP I in the mouse limb bud (Dollé *et al.* 1989) or the *in situ*s showing CRABP II in the mouse nervous system (Ruberte *et al.* unpublished data). Thus there are still several unresolved issues concerning the relation between mRNA and protein distributions, which must be borne in mind with regard to the following descriptions.


CRABP IN THE CHICK CENTRAL NERVOUS SYSTEM

When the chick embryo was dissected into its component parts, CRABP was detected at highest levels in the limb buds, followed by spinal cord, followed by brain and it was not detected in other regions of the embryo, such as the trunk or eye (Momoi *et al.* 1989) confirming the importance of this molecule in CNS development.

In the stage 12 chick embryo (16 somites, 4 days of development), CRABP transcripts were detected by *in situ* hybridization in the cells in the outer layer of the mesencephalon, rhombencephalon, and spinal cord, but not in the prosencephalon (Vaessen *et al.* 1990). By stage 22 these areas were still labelled; in addition, labelling was present in the cranial and dorsal root ganglia and in the dorsolateral areas of the spinal cord. Outside the CNS, CRABP expression was noted in the otic vesicle and particularly in neural crest derivatives, the craniofacial mesenchyme and the mesenchyme of the branchial arches.

We have studied the distribution of CRABP protein in the CNS during chick development (Maden *et al.* 1989*a*; Maden *et al.* 1991). CRABP-positive cells first appeared in the neural tube at stage 9 (7 somites, 3 days of development) and were located in the rhombencephalic region, appearing very soon after closure of the neural folds (at stage 8). As the neural tube continued to develop and somites were added in a posterior direction, CRABP immunoreactivity spread posteriorly within the neural tube to remain about 6–8 somites anterior to the last formed somite. By stage 11, a remarkable pattern of labelling appeared in the newly formed rhombomeres, such that a single stripe was created. Immunoreactivity was present from more caudal regions up to and including rhombomere 6, was absent from rhombomere 5, present in rhombomere 4 and absent again in the rest of the hindbrain. This pattern remained until stages 14–16, by which time rhombomeres 1 and 2 had become labelled thus creating two unlabelled stripes. Following this the entire hindbrain became labelled and the striping pattern was lost. Over this period, the mesencephalon was also CRABP positive, especially the roof, whereas the prosencephalon remained unlabelled. These temporal patterns are summarized in Fig. 10.1.

The CRABP-positive neuroblasts in the developing spinal cord increased in number with time and eventually formed the commissural neurons in the lateral mantle layer (Maden *et al.* 1989*a*; Momoi *et al.* 1989). As development of the spinal cord proceeded, CRABP-positive commissural axons projected lateral and medial to the unlabelled motoneurons and then underneath the ventral floor plate.

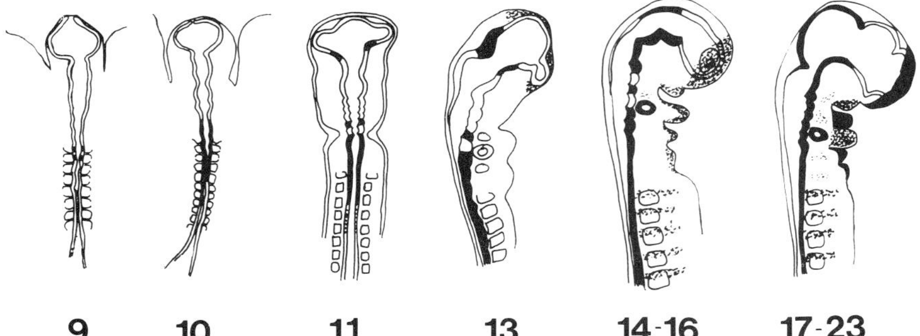



**Fig. 10.1** Drawings of the distribution of CRABP protein (shown as dark or stippled areas, stippled being lighter staining) at various stages of chick embryogenesis. Stages are marked below the drawings. Taken from Maden *et al.* (1991) and published with permission from Company of Biologists Ltd.

Outside the neural tube the major population of cells that expressed CRABP was the neural crest. Crest cells were seen from stage 13 onwards migrating out from the cephalic regions to populate the craniofacial mesenchyme and the branchial arches (Maden *et al.* 1991). In the branchial arches CRABP-positive cells were found peripherally, underneath the unlabelled epidermis and as found in the *in situ* hybridization study (Vassen *et al.* 1990) there was a variation in intensity of immunoreactivity between different arches. Arch 1 was the most intense followed by the frontonasal mesenchyme and arches 3 and 4 with the maxillary prominence and arch 2 the faintest. The only other cranial tissues to show immunoreactivity were the otic vesicle and some of the cranial ganglia.

In the trunk, CRABP-positive neural crest cells were seen migrating through the anterior halves of the somites from stage 14 onwards. As a continuation of this cell-type-specific localization, the neural crest derivatives, i.e. dorsal root ganglia, sensory axons, sympathetic ganglia, and enteric ganglia, retain CRABP antigenicity (Maden *et al.* 1989*a*).

## CRBP IN THE CHICK CENTRAL NERVOUS SYSTEM

We have also examined the chick embryo for the presence of retinol-binding protein, CRBP. Surprisingly, considering its widespread distribution in the mouse embryo (see below), immunocytochemical studies have identified only one specific part of the CNS in which CRBP can be detected, the ventral floor plate (Maden *et al.* 1989*b*). This showed a fascinating complementarity with the CRABP distribution within the commissural neurons of the mantle layer and in the commissural axons, which pass *below* the CRBP-positive floor plate cells. It is known, largely from the work of Jessell and colleagues (Dodd and Jessell 1988; Tessier-Lavigne *et al.* 1988; Placzek *et al.* 1990), that the commissural axons of the developing neural tube are attracted towards the ventral floor plate by the action of a floor plate chemoattractant. The axons then make abrupt right-angled turns to project in the rostrocaudal axis along the lateral surface of the floor plate.

These observations, taken together, have led us to the following speculation for the role of CRBP, CRABP, and their ligands in this aspect of neural development (Maden *et al.* 1989b). As the floor plate is rich in CRBP it could take up retinol from the blood, metabolize it to RA and release it to act as a chemoattractant. RA will thus diffuse dorsally within the neural tube and establish a gradient. As only commissural fibres possess CRABP, only they are able to respond to the gradient of RA, and they do so by growing up the gradient and thus arrive at the floor plate (Fig. 10.2). Recently, Wagner *et al.* (1990) inserted grafts of floor plate into the anterior margin of the chick limb bud, thereby inducing limb duplications exactly as one would obtain after grafting RA soaked beads. These results provide strong support for our hypothesis.

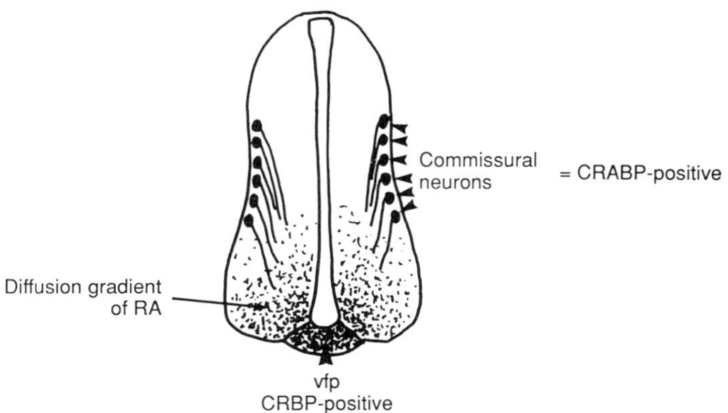


Fig. **10.2** Drawing representing a transverse section of the developing chick neural tube showing the distribution of CRABP and CRBP and a hypothesis of their function. The only CRBP-positive area is the ventral floor plate (vfp). It is suggested that the vfp takes up retinol, metabolizes it to RA and releases it into the neural tube where it will diffuse out forming a gradient (represented by shading). The CRABP-positive neurons are those of the commissural system in the lateral mantle layer. Being the only cells with CRABP, they alone respond to the RA gradient by growing their axons up the gradient and thus arrive at the vfp.

## CRABP IN THE MOUSE CENTRAL NERVOUS SYSTEM

From the available data, the distribution of CRABP-positive cell types in the mouse embryo seems to be very similar to that described above for the chick embryo. The most notable difference is that expression begins at an earlier stage in the mouse embryo. Like the chick embryo, CRABP is expressed at high levels in early to mid-gestation stages and expression drops markedly between days 16 and 19, as shown by developmental Northerns (Vaessen *et al.* 1989) and immunoblots (Dencker *et al.* 1990).

In the presomite mouse embryo (day 7.5–8.0) transcription of the CRABP gene as revealed by *in situ* hybridization occurred throughout the early mesoderm, including the allantois, and continued at later stages in the undifferentiated mesenchyme lateral to the primitive streak (Ruberte *et al.* 1991). As differentiation of the cranial mesenchyme occurred in early somite stage embryos (day 8), hybridization declined, except in expected areas of neural crest accumulation, for example the first branchial arch. In the neural epithelium, transcription began on day 8 caudal to the preotic sulcus and then spread (day 8.5) rostrally to incorporate the midbrain, hindbrain, and part of the cervical neural tube. On day 9 (Vaessen *et al.* 1989) CRABP transcripts were located in the outer layer of neural epithelium in the dorsal midbrain and in the anterior hindbrain. CRABP transcripts were noticably absent from any forebrain region. Outside the neural tube CRABP transcription continued in undifferentiated mesenchyme lateral to the primitive streak and, in cephalic regions, was concentrated in areas of neural crest accumulation, such as the frontonasal mesenchyme, branchial arches and cranial ganglia, and in the otic vesicle (Ruberte *et al.* 1991).

In the developing mouse spinal cord, Perez-Castro *et al.* (1989) have shown CRABP transcripts in the mantle layer of the alar plate and intermediate horn, suggesting an association with sensory neurons and commissural neurons.

We have studied the distribution of CRABP protein in the developing spinal cord (Maden *et al.* 1990) and find an identical pattern to that of the chick embryo. At around day 9.5, CRABP-positive neuroblasts appeared in the mantle layer; one day later CRABP-positive axons projecting from these neuroblasts could be seen passing through the motor neurons on their way to the ventral floor plate. Thus these were commissural neurons. Outside the neural tube, CRABP immunoreactivity was present in neural crest cells passing through the anterior halves of the somites and accumulating in specific areas such as the dorsal root ganglia, Schwann cells along the nerves and enteric ganglia in the gut wall.

Studies on the distribution of CRABP protein in cephalic regions are the subject of the results described below, but one report will be mentioned here. Dencker *et al.* (1990) showed CRABP immunoreactivity in the roof of the midbrain and in the floor, roof, and walls of the hindbrain. Within the hindbrain the staining intensity of the rhombomeres appeared to vary. Outside the neural tube immunoreactivity was present in neural crest and its derivatives, namely, the frontonasal mesenchyme, the branchial arches, and in trails of crest cells entering the first and second arches.


## CRBP IN THE MOUSE CENTRAL NERVOUS SYSTEM

In contrast to CRABP, the distribution of CRBP in the mouse embryo is completely different from that of the chick embryo as described above, although it must be said that chick CRBP *in situ* hybridizations have not yet been done to confirm that the difference is not due to a comparison between RNA transcript localization in the mouse with protein localization in the chick (but see the amphibian CNS below).

In the presomite mouse embryo CRBP transcripts were mainly confined to the primitive streak region, where they were present throughout the epiblast, mesenchyme, and proximal allantois (Ruberte *et al.* 1991). In the neuroepithelium of the presomite and early somite embryo (day 8) CRBP transcripts were most abundant at the apical and basal surfaces of the epithelium, except in the forebrain region where they were undetectable. CRBP was also present in the neural crest migrating to, and within, the frontonasal mesenchyme and first branchial arch.

In the early spinal cord, CRBP transcripts were present only in the ventral part of the cord, which will give rise to the motoneurons (Perez-Castro *et al.* 1989). They were also present in the spinal nerves, dorsal root ganglia of the cervical region, and some cranial ganglia.

By immunocytochemistry we have also shown the ventral localization of CRBP protein in the developing spinal cord of the mouse embryo (Maden *et al.* 1990). It appeared on day 10 as a ventrally located stripe in the early cord, before any obvious differentiation of motor neurons. When they grew out, the motor axons also contained CRBP protein, resulting in a striped pattern in the brachial plexus where labelled motor nerves mixed with unlabelled sensory nerves. This suggests that retinol may play some role in differentiation of the motor neurons. The other areas that contained CRBP were the notochord and sympathetic ganglia. Thus, in contrast to the chick embryo, the ventral floor plate is not a site of CRBP localization, implying that the scheme described above for the guidance of commissural axons to the floor plate (Fig. 10.2) does not operate on the mouse. Perhaps this function is taken over by the CRBP-positive notochord.

We now describe two sets of experiments undertaken to fill in some of the gaps in our knowledge of CRABP distribution in the CNS. First, we have examined the distribution of CRABP protein in the cephalic regions of the developing mouse embryo. Second, we describe our experiments to determine where and when CRABP appears during CNS development in amphibians, the other vertebrate group which is currently used for RA studies on the nervous system.


## CRABP IN THE MOUSE CEPHALIC REGION

We have studied the distribution of CRABP protein in day 8–day 11 mouse embryos using the same antibody and immunocytochemical techniques as for the chick (Maden *et al.* 1991).

In the presomite (d8) embryo, CRABP immunoreactivity was seen in a thin stripe of mesenchyme in the region of the future hindbrain, posterior to the head fold. The staining was present only in the mesenchyme, not in the neuroepithelium or endoderm and could conceivably represent the initial establishment of a pattern that is later transferred to the neuroepithelium. This is a much more restricted localization than that found in CRABP *in situ* studies (see p. 124) where transcripts were found throughout the mesoderm and allantois. Possible reasons for the differences are discussed below.

In early somite embryos (d8.5) CRABP immunoreactivity was located in a stripe of neuroepithelium located in the future hindbrain region and in migrating neural crest cells. The hindbrain labelling was discrete and the labelled segment was located adjacent to the developing otic placode (Fig. 10.3(*a*)). Three areas of slight concavity were present in sections of this stage (Fig. 10.3(*a*)), representing the first appearance of rhombomeric subdivisions within the hindbrain, and the caudal two of these three areas were the ones that were labelled.

In day 9 embryos, CRABP-labelled neural crest cells were present scattered in the mesenchyme around the diencephalon and optic vesicle and as a tongue of cells leaving the most rostral labelled rhombomere (Fig. 10.3(*b*)). The immunoreactive hindbrain segments remained at the caudal end of the rhombencephalon and, as more rhombomeres were formed, three of them were labelled (Fig. 10.3(*b*)). These could now be clearly identified as rhombomeres 4, 5, and 6 due to their location adjacent to the otic vesicle.

In day 9.5 embryos, rhombomeres 4, 5, and 6 were still intensely CRABP positive, although the caudal boundary at rhombomere 6 was not so sharply delineated as at earlier stages and faint labelling appeared to tail off into rhombomere 7. Expression of CRABP in these three rhombomeres was transient because labelling was fading by day 10.5 and had disappeared altogether on day 11, to be replaced by labelling of the axons of the mantle layer throughout the rhombencephalon. This axonal immunoreactivity seemed to be a uniform response to the initiation of outgrowth within the mesencephalon, rhombencephalon and spinal cord (Maden *et al.* 1990) and is also seen in the chick embryo (Maden *et al.* 1991). Thus there are two phases of mouse neural development in which CRABP is present. First, an early, transient phase in the neuroepithelium throughout rhombomeres 4, 5, and 6

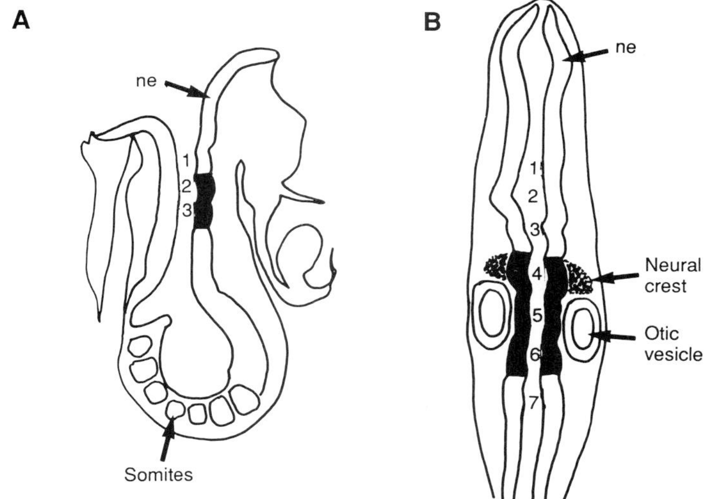



**Fig. 10.3** Drawings of the distribution of CRABP in the early mouse embryo. (*a*) Seven-somite embryo with the neural epithelium (ne) still open and three bulges, the early rhombomeres (1, 2, 3), present in the future hindbrain region. Only the caudal two of these are CRABP-positive (shown in dark shading). (*b*) Day 9 embryo sectioned through the hindbrain showing the rhombomeres (numbered 1, 2, 3, 4, 5, 6, 7) of the neural epithelium (ne) of which only three (numbers 4, 5 and 6) are CRABP-positive (shown in dark shading). The numbers of the rhombomeres can be assigned by reference to the otic vesicle. Also CRABP-positive are the neural crest cells leaving rhombomere 4 as a tongue of cells.

and second, a later phase in a wider population of cells as axogenesis commences.

CRABP immunoreactivity in the neural crest and the branchial arches showed an interesting variation related to the rhombomeric patterning. In day 9 embryos the stream of CRABP-positive crest cells from rhombomere 4 was directed towards arch 2, and that from rhombomere 6 towards arch 3. This arrangement left arch 1 CRABP-negative in contrast to the CRABP-positive frontonasal mass, maxillary prominence, arch 2 and arch 3.

In comparison with the distribution of CRABP I transcripts determined by *in situ* hybridization (Ruberte *et al.* 1991) CRABP protein is clearly much more restricted. The most likely explanation therefore is that CRABP mRNA is subject to post-transcriptional regulation.

CRABP IN THE AMPHIBIAN CENTRAL NERVOUS SYSTEM

It is clear that, from an early stage of development, the CNS of birds and mammals contains CRABP in particular areas, which may go some way towards explaining the observation described in the first section of this chapter (see p. 119) that the nervous system is a primary target of RA action during development. The other vertebrate nervous system that has been a focus of study for RA action on the nervous system is the amphibian, *Xenopus*.

In *Xenopus*, RA administration at early gastrulation stages (stage 10) results in a progressive truncation of the anteroposterior axis starting with the forebrain (Durston *et al.* 1989), a repression of anterior-specific genes and induction of a posterior-specific gene (Sive *et al.* 1990), and a disruption of rostral hindbrain segmentation (Papalopulu *et al.* 1991). This effect is temporally restricted to stages 9–14 (Sive *et al.* 1990). We can therefore ask: is CRABP present during these stages?

We have answered this question with immunocytochemistry by studying the distribution of CRABP on sections of embryos from early stages to post hatching stages. Axolotl embryos (*Ambystoma mexicanum*) were used for these experiments because the anti-CRABP antibody does not cross-react with *Xenopus* but does cross-react with axolotls, producing a single band of the appropriate molecular weight on immunoblots of axolotl CNS (Hunter *et al.* 1991).

As a preliminary, therefore, we had to show that axolotl embryos react to RA administration in the same way as *Xenopus* embryos. In the first experiment, axolotl embryos at the early gastrula stage were treated continuously with increasing concentrations of all-*trans*-RA. At concentration of $1.6 \times 10^{-7}$ M and above we noted a graded truncation of the anteroposterior axis similar to that described by Durston *et al.* (1989) for *Xenopus* (Fig. 10.4(*a*)). Interestingly, parallel experiments with *Xenopus* and axolotl embryos suggested that the axolotl is more sensitive to RA than *Xenopus*. As with *Xenopus*, however, the maximum period of sensitivity of axolotl embryos to RA is during gastrulation and neurulation; exposure during early cleavage stages or after neural tube closure has less effect. There was also disruption of tail formation in the axolotl, as reported to occur in *Xenopus* by Papalopulu *et al.* (1991) (Fig. 10.4(*a*)).

The anteroposterior reduction of axolotl development was particularly marked in neural structures and the histology of typical embryos is shown in Fig. 10.4(*b–f*). For example, at $8 \times 10^{-7}$ M RA, forebrain and midbrain structures are absent and there are no nasal placodes or eyes. However, head mesenchyme is present, together with a small heart, apparently normal notochord and mytomes. By $4 \times 10^{-6}$ M RA no brain structures are visible, but axial mesodermal structures appear relatively normal, as does the spinal cord (Fig. 10.4(*f*)). Overall, these results are very similar to those observed by Durston *et al.* (1989).

Having established the similarity of the effects of RA on the *Xenopus* and axolotl CNS, we then examined serially sectioned embryos at 14 different stages of axolotl development, from early gastrulation to a 12 mm hatched larva, for the presence of CRABP protein in the CNS. No immunoreactivity

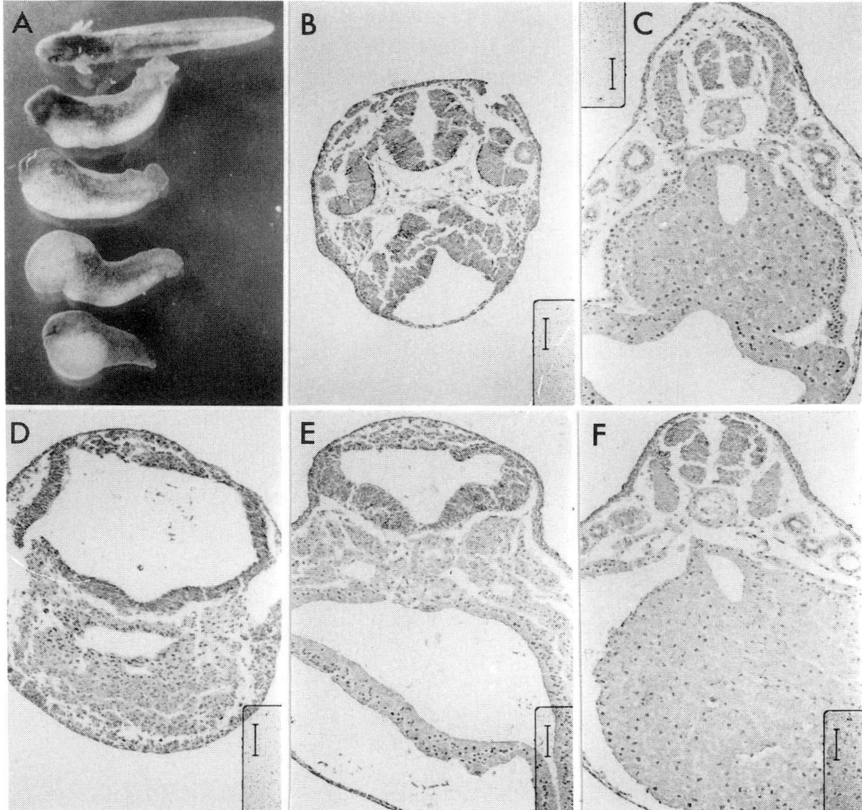

**Fig. 10.4** The effects of retinoic acid on early axolotl development. (*a*) Examples of axolotl larvae after continuous treatment with RA from the late blastula stage. Top, control embryo. Underneath, embryos treated with $8 \times 10^{-7}$, $4 \times 10^{-6}$, $2 \times 10^{-5}$ and $10^{-4}$ M RA, respectively. (*b*) Head of normal embryo showing eye cups and lenses. (*c*) Trunk of normal embryo. (*d*) Anterior region of embryo treated with $2 \times 10^{-5}$ M RA at a level where the nervous system can first be recognized. Notice disorganized appearance of the brain. (*e*) Further posterior in the same embryo as shown in (*d*). Notice expanded hindbrain. (*f*) The trunk region of the embryo shown in (*d*) and (*e*) appears relatively normal. Scale bars in (*b*)–(*f*) are 100 μm.

could be detected until very late in development, at stage 39. By this time the larva is fully formed, has begun to twitch its body and will soon hatch. In other words, the CNS is well developed. Thus, we could not detect any CRABP at the time of RA administration or during the period of neuronal specification.

The period when CRABP begins to be synthesized, however, corresponds to axogenesis and the development of axon pathways. CRABP is localized in very specific regions of the CNS. In the spinal cord it is found in the axons of the white matter and was evident in commissural neurons running below the ventral floor plate (Fig. 10.5(*a*)). In the brain it was evident that not all axons were CRABP positive. For example, in the hindbrain, axons in the dorsal tracts were unlabelled (Fig. 10.5(*b*)).

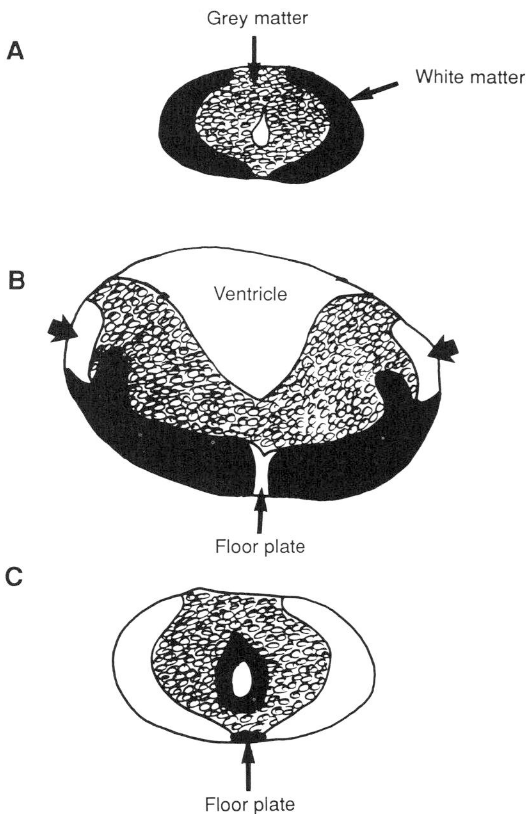


**Fig. 10.5** Drawings of the distribution of CRABP and CRBP in the axolotl spinal cord (after stage 40, 1–2 cm in length). Immunoreactivity is represented as black shading. (*a*) CRABP is seen in the axons of the white matter rather than in any of the cells (represented as open circles) of the grey matter. (*b*) The distribution of CRABP in the hindbrain showing two additional features: (i) not all axon tracts are CRABP-positive, those in the dorsal region (arrows) being unlabelled; (ii) the absence of CRABP reactivity in the floor plate region can be clearly seen here. (*c*) CRBP localization in the cord shows a reciprocal distribution to that of CRABP with immunoreactivity present in the floor plate region and occasionally in ependymal cells surrounding the neural canal.

We have also examined the axolotl CNS for the presence of CRBP and it shows an interesting complementarity to CRABP, remarkably similar to that described for the chick embryo (Maden *et al.* 1989*b* and see p. 123). CRBP is detected in a subpopulation of cells in contact with the neural canal and in the dorsal and ventral midline (Fig. 10.5(*c*)). In the axolotl these cells were predominantly radial glia, which have processes running through the white matter to the periphery of the cord (Holder *et al.* 1990). Immunoreactivity was particularly clear in the processes of the ventral floor plate glia. No immunoreactivity was detected in the white matter of the spinal cord.

Finally, we have confirmed the relevance of these results to the *Xenopus* embryo in the absence of the appropriate antibodies by performing sucrose density gradient centrifugation assays on *Xenopus* embryos from oöcyte stages to stage 40. No CRABP was detectable until stage 24 somewhat earlier than in the axolotl. Thus, we may conclude that the effects of RA on the amphian CNS cannot be mediated by CRABP.


CONCLUSIONS

The development of the nervous system may be seen in terms of three phases: (i) neuronal specification; (ii) neurite outgrowth; and (iii) maintenance. There is now considerable evidence to suggest that RA is involved in each of these phases.

The first phase—neuronal specification—is grossly altered by either the administration of excess RA or by a deficiency of vitamin A in the maternal diet, as described on page 119. Precise areas of the CNS, such as the cerebellum or rostral hindbrain segments, can be deleted (Morriss and Thorogood 1978; Holder and Hill 1991; Papalopulu *et al.* 1991). As we describe here, in the mouse embryo CRABP is present in specific domains of the early CNS, such as the roof of the midbrain, in specific rhombomeres (r4, r5, and r6) and in the neural crest that derives from these rhombomeres. Similar results have been obtained with the chick embryo (see p. 121). The RARs are also present in specific domains of the developing CNS (Ruberte *et al.* 1991; Smith and Eichele 1991). Thus, it is quite conceivable that precise levels of RA in precise regions of the developing CNS could play a part in neuronal specification in the vertebrate CNS.

With regard to the interesting expression domains of CRABP, the same does not seen to be true for the amphibian CNS. Although specification of the amphibian CNS can be altered (Durston *et al.* 1989) we could not detect the presence of CRABP in *Xenopus* or the axolotl until much later in development, when axons were forming their pathways in the spinal cord, that is, during neurite outgrowth.

Thus, during neurite outgrowth—the second phase of CNS development— RA is also likely to be involved. In each of the vertebrate species examined or

132  *M. Maden* et al.

described here–axolotl, chick and mouse—CRABP was present in a subset of spinal cord neurons—the commissural neurons. Moreover, the binding protein for retinol, CRBP, was present in axolotls and chicks in a complementary location in the ventral floor plate. As a result of these localizations we have proposed a specific hypothesis for the function of CRABP, CRBP, retinol, and RA in the spinal cord (See Fig. 10.2). The mouse embryo, however, does not have CRBP in the floor plate of the neural tube, but in a vertical stripe in cells that will become motor neurons. Motor axons subsequently remain CRBP-positive, suggesting a role for retinol in determination of motor neurons and outgrowth of their axons.

In the third phase of neural development—maintenance–retinoids may still play a role because the binding proteins continue to be expressed in the adult CNS (Ong *et al.* 1982). Thus, from the earliest moment of specification the levels of retinoids may have to be strictly regulated, perhaps in part by the binding proteins, to enable proper functioning of the CNS.


REFERENCES

Bailey, J. S. and Siu, C-H. (1988). Purification and partial characterisation of a novel binding protein for retinoic acid from neonatal rat. *Journal of Biological Chemistry*, **263**, 9326–32.
Bashor, M. M., Toft, D. O., and Chytil, F. (1973). In vitro binding of retinol to rat tissue components. *Proceedings of the National Academy of Science USA*, **70**, 3483–7.
Crow, J. A. and Ong, D. E. (1985). Cell-specific immunohistochemical localisation of a cellular retinol-binding protein (type two) in the small intestine of the rat. *Proceedings of the National Academy of Science USA*, **82**, 4707–47.
Dencker, L., Annerwall, E., Busch, C., and Eriksson, U. (1990). Localisation of specific retinoid-binding sites and expression of cellular retinoic acid-binding protein (CRABP) in the early mouse embryo. *Development*, **110**, 343–52.
Dodd, J. and Jessell, T. M. (1988). Axon guidance and the patterning of neuronal projections in vertebrates. *Science*, **242**, 692–99.
Dollé, P., Ruberte, E., Kastner, P., Petkovich, M., Stoner, C., Gudas, L., and Chambon, P. (1989). Differential expression of the genes encoding the retinoic acid receptors α, β, γ and CRABP in the developing limbs of the rat. *Nature*, 342, 702–5.
Dollé, P., Ruberte, E., Leroy, P., Morriss-Kay, G., and Chambon, P. (1990). Retinoic acid receptors and cellular retinoid binding proteins. I. A systematic study of their differential pattern of transcription during mouse organogensis. *Development*, **110**, 1133–52.
Durston, A. J., Timmermans, J. P. M., Hage, W. J., Hendricks H. F. J., de Vries, N. J., Heideveld, M., and Neiuwkoop, P. D. (1989). Retinoic acid causes an anteroposterior transformation in the developing central nervous system. *Nature*, **340**, 140–4.
Giguère, V., Lyn, S., Yip, P., Siu, C-H., and Amin, S. (1990). Molecular cloning of cDNA encoding a second cellular retinoic acid-binding protein. *Proceedings of the National Academy of Science USA*, **87**, 6233–7.

Holder, N., Clarke, J. D. W., Kamalati, T., and Lane, E. B. (1990). Heterogeneity in spinal radial glia demonstrated by intermediate filament expression and HRP labelling. *Journal of Neurocytology*, **19**, 915–28.

Holder, H. and Hill, J. (1991). Effects of retinoic acid on zebrafish embryos: modulation of engrailed protein expression at the midbrain-hindbrain border and development of cranial ganglia. *Development*, (in press).

Hunter, K., Maden, M., Summerbell, D., Eriksson, U., and Holder, N. (1991). Retinoic acid stimulates neurite outgrowth in the amphibian spinal cord. *Proceedings of the National Academy of Science USA*, **88**, 3666–70.

Kalter, H. and Warkany, J. (1959). Experimental production of congenital malformations in mammals by metabolic procedure. *Physiological Review*, **39**, 69–115.

Kitamoto, T., Momoi, T., and Momoi, M. (1988). The presence of a novel cellular retinoic acid-binding protein in chick embryos: purification and partial characterisation. *Biochemical Biophysical Research Communications*, **157**, 1302–8.

Kitamoto, T., Momoi, M., and Momoi, T. (1989). Expression of cellular retinoic acid–binding protein in chick embryos: purification and partial charcterisation. *Biochemical and Biophysical Research Communications*, **164**, 531–6.

Lammer, E. J., Chen, D. T., Hoar, R. M., Agnisti, N. D., Benke, P. J., Braun, J. T., Curry, C. J., Fernhoff, P. M., Grix, A. W., Lott, I. T., Richard, J. M., and Shyan, C. S. (1985). Retinoic acid embryopathy. *New England Journal of Medicine*, **313**, 837–41.

Maden, M., Ong, D. E., Summerbell, D., Chytil, F., and Hirst, E. A. (1989*a*). Cellular retinoic acid-binding protein and the role of retinoic acid in the development of the chick embryo. *Developmental Biology*, **135**, 124–32.

Maden, M., Ong, D. E., Summerbell, D., and Chytil, F. (1989*b*). The role of retinoid-binding proteins in the generation of pattern in the developing limb, the regenerating limb and the nervous system. *Development*, (Suppl), 109–19.

Maden, M., Ong, D. E., and Chytil, F. (1990). Retinoid-binding protein distribution in the developing mammalian nervous system. *Development*, **109**, 75–80.

Maden, M., Hunt, P., Eriksson, U., Kuriowa, A., Krumlauf, R., and Summerbell, D. (1991). Retinoic acid-binding protein, rhombomeres and the neural crest. *Development*, **111**, 35–44.

Momoi, T., Kitamoto, T., Kasuya-Sato, J., Seno, H., and Momoi, M. (1989). Developmental changes in the expression and distribution of cellular retinoic acid binding protein (CRABP) in the central nervous system of the chick embryo. *Biomedical Research*, **10**, 43–8.

Morriss, G. M. and Thorogood, P. V. (1978). An approach to cranial neural crest cell migration and differentiation in mammalian embryos. In *Development in mammals*, vol. 3, (ed. M. Johnson), pp. 363–412. Cambridge University Press, Cambridge.

Ong, D. E. and Chytil, F. (1975). Retinoic acid-binding protein in rat tissue. Partial purification and comparison to rat tissue retinol-binding protein. *Journal of Biological Chemistry*, **250**, 6113–17.

Ong, D. E., Crow, J. A., and Chytil, F. (1982). Radioimmunochemical determination of cellular retinol- and cellular retinoic acid-binding proteins in cytosols of rat tissues. *Journal of Biological Chemistry*, **257**, 13385–9.

Papalopulu, N., Clarke, J. D. W., Wilkinson, D., Krumlauf, R., and Holder, N. (1991). Retinoic acid causes abnormal development and segmental patterning of the anterior hindbrain in *Xenopus* embryos. *Development*, (in press).

134   *M. Maden* et al.

Perez-Castro, A. V., Toth-Rogler, L. E., Wei, L-N., and Nguyen-Huu, M. C. (1989). Spatial and temporal pattern of expression of the cellular retinoic acid-binding protein and the cellular retinol-binding protein during mouse embryogenesis. *Proceedings of the National Academy of Science USA*, **86**, 8813–17.

Placzek, M., Tessier-Lavigne, M., Yamada, T., Jessell, T., and Dodd, J. (1990). Mesodermal control of neural cell identity; floor plate induction by the notochord. *Science*, **250**, 985–8.

Pratt, R. M., Goulding, E. H., and Abbott, B. D. (1987). Retinoic acid inhibits migration of cranial neural crest cells in cultured mouse embryo. *Journal of Craniofacial Genetics and Developmental Biology*, **7**, 205–17.

Ruberte, E., Dollé, P., Chambon, P., and Morriss-Kay, G. (1991). Retinoic acid receptors and cellular retinoid binding proteins. II. Their differential pattern of transcription during early morphogenesis in mouse embryos. *Development*, **111**, 45–60.

Rowe, A., Eager, N. S. C., and Brickell, P. M. (1991). A member of the RXR nuclear receptor family is expressed in neural-crest-derived cells of the developing chick peripheral nervous system. *Development*, **111**, 771–8.

Sive, H. L., Draper, B. W., Harland, R. M., and Weintraub, H. (1990). Identification of a retinoic acid-sensitive period during primary axis formation in *Xenopus laevis*. *Genes and Development*, **4**, 932–42.

Smith, S. M. and Eichele, G. (1991). Temporal and regional differences in the expression pattern of distinct retinoic acid receptor- transcripts in the chick embryo. *Development*, **111**, 245–52.

Tessier-Lavigne, M., Placzek, M., Lumsden, A. G. S., Dodd, J., and Jessell, T. M. (1988). Chemotropic guidance of developing axons in the mammalian central nervous system. *Nature*, **336**, 775–8.

Thorogood, P., Smith, L., Nicol, A., McGinty, R., and Garrod, D. (1982). Effects of vitamin A on the behaviour of migratory neural crest cells *in vitro*. *Journal of Cell Science*, **57**, 331–50.

Vaessen, M-J., Kootwijk, E., Mummery, C., Hilkens, J., Bootsma, D., and van Kessel, A. G. (1989). Preferential expression of cellular retinoic acid binding protein in a subpopulation of neural cells in the developing mouse embryo. *Differentiation*, **40**, 99–105.

Vaessen, M-J., Meijers, J. H. C., Bootsma, D., and van Kessel, A. G. (1990). The cellular retinoic-acid-binding protein is expressed in tissues associated with retinoic-acid-induced malformations. *Development*, **110**, 371–8.

Wagner, M., Thaller, C., Jessell, T., and Eichele, G. (1990). Polarizing activity and retinoid synthesis in the floor plate of the neural tube. *Nature*, **345**, 819–22.

# PART III

# Retinoids and developing epithelia

# 11

# Interrelationships between two families of multifunctional effectors: retinoids and TGF-$\beta$

Anita B. Roberts, Adam B. Glick, and Michael B. Sporn

## INTRODUCTION

It has been recognized for quite some time that retinoids are important regulators of cell growth and differentiation. As early as 1925, Wolbach and Howe studied histological changes in tissues of vitamin A-deficient rats and suggested that there were important effects of retinoid status on both the differentiation and proliferation of cells. Later, Fell and Mellanby (1953) demonstrated that retinoids could control the phenotype of epithelial cells in tissue culture by showing that either retinol or retinyl acetate could induce keratinized embryonic chicken skin to begin secreting mucus and to elaborate ciliated cells; the opposite occurs in retinoid deficiency, where nasal or tracheal epithelia that are normally mucous-secreting and ciliated undergo a change to a squamous, keratinizing epithelium (Wolbach and Howe 1925). Approaching the problem somewhat differently, Lasnitzki (1955) demonstrated that the premalignant phenotype of mouse prostate glands that had been treated with the carcinogen 3-methyl cholanthrene could be altered by retinoid treatment in organ culture. The atypical cells were replaced by cells with more normal morphology, demonstrating that retinoids could suppress abnormal cellular differentiation that had been induced by a carcinogen and restore a more normal pattern of cellular differentiation. Collectively, these data underscore the importance of retinoids in the maintenance of epithelial homeostasis.

For the past 10 years, our laboratory has been studying a family of growth factors called transforming growth factor-$\beta$ (TGF-$\beta$; for a recent review see Roberts and Sporn 1990). Three distinct isoforms of this peptide, TGF-$\beta$s 1, 2, and 3, are found, often differentially expressed, in mammalian species. Although TGF-$\beta$s are produced by and act on most cell types, these peptides, like the retinoids, have profound effects on epithelia, strongly inhibiting the growth of most epithelial cells *in vitro* and exhibiting a localization pattern *in vivo* that strongly implicates them in differentiation and maintenance of epithelial tissues, and especially in interactions of epithelial and mesenchymal cells (Heine *et al.* 1990). The data of Barnard *et al.* (1989) suggest that endogenous TGF-$\beta$ might function to suppress growth and regulate differentiation of epithelial cells *in vivo*; they have demonstrated an inverse relationship between mitotic labelling and expression of TGF-$\beta$1 mRNA in isolated intestinal enterocytes. Moreover, Silberstein and Daniel (1987) demonstrated inhibitory effects of exogenous TGF-$\beta$ on growth and morphogenesis of mammary epithelium into end-buds; the inhibition was reversible and localized to regions adjacent to implanted slow-release pellets of the peptide.

Both retinoids and the TGF-$\beta$s are multifunctional, that is, the nature of their action on a target cell is critically dependent on many parameters, including the cell type and its state of differentiation, and the particular set of growth factors and hormones acting on the cell (Sporn and Roberts 1988). Many epithelial tissues exist in a dynamic state in which highly regulated mechanisms exist to control cellular differentiation and proliferation. That control depends, in part, on the immediate environment of the cell, including extracellular effector peptides, cell–cell contacts, and cell–matrix interactions (Nathans and Sporn 1991).

Ultimately, retinoids are thought to regulate the phenotype of cells by altering their pattern of gene transcription and expression of proteins (Sporn and Roberts 1991). The important question then becomes one of identifying which genes are regulated by the retinoids and which specific mechanisms are utilized. Although important new discoveries have been made, particularly in identification of nuclear receptors that mediate the effects of retinoids on gene transcription (Krust *et al.* 1989; Mangelsdorf *et al.* 1990; O'Malley 1990), the totality of the mechanism of action of retinoids in control of cellular differentiation and proliferation remains to be elucidated. To begin to address this problem, we have investigated whether retinoids—hormone-like molecules that act through a set of nuclear receptors—might, in some instances, control the expression of TGF-$\beta$—peptides that act principally through cell membrane receptors—and whether, as a consequence, certain of the actions of retinoids on epithelial tissues might be mediated by one or more of the TGF-$\beta$ isoforms. We demonstrate that retinoids selectively control expression of the TGF-$\beta$ isoforms not only *in vitro*, in cultured epithelial cells, but also *in vivo*, suggesting that the TGF-$\beta$s could play an important role in the mechanism of action of the retinoids.

RETINOIDS CONTROL EXPRESSION OF TGF-$\beta$2 IN CELLS *IN VITRO*

Retinoic acid (RA) and TGF-$\beta$ each inhibit DNA synthesis in cultured keratinocytes (Glick *et al.* 1989; Coffey *et al.* 1988*a,b*). For this reason, we investigated whether RA might regulate expression of TGF-$\beta$ mRNAs or proteins in these cells. In primary murine keratinocytes treated with 3 $\mu$M RA for 3–4 days, levels of secreted TGF-β were elevated greater than 200-fold (Glick *et al.* 1989). The amount of TGF-β secreted was dependent on the dose of RA; the concentration of RA half-maximal for induction of TGF-β secretion was 0.3–0.4 $\mu$M, similar to the $ED_{50}$ for inhibition of DNA synthesis by RA. Use of antisera specific for the different TGF-β isoforms demonstrated that the secreted TGF-β was nearly all the type 2 isoform.

Treatment of murine keratinocytes with retinoic acid elevated mRNA levels for both TGF-βs 1 and 2, although the increase in the level of TGF-β2 mRNA was substantially greater than that of TGF-β1 mRNA. However, nuclear run-on experiments demonstrated that the increase in steady-state level of TGF-β2 mRNA was not transcriptional, but rather must result from post-transcriptional effects (Glick *et al.* 1989).

A highly significant biological finding is that the TGF-β2 secreted in response to treatment of the keratinocytes with RA is in a biologically active form (Glick *et al.* 1989). Most cells in culture, as well as platelets induced to degranulate, release TGF-β in a biologically inactive, latent form in which the mature, active TGF-β is complexed non-covalently with the remainder of its own precursor, called the latency-associated protein (LAP; Wakefield *et al.* 1989). Interestingly, the ability of RA to induce secretion of activated TGF-β is shared with three other members of the steroid/retinoid family of nuclear receptors (O'Malley 1990), namely oestrogen, tamoxifen, and gestodene, a synthetic progestin (Knabbe *et al.* 1987; Colletta *et al.* 1990, 1991). The pharmacokinetics of the latent and active forms of TGF-β *in vivo* differ markedly, with the latent form having a substantially prolonged half-life and larger volume of distribution than active TGF-β (Wakefield *et al.* 1990). This could have important implications for the local action of the TGF-β induced by retinoids or steroid hormones. Use of blocking antibodies to TGF-β2 demonstrated that approximately 30 per cent of the inhibition of keratinocyte DNA synthesis resulting from treatment with retinoic acid was mediated by the secreted TGF-β2 (Glick *et al.* 1989).

Danielpour *et al.* (1991) have demonstrated that RA induces secretion of TGF-β in other cells as well, including human lung carcinoma A-549 cells and normal rat kidney (NRK) fibroblasts. As in the case of the keratinocytes, the isoform of TGF-β induced is TGF-β2. These cells are at least 100-fold more sensitive to induction by RA than are the keratinocytes, with $ED_{50}$s of 1–10 nM, but the magnitude of the induction of TGF-β2 was only 30–50 per cent that of the keratinocytes. The biological significance of the increase in secretion in TGF-β2 in these cells is not known.


RETINOIDS INDUCE SYNTHESIS OF TGF-β IN MOUSE EPIDERMIS *IN VIVO*

To ascertain that the observed induction of TGF-β2 secretion by RA was not restricted to *in vitro* cell systems, RA was applied to the shaved backs of adult mice and the skin processed for immunohistochemical analysis 48 hours later. As in the *in vitro* systems, there was a specific and selective induction of TGF-$\beta$2 in the epidermis (Glick *et al.* 1989). Use of antibodies specific for either TGF-$\beta$1 or TGF-$\beta$2 demonstrated that staining for TGF-$\beta$1 in the epidermis is minimal both before and after RA treatment. In contrast, staining for TGF-$\beta$2 was localized in the differentiated secretory cells of the sebaceous glands of the control epidermis. Following treatment of the skin with RA, staining for TGF-$\beta$2 increased significantly and was found in the follicular and inter-follicular epithelium, as well as in the sebaceous glands. These data strongly suggest that the effects of retinoic acid on TGF-$\beta$2 secretion by keratinocytes *in vitro* are relevant to the action of retinoic acid on epidermis *in vivo*.

RETINOID DEFICIENCY AND REPLETION HAVE PROFOUND TISSUE-SPECIFIC EFFECTS ON EXPRESSION OF THE TGF-$\beta$ ISOFORMS IN EPITHELIAL TISSUES *IN VIVO*

To strengthen the mechanistic interrelationships between retinoids and TGF-$\beta$, a study was undertaken to determine the immunohistochemical localization of TGF-$\beta$s 1, 2, and 3 in epithelial tissues of normal rats, vitamin A-deficient rats, and deficient rats supplemented with RA. As vitamin A deficiency is known to have profound effects on a variety of epithelial tissues, it was expected that, if TGF-$\beta$s were critical elements in control of cellular differentiation by retinoids, then the epithelia of the animals in each of the groups would exhibit corresponding changes in the localization patterns of TGF-$\beta$. The results demonstrate tissue-specific expression of the TGF-$\beta$ isoforms, which changes with the retinoid status, is consistent with the normal state of differentiation of each of the particular epithelial tissues and provides strong support for a prominent role of the TGF-$\beta$s in the mechanism of action of retinoids in control of epithelial cell differentiation (Glick *et al.* 1991).

The tissues examined included epidermis and bronchiolar, tracheal, intestinal, colonic, and vaginal epithelium (Glick *et al.* 1991). The specific findings in each tisue are summarized in Table 11.1. Overall, as had been observed *in vitro*, the levels of the type 2 isoform of TGF-$\beta$ were most sensitive to changes in retinoid levels. However, strong induction of TGF-$\beta$3 was also seen in most epithelia. In contrast, the levels of TGF-$\beta$1 were altered by the retinoid status in only the epidermis and vaginal epithelium (Table 11.1). In every case, the effect of the retinoid status on the expression of the TGF-$\beta$ isoforms was apparent within 24–48 hours of an oral dose of RA and was reversible, suggesting an important interaction between the two families of effectors.

Although the particular target cell types differed in the various epithelial tissues, a common pattern emerged with respect to the effects of retinoids on

**Table 11.1** Effects of vitamin A-deficiency and repletion with retinoic acid on the expression of the TGF-β isoforms in epithelial tissue of rats

| Tissue/retinoid status | TGF-β staining pattern[a] |
|---|---|
| **Epidermis** | |
| control | Little or no staining |
| deficient | Little or no staining |
| +RA, 1d | Increased staining for all three isoforms |
| **Intestinal mucosa** | |
| control | Staining for TGF-β2 in lamina propria and epithelial cells lining the villi (increasing toward the tip) |
| deficient | Loss of TGF-β2 staining |
| +RA, 1d | Increase in TGF-β2 staining in lamina propria |
| +RA, 2d | Increase in TGF-β2 staining in villus crypt, moves up to midway point by 4d |
| +RA, 8d | Increased staining lost |
| **Colonic mucosa** | |
| control | Slight staining for TGF-β2 in surface epithelial cells |
| deficient | Loss of TGF-β2 staining |
| +RA, 1d | Increase of TGF-β2 staining in surface epithelium; staining for TGF-βs 1 and 3 in lamina propria |
| +RA, 2d | Intense staining for TGF-βs 2 and 3 in surface epithelia |
| +RA, 4d | Increase in staining for TGF-βs 2 and 3 in lamina propria and mucosa |
| +RA, 8d | Staining lost |
| **Bronchial epithelium** | |
| control | TGF-β2 staining in ciliated cells |
| deficient | Ciliated cells lost, TGF-β2 staining lost |
| +RA, 2d | Increase in TGF-β2 staining in basal cells |
| +RA, 4d | Cilia intensely stained for TGF-βs 2 and 3 |
| +RA, 8d | All staining lost |
| **Tracheal epithelium** | |
| control | Faint staining for TGF-β2 |
| deficient | No staining for any TGF-β isoforms |
| +RA, 1d | Metaplastic lesions strongly stained for TGF-βs 2 and 3; no staining in tracheal epithelium |
| +RA, 2d | Staining in squamous lesions lost |
| +RA, 4d | Strong staining for TGF-βs 2 and 3 in epithelium surrounding squamous lesions |
| +RA, 8d | All staining lost |
| **Vaginal epithelium** | |
| control | No staining for any TGF-β isoforms |
| deficient | Intense staining for TGF-β2 > TGF-βs 1 and 3 in cell layers analogous to granular layer of epidermis |
| +RA, 1d | All staining lost |
| +RA, 4d | Increase in staining for all three isoforms in newly keratinized vaginal epithelium |

[a] Data adapted from immunohistochemical staining in Glick *et al.* 1991.

142    *A. B. Roberts, A. B. Glick, and M. B. Sporn*

TGF-$\beta$ expression in respiratory and intestinal epithelium (Glick *et al.* 1991).
In each of these tissues, expression of the TGF-$\beta$ isoforms in the control rats
was limited to TGF-$\beta$2 and this staining disappeared in the deficient tissues.
Treatment of the deficient rats with an oral dose of 100 $\mu$g RA, which
transiently reverses the deficiency, resulted in a rapid induction of TGF-$\beta$2
and TGF-$\beta$3 expression, which then disappeared 4–8 days later as the animal
again became deficient. In contrast, in epidermis, RA treatment resulted in
induction of all three isoforms of TGF-$\beta$. The time course of return to the
deficient state is consistent with previous data showing that approximately
90 per cent of either a physiological or pharmacological dose of RA given to
deficient animals is eliminated within 5 days (Roberts and DeLuca 1967).

An opposite pattern of TGF-$\beta$ expression was seen in vaginal epithelium
(Glick *et al.* 1991). This tissue is normally stratified and squamous in rats and
keratinizes in vitamin A-deficient animals (Wolbach and Howe 1925). While
no staining for any of the TGF-$\beta$ isoforms was detected in control rat tissues,
there was intense staining for all three isoforms in the vaginal epithelium of the
vitamin A-deficient animals, with that for TGF-$\beta$2 being most prominent.
Within 24 hours after dosing with RA, the normal stratified, squamous
epithelium had been partially restored and all staining for TGF-$\beta$ was lost.
Four days later, as the animals again became deficient and the vaginal
epithelium again began to keratinize, there was a marked increase in the
staining of all three isoforms of TGF-$\beta$.

OTHER DIFFERENTIATING AGENTS ALSO INDUCE EXPRESSION OF
TGF-$\beta$2

Although the above data solidly implicate TGF-$\beta$ in epithelial cell differentiation resulting from changes in the tissue retinoid status, they do not implicate
TGF-$\beta$s in the direct mechanism of action of retinoids *per se*. That is, the
changes in expression of the TGF-$\beta$s could be a consequence of the sequence of
events initiated by retinoid treatment, rather than a direct mediator of those
effects. In cultures of murine keratinocytes, induction of differentiation
brought about by an increase in the $Ca^{2+}$ concentration of the medium also
selectively increases expression of TGF-$\beta$2, analogous to that seen following
treatment of the cells with retinoic acid (Glick *et al.* 1990; Table 11.2). In both
cases, the TGF-$\beta$2 is thought to mediate the inhibition of DNA synthesis
associated with the differentiation (Glick *et al.* 1989, 1990).

DISCUSSION

**Multifunctional nature of the retinoids and TGF-$\beta$**

In 1984, in a description of the cellular biology of the retinoids, we suggested
that:

**Table 11.2** Comparison of the effects of retinoids and $Ca^{2+}$ on induction of TGF-$\beta$2 secretion in mouse keratinocytes

| Time (days) | Treatment[a] | |
|---|---|---|
| | 1.4 mM $Ca^{2+}$ | 3 $\mu$M retinoic acid |
| | TGF-$\beta$ (pM) | |
| 1 | 19 | $<1.0$ |
| 2 | 54 | 28 |
| 3 | — | 102 |
| 4 | — | 172 |

[a] Data from Glick *et al.* 1989, 1990.

If one picks and chooses appropriate cellular systems, appropriate concentrations of retinoids, and appropriate time points, then one can eventually find a set of data that will corroborate almost any preconceived notion relating to the effects of retinoids on cell proliferation and differentiation (Roberts and Sporn 1984).

Is it fate or simply chance that our laboratory should now also be involved in research on another set of molecules, the TGF-$\beta$s, which are equally pleiotropic and multifunctional? Answer as you will, it is now clear that, in many instances, the biological functions and mechanisms of action of these two sets of molecules and their receptors are closely linked. RA and the TGF-$\beta$s are both acknowledged to be key regulators of the growth and differentiation of cells and, while not all of their actions are interrelated, their effects on growth and differentiation of epithelial tissue are likely to involve certain common elements. Moreover, the demonstrated effects of RA on expression of the TGF-$\beta$ isoforms, placed in the context of the large volume of literature describing effects of retinoids not only on epithelial tissue homeostasis, but also on embryonic development and the chemoprevention of cancer, suggest that the mechanisms of action of the retinoids and TGF-$\beta$ will be interrelated in those processes as well (Glick *et al.* 1991; Sporn and Roberts 1991).

## Mechanism of effects of retinoids on TGF-$\beta$ expression

The exact mechanism underlying the induction of TGF-$\beta$ expression by retinoids is still elusive. *In vitro* studies with murine keratinocytes rule out effects of retinoids on transcription of TGF-$\beta$2 in that system, despite a nearly 200-fold induction of secreted peptide (Glick *et al.* 1989). Interestingly, two other ligands of the steroid hormone family of nuclear receptors, tamoxifen and the synthetic progestin, gestodene, which result in increased secretion of TGF-$\beta$1 by target cells, also do so principally by post-transcriptional mechanisms (Coletta *et al.* 1990, 1991). Whether the ability of these agents to release the activated form of TGF-$\beta$ is a result of changes in the intracellular processing of the TGF-$\beta$s or in activation by an extracellular protease-dependent mechanism, is also not understood. These limited observations suggest that the TGF-$\beta$ genes themselves are *not* direct targets of retinoid action and that other, less well understood, mechanisms are operative. None the less, use of antibodies that block the biological activity of the TGF-$\beta$s suggests that autocrine or paracrine action of the secreted, active TGF-$\beta$ on the target cell does account, in part, for the inhibition of growth observed following treatment with either tamoxifen, gestodene, or RA (Knabbe *et al.* 1987; Glick *et al.* 1989; Coletta *et al.* 1990, 1991).

**Possible role of retinoids in control of TGF-$\beta$ expression in embryogenesis**

There is now a wealth of literature based on both *in situ* hybridization and immunolocalization techniques describing specific and selective expression of the TGF-$\beta$ isoforms in embryogenesis (Heine *et al.* 1987, 1990; Lehnert and Akhurst 1988; Schmid *et al.* 1991). The results suggest that as development enters a phase of morphogenesis characterized by more complex differentiation events, the three TGF-$\beta$ isoforms are expressed in differential patterns with TGF-$\beta$1 preferentially localized in mesenchyme and TGF-$\beta$s 2 and 3 in epithelia (Schmid *et al.* 1991). As an example, in the developing mouse lung, TGF-$\beta$1 transcripts are localized in mesenchymal cells, while transcripts of TGF-$\beta$s 2 and 3 are localized in the epithelial linings of the bronchi (Schmid *et al.* 1991). In agreement with these findings, Heine *et al.* (1990) demonstrated colocalization of TGF-$\beta$1 protein and matrix proteins in mesenchymal cells lining the bronchiolar ducts at the time of branching morphogenesis, while staining for TGF-$\beta$2 is localized to the epithelial cells (K. Flanders, personal communication).

It is noteworthy that, in both *in vitro* systems and the studies on mature rats reported here, the major effects of retinoic acid and of retinoid deficiency were on TGF-$\beta$2 and, to a lesser extent, TGF-$\beta$3 expression (Glick *et al.* 1989, 1991; Danielpour *et al.* 1991). This suggests that the expression of these two TGF-$\beta$ isoforms in epithelia during development may also be regulated by RA that, based on its localization in certain embryonic structures, has been termed an endogenous morphogen (Thaller and Eichele 1987; Simeone *et al.* 1990). The existence of cellular binding proteins for retinol and RA and the recent identification of as many as six different nuclear RA receptors (Krust *et al.* 1989; Mangelsdorf *et al.* 1990), each of which are also expressed in a distinct tissue- and time-dependent manner during embryogenesis (Dollé *et al.* 1990; Ruberte *et al.* 1991) assures us that the problem of defining specific relationships between these binding proteins and receptors and induction of specific TGF-$\beta$ isoforms will be extremely complex. Lastly, based on the probable relationship of retinoids and TGF-$\beta$ in normal development and vitamin A-deficiency, it is likely that teratogenic effects of retinoids may also be mediated, in part, by aberrant or excessive expression of the TGF-β isoforms.

## Interrelationships between retinoids and TGF-β in chemoprevention of cancer

Epithelial tissues that depend on retinoids for appropriate cell differentiation and growth account for more than three-quarters of the total primary cancer in both men and women, including organ sites and tissues such as the bronchi and trachea, stomach, intestine, uterus, bladder, testis, breast, prostate, pancreas, and skin (Sporn and Roberts 1991). Extensive data have been obtained on the use of retinoids for prevention or treatment of epithelial cancer in experimental animals (see Moon and Itri 1984 for a review) and significant clinical results have now been obtained in prevention of head and neck cancer, as well as skin cancer, with 13-*cis*-retinoic acid (Kraemer *et al.* 1988; Hong *et al.* 1990). The ability of retinoids to delay the preneoplastic progression of carcinogen-treated cells *in vitro* and to modify the differentiation and proliferation of both preneoplastic and neoplastic cells was an important conceptual advance in our understanding of the mechanism of action of retinoids in the suppression of carcinogenesis (Lasnitzki 1955; Merriman and Bertram 1979; Breitman *et al.* 1980). As, like the retinoids, the TGF-βs are also potent inhibitors of epithelial cell proliferation, it should not be surprising that numerous studies now suggest that loss of responsiveness of cells to TGF-β is often associated with neoplastic progression (Wakefield and Sporn 1989). The demonstration that retinoids can induce expression of TGF-β as well as TGF-β receptors (Ruscetti *et al.* 1991) suggests that local enhancement of responsiveness to TGF-β may play a role in the mechanism of chemoprevention by retinoids. In a broader sense, the data on the retinoids, together with the observed local induction of TGF-β activity by two other members of the steroid/retinoid family of nuclear receptors, tamoxifen and gestodene, open new horizons in therapy and prevention of epithelial cancers (Colletta *et al.* 1990, 1991). In fact, in a rat tumour model for human breast cancer, studies now show that combined treatment with a retinoid and tamoxifen, following removal of the primary tumour, was consistently more effective than either agent alone in suppressing the postsurgical appearance of new mammary tumours (Ratko *et al.* 1989); our data would suggest that multiple isoforms of TGF-β might be induced by this regimen.


CONCLUSIONS

Although much is still to be learned about the mechanisms responsible for induction of TGF-β activity by retinoids, about the involvement of specific retinoid receptors, and about the factors regulating the differential effects of retinoids on the various isoforms of TGF-$\beta$, it is clear that the identification of the interrelationships between these two super-families of multifunctional effectors has heralded a new era of discovery of the underlying regulatory events controlling cellular proliferation and differentiation.


REFERENCES

Barnard, J. A., Beauchamp, R. D., Coffey, R. J., and Moses, H. L. (1989). Regulation of intestinal epithelial cell growth by transforming growth factor-beta. *Proceedings of the National Academy of Sciences USA*, **86**, 1578–82.

Breitman, T. R. (1982). Induction of terminal differentiation of HL-60 and fresh leukemic cells by retinoic acid. In *Expression of differentiated functions of cancer cells*, (eds R. P. Revoltella and G. Pontieri), pp. 257–73. Raven Press, New York.

Breitman, T. R., Selonick, S. E., and Collins, S. J. (1980). Induction of differentiation of the human promyelocytic leukemia cell line (HL-60) by retinoic acid. *Proceedings of the National Academy of Sciences of the USA*, **77**, 2936–40.

Coffey, R. J., Bascom, C. C., Sipes, N. J., Graves-Deal, R., Weissman, B. E., and Moses, H. L. (1988*a*). Selective inhibition of growth-related gene expression in murine keratinocytes by transforming growth factor-$\beta$. *Molecular and Cellular Biology*, **8**, 3088–93.

Coffey, R. J., Sipes, N. J., Bascom, C. C., Graves-Deal, R., Pennington, C. Y., Weissman, B. E., and Moses, H. L. (1988*b*). Growth modulation of mouse keratinocytes by transforming growth factors. *Cancer Research*, **48**, 1596–602.

Colletta, A. A., Wakefield, L. M., Howell, F. V., Danielpour, D., Baum, M., and Sporn, M. B. (1991). The growth inhibition of human breast cancer cells by a novel synthetic progestin involves the induction of transforming growth factor-$\beta$. *Journal of Clinical Investigation*, **87**, 277–83.

Colletta, A. A., Wakefield, L. M., Howell, F. V., van Roozendaal, K. E. P., Danielpour, D., Ebbs, S. R., Sporn, M. B., and Baum, M. (1990). Anti-oestrogens induce the secretion of active transforming growth factor-$\beta$ from human fetal fibroblasts. *British Journal of Cancer*, **62**, 405–9.

Danielpour, D., Kim, K. Y., Winokur, T. S., and Sporn, M. B. (1991). Differential regulation of the expression of transforming growth factor-$\beta$s 1 and 2 by retinoic acid, epidermal growth factor and dexamethasone in NRK-49F and A549 cells. *Journal of Cellular Physiology*, **148**, 235–44.

Dollé, P., Ruberte, E., Leroy, P., Morriss-Kay, G., and Chambon, P. (1990). Retinoic acid receptors and cellular retinoid binding proteins I. a systematic study of their differential pattern of transcription during mouse organogenesis. *Development*, **110**, 1133–51.

Fell, H. B. and Mellanby, E. (1953). Metaplasia produced in cultures of chick endoderm produced by high vitamin A. *Journal of Physiology*, **119**, 470–88.

Glick, A. B., Danielpour, D., Morgan, D., Sporn, M. B., and Yuspa, S. H. (1990). Induction and autocrine receptor binding of transforming growth factor-$\beta$2 during terminal differentiation of primary mouse keratinocytes. *Molecular Endocrinology*, **4**, 46–52.

Glick, A. B., Flanders, K. C., Danielpour, D., Yuspa, S. H., and Sporn, M. B. (1989). Retinoic acid induces transforming growth factor-β2 in cultured keratinocytes and mouse epidermis. *Cell Regulation*, **1**, 87–97.

Glick, A. B., McCune, B. K., Abdulkarem, N., Flanders, K. C., Lumadue, J. A., Smith, J. M., and Sporn, M. B. (1991). Complex regulation of TGF-β expression by retinoic acid in the vitamin A-deficient rat. *Development*, **111**, 1081–6.

Heine, U. I., Munoz, E. F., Flanders, K. C., Ellingsworth, L. R., Lam, H.-Y.P.P., Thompson, N. L., Roberts, A. B., and Sporn, M. B. (1987). Role of transforming growth factor-β in the development of the mouse embryo. *Journal of Cell Biology*, **105**, 2861–76.

Heine, U. I., Munoz, E. F., Flanders, K. C., Roberts, A. B. and Sporn, M. B. (1990). Colocalization of TGF-beta 1 and collagen I and III, fibronectin and glycosamino-glycans during lung branching morphogenesis. *Development*, **109**, 29–36.

Hong, W. K., Lippman, S. M., Itri, L. M., Karp, D. D., Lee, J. S., Byers, R. M. *et al.* (1990). Prevention of second primary tumors in squamous cell carcinoma of the head and neck with 13-*cis*-retinoic acid. *New England Journal of Medicine*, **323**, 795–801.

Knabbe, C., Lippman, M. E., Wakefield, L. M., Flanders, K. C., Kasid, A., Derynck, R., and Dickson, R. B. (1987). Evidence that transforming growth factor-β is a hormonally regulated negative growth factor in human breast cancer cells. *Cell*, **48**, 417–28.

Kraemer, K. H., DiGiovanna, J. J., Moshell, A. N., Tarone, R. E., and Peck, G. L. (1988). Prevention of skin cancer in xeroderma pigmentosa with the use of oral isotretinoin. *New England Journal of Medicine*, **318**, 1633–7.

Krust, A., Kastner, P. H., Petkovich, M., Zelent, A., and Chambon, P. (1989). A third human retinoic acid receptor, hRAR-γ. *Proceedings of the National Academy of Sciences USA*, **86**, 5310–14.

Lasnitzki, I. (1955). The influence of A-hypervitaminosis on the effect of 20-methyl cholanthrene on mouse prostate glands grown *in vitro*. *British Journal of Cancer*, **9**, 434–41.

Lehnert, S. A. and Akhurst, R. J. (1988). Embryonic expression pattern of transforming growth factor-β type 1 RNA suggests both paracrine and autocrine mechanisms of action. *Development*, **104**, 263–73.

Mangelsdorf, D. J., Ong, E. S., Dyck, J. A., and Evans, R. M. (1990). Nuclear receptor that identifies a novel retinoic acid response pathway. *Nature*, **354**, 224–9.

Merriman, R. L. and Bertram, J. S. (1979). Reversible inhibition by retinoids of 3-methylcholanthrene-induced neoplastic transformation in C3H/10T$\frac{1}{2}$ clone 8 cells. *Cancer Research*, **39**, 1661–6.

Moon, R. C. and Itri, L. M. (1984). Retinoids and cancer. In *The retinoids*, vol. 2, (eds M. B. Sporn, A. B. Roberts, and D. S. Goodman), pp. 327–71. Academic Press, New York.

Nathans, C. and Sporn, M. B. (1991). Cytokines in context. *Journal of Cell Biology*, **113**, 981–6.

O'Malley, B. (1990). The steroid receptor superfamily: more excitement predicted for the future. *Molecular Endocrinology*, δ, 363–9.

Ratko, T. A., Detrisac, C. J., Dinger, N. M., Thomas, C. F., Kelloff, G. J., and Moon, R.C. (1989). Chemopreventive efficacy of combined retinoid and tamoxifen treatment following surgical excision of a primary mammary cancer in female rats. *Cancer Research*, **49**, 4472–6.

Roberts, A. B. and DeLuca, H. F. (1967). Pathways of metabolism of retinol and retinoic acid in the rat. *Biochemical Journal*, **102**, 600–5.

Roberts, A. B. and Sporn, N. B. (1984). Cellular biology and biochemistry of the retinoids. In *The retinoids*, vol. 2, (eds M. B. Sporn, A. B. Roberts, and D. S. Goodman), pp. 209–86. Academic Press, New York.

Roberts, A. B. and Sporn, M. B. (1990). The transforming growth factors-$\beta$. In the *Handbook of Experimental Pharmacology, Peptide growth factors and their receptors*, vol. 95/1, (eds. M. B. Sporn and A. B. Roberts), pp. 419–72. Springer-Verlag, Heidelberg.

Ruberte, E., Dollé, P., Chambon, P., and Morriss-Kay, G. (1991) Retinoic acid receptors and cellular retinoid binding proteins: II. Their differential pattern of transcription during early morphogenesis in mouse embryos. *Development*, **111**, 45–60.

Ruscetti, F., DuBois, C., Falk, L., Jacobsen, E., Sing, G., Longo, D., Wiltrout, R., and Keller, J. (1991). *In vivo* and *in vitro* effects of TGF-$\beta$1 on normal and leukemic hematopoiesis. *The Ciba Foundation Symposium*, **157**, 212–27.

Schmid, P., Cox, D., Bilve, G., Maier, R., and McMaster, G. K. (1991). Differential expression of TGF-$\beta$1, $\beta$2, and $\beta$3 genes during mouse embryogenesis. *Development*, **111**, 117–30.

Silberstein, G. B. and Daniel, C. W. (1987). Reversible inhibition of mammary gland growth by transforming growth factor-$\beta$. *Science*, **237**, 291–3.

Simeone, A., Acampora, D., Arcioni, L., Andrews, P. W., Boncinelli, E., and Mavilio, F. (1990). Sequential activation of HOX2 homeobox genes by retinoic acid in human embryonal carcinoma cells. *Nature*, **346**, 763–6.

Sporn, M. B. and Roberts, A. B. (1988) Peptide growth factors are multifunctional. *Nature*, **332**, 217–19.

Sporn, M. B. and Roberts, A. B. (1991). Interactions of retinoids and transforming growth factor-$\beta$ in regulation of cell differentiation and proliferation. *Molecular Endocrinology*, **5**, 3–7.

Thaller, C. and Eichele, G. (1987). Identification and spatial distribution of retinoids in the developing chick limb bud. *Nature*, **327**, 625–8.

Wakefield, L. M., Smith, D. M., Broz, S., Jackson, M., Levinson, A. D., and Sporn, M. B. (1989). Recombinant TGF-$\beta$1 is synthesized as a two component latent complex that shares some structural features with the native platelet latent TGF-$\beta$1 complex. *Growth Factors*, **1**, 203–18.

Wakefield, L. M. and Sporn, M. B. (1989). Suppression of carcinogenesis: a role for TGF-$\beta$ and related molecules in prevention of cancer. In *Tumor suppressor Genes*, (ed. G. Klein). Marcel Dekker Inc., New York.

Wakefield, L. M., Winokur, T. S., Hollands, R. S., Christopherson, K., Levinson, A. D., and Sporn, M. B. (1990) Recombinant latent transforming growth factor $\beta$1 has a longer plasma half-life in rats than active transforming growth factor $\beta$1, and a different tissue distribution. *Journal of Clinical Investigation*, **86**, 1976–84.

Wolbach, S. B. and Howe, P. R. (1925). Tissue changes following deprivation of fat-soluble vitamin A. *Journal of Experimental Medicine*, **43**, 753–77.

# 12

# Palatal expression of TGF-$\beta$ isoforms in normal and retinoic acid-treated mouse embryos

David R. FitzPatrick, Barbara D. Abbott, and Rosemary J. Akhurst

INTRODUCTION

The transforming growth factors type beta (TGF-$\beta$s) are a fascinating family of closely related dimeric proteins that have tissue-specific effects on the growth, differentiation, and extracellular matrix deposition of cells in culture (for review see Massague 1990; Roberts and Sporn 1990). Although there are at least eight isoforms of TGF-$\beta$ in vertebrates (Thomsen and Melton 1990), only three have been characterized in mammalian species, TGF-$\beta$1 (Moses *et al.* 1981; Roberts *et al.* 1981; Derynck *et al.* 1985), TGF-$\beta$2 (Cheifetz *et al.* 1987; Seyedin *et al.* 1987; Wrann *et al.* 1987; Hanks *et al.* 1988), and TGF-$\beta$3 (ten Dijke *et al.* 1988; Miller *et al.* 1989; Denhez *et al.* 1990).

Although the reported *in vitro* properties of each isoform show striking similarities, differences up to 50-fold have been demonstrated in the potency of their growth inhibitory effects on cultured cells (Ohta *et al.* 1987). Cheifetz *et al.* (1990) have shown that TGF-$\beta$ receptor subsets determine these differences in cellular responsiveness.

The first suggestion that TGF-$\beta$ may be involved in developmental processes came from the multifunctional nature of its biological activity *in vitro*. The TGF-$\beta$s display considerable homology in amino acid sequence to several key developmental genes, such as the decapentaplegic gene complex in *Drosophila* (Padgett *et al.* 1987), *Xenopus* mesoderm-inducing factor, XTC-MIF (activin) (Smith 1989) and Müllerian inhibitory substance (Cate *et al.* 1986). However, the most convincing evidence that the TGF-$\beta$s might endogenously regulate developmental processes in mammals is provided by the embryonic localization of TGF-$\beta$ protein and RNA within mammalian embryos (Heine *et al.* 1987; Lehnert and Akhurst 1988; Wilcox and Derynck 1988; Pelton *et al.* 1989; Akhurst *et al.* 1990*a*; FitzPatrick *et al.* 1990; Gatherer *et al.* 1990; Millan *et al.* 1991). As yet, no unequivocal data have demonstrated a specific function for any of the TGF-$\beta$ isoforms in developmental events *in vivo*.

The first evidence for the involvement of the TGF-$\beta$s in retinoid-induced effects was provided by Glick *et al.* (1989), who demonstrated retinoic acid

(RA)-induced expression of biologically-active TGF-$\beta$2 in cultured mouse keratinocytes. In this case the induction was correlated with inhibition of growth of the keratinocytes; by blocking the availability of TGF-$\beta$2 with specific antibodies, the authors also found that the biological response to RA was, in part, mediated by TGF-$\beta$2.

The specificity of expression of all three TGF-$\beta$ isoforms, both spatially and temporally, during mammalian palatogenesis, and the suggestion that TGF-$\beta$s may mediate the effects of RA, led us to investigate alterations in the normal expression patterns of these growth factors in RA-induced cleft palate. In this review we will discuss these studies and place them in context with the work of others on both descriptive and functional studies of TGF-$\beta$ in development.

RETINOID-INDUCED CLEFT PALATE

**Normal palatogenesis**
Development of the mammalian secondary palate can be viewed in four stages:

1. Initiation of the primordial palatal shelves from the medial aspect of the maxillary processes within the primitive oral cavity.
2. Rapid vertical downgrowth of the palatal shelves lateral to the tongue.
3. Reorientation of the shelves to come into apposition above the tongue, with coincidental cell death and subsequent sloughing of the outer layer (periderm) of the medial edge epithelia (MEE) of both palatal shelves.
4. Fusion of the MEE to form the midline epithelial seam (MES), which subsequently disrupts, thus establishing continuity in the palate (Greene and Pratt 1976; Ferguson 1988).

Palatogenesis represents an attractive system in which to study both normal and aberrant development because of its relatively late occurence in organogenesis (11.5–14.5 gestational days (GD) in the mouse) and the ease of visualizing phenotypic alterations.

**Stage-specific effects of retinoids**

Cleft palate was noted as a feature of retinoid embryopathy by Cohlan (1953), who studied the effects of excessive vitamin A on rat embryos. Since then retinoid-induced cleft palate has been noted in simian species (Fantel *et al.* 1977), humans (Rosa *et al.* 1986) and some rodent species (Shenefelt 1972; Newall and Edwards 1981). The use of the vitamin A analogue RA, which is not stored in maternal fat or liver and, therefore, has a short half-life (Kochhar 1967), has enabled the study of retinoid exposure at precise developmental stages. It is now clear that the aetiology of the palatal cleft varies with the gestational timing of the RA exposure (Newall and Edwards 1981; Sulik *et al.* 1987; Abbott *et al.* 1989; Sulik *et al.* 1989).

In C57BL mice, treatment with a single dose of RA at or before 10 days of gestation (GD) causes severe hypoplasia of the palatal shelves, which then fail to meet in the midline resulting in cleft palate. This has been termed mesenchymal failure (Abbott *et al.* 1989). In mouse embryos treated on 12 GD or later, however, the aetiology of the cleft palate seems to be the result of changes in the MEE. These embryos produce normal-sized palatal shelves that meet in the midline but fail to fuse (Abbott *et al.* 1989), hence the term epithelial failure.

**Effects of retinoids on palatal mesenchyme**

Several authors have suggested that the RA-induced hypoplasia of the palatal shelves is due to the effect of RA on migration and proliferation of cranial neural crest (NC)-derived cells (Webster *et al.* 1986; Pratt *et al.* 1987; Davis *et al.* 1990). Although the cranial NC is known to contribute significantly to the mesenchyme of the maxillary processes of both chick (Le Douarin 1982; Noden 1986) and mouse (Nichols 1986; Pratt *et al.* 1987) embryos, no direct evidence is available that it contributes to mammalian palatal mesenchyme. Several cranial NC-derived ectomesenchymal phenotypes are, however, observed in the course of palate development (Slavkin 1984) and it would seem likely that NC cells are involved in this process.

Evidence supporting the hypothesis that migrating cranial NC cells are important sites of action in RA embryopathy comes from both *in vivo* and *in vitro* studies. There are now several reports of retinoid-induced NC cell death and also of delay or disorganization of the normal NC migration pathways *in vivo* (Morriss 1973; Wiley *et al.* 1983; Webster *et al.* 1986). Thorogood *et al.* (1982) have demonstrated behavioural changes in cultured primary mesenchymal and NC cells after treatment with retinol, which include increased surface membrane blebbing and loss of cell-substratum adhesiveness. Davis *et al.* (1990) have further shown that retinoid-treated cranial NC cells in culture generate cytotoxic radical oxygen species and have suggested this as a major factor in retinoid-induced cell death *in vivo*.

Migration of the cranial NC is thought to be complete by 9 GD in the mouse (Nichols 1986). However, mouse embryos exposed to RA at gestation stages later than this show palatal shelf hypoplasia (Sulik *et al.* 1987; Abbott *et al.* 1989; Sulik *et al.* 1989). This suggests that there may be sites of retinoid action on the developing palatal mesenchyme other than migrating NC cells. In an attempt to identify these sites, Sulik *et al.* (1987, 1989) have correlated the degree of RA-induced hypoplasia of the palatal shelves with areas of excessive cell death in the trigeminal placode, an area of thickened ectoderm on the proximal aspect of the first visceral arch, thought to contribute cells that form neurons in the trigeminal ganglion (for review see Verwoerd and van

Oostrom 1979). The mechanism by which terminal differentiation in the trigeminal placode could influence proliferation of cells destined to become palatal mesenchyme, however, remains unclear.

**The effects of retinoids on palatal epithelia**

Using tissue recombination techniques, Ferguson and Honig (1984) have shown that the regionally-specific epithelial phenotypes of normal palatal shelves (oral, nasal, and MEE) are determined by signals from the underlying mesenchyme. Normal differentiation of the MEE is, however, diverted to that of squamous oral-type epithelium after RA treatment at 10 GD and to a pseudostratified nasal epithelium after RA treatment on 12 GD (Abbott *et al.* 1989). It can thus be postulated that RA may in some way change the nature of the mesenchymal signal. There are, as yet, few clues to the physiochemical nature of this signalling. It is, however, of interest that RA treatment on 12 GD alters the expression of epidermal growth factor (EGF) receptor in the MEE (Abbott *et al.* 1988), suggesting a role for peptide growth factors in region-specific epithelial differentiation.

TGF-$\beta$S DURING MAMMALIAN DEVELOPMENT

**Global localization studies of TGF-$\beta$ isoforms**

Ideally, an *in vivo* study to address the role of any gene in morphogenetic processes would employ sensitive and specific techniques for the localization of both the transcribed RNA and its translated product in the embryo. However, the amino acid and nucleotide sequence homology shared between the TGF-$\beta$ isoforms has posed a particular problem in the study of these gene products.

The generation of isoform-specific antibodies for immunohistochemical localization of TGF-$\beta$ peptides has proved particularly difficult, with cross-reactivity between the isoforms often occurring in one assay but not in another (Flanders *et al.* 1988) and further complicated by the fact that antibodies raised to identical peptides can recognize different epitopes (Flanders *et al.* 1989). However, by using *in situ* hybridization to cellular RNA the problem of cross-reactivity can be minimized by the use of species- and isoform-specific riboprobes made from the relatively divergent 5' (or precursor) areas of the gene and employing highly stringent hybridization conditions.

Protein, and to a lesser extent, RNA localization studies of TGF-$\beta$ expression in the embryo must, therefore, be approached with a healthy scepticism, not least because there may be other, as yet, undiscovered mammalian TGF-$\beta$ isoforms to further cloud already muddy waters.

## TGF-β during normal palatogenesis

Heine *et al.* (1987), in the first extensive protein localization study of TGF-β1 in the mouse embryo, reported immunohistochemical staining in the mesenchyme of the developing palatal shelves. In other studies, TGF-β1 peptide has been found to be uniformly distributed between palatal epithelium and mesenchyme (Abbott and Birnbaum 1990) while TGF-β1 RNA has been localized to the MEE (Akhurst *et al.* 1990*a*). TGF-β2 polypeptide has been detected immunohistochemically in palatal epithelia with regional mesenchymal expression on 14 GD (Abbott and Birnbaum 1990).

Few *in vitro* studies exist on the effects of the exogenous TGF-βs on palatal tissues, although Sharpe and Ferguson (1988) have shown TGF-β1 to have a profound growth inhibitory effect on cultured murine palatal mesencyme. It is also known that embryonic murine palates exposed to TGF-β1 in organ culture show a concentration-dependent inhibition of MEE peridermal degeneration and inhibition of basal epithelial transformation (B. D. Abbott, unpublished data).

We have recently studied the expression of all three mammalian TGF-β genes during the critical stages of secondary palatogenesis using *in situ* hybridization of radiolabelled, isoform-specific complementary RNA probes (riboprobes) (FitzPatrick *et al.* 1990). The temporal and spatial distribution of transcripts from these three genes is summarized in Fig.12. 1. The first evidence for expression of any of these genes during murine palatogenesis at 13.5 GD, with epithelial expression of TGF-β3 in the vertical palatal shelves and in the epithelium of the nasal septum that will go on to fuse with the palatal shelves. The expression of TGF-β3 appears to be restricted to the areas that will form the medial edge epithelium (MEE) of the palatal shelf after reorientation (Fig. 12.2(*c*). This remarkable restriction of TGF-β3 RNA expression to the MEE is also seen after shelf elevation and continues in the epithelial seam (Fig. 12.3(*b,d*)) until loss of epithelial expression as the seam disrupts by epithelial to mesenchymal cell transformation.

TGF-β1 RNA shows a similar epithelial pattern of expression to that of TGF-β3, but is not detectable in the MEE until after palatal shelf elevation and at this stage apparently has a lower transcript prevalence than TGF-β3. Previous work in mouse embryos (Lehnert and Akhurst 1988) has localized TGF-β1 RNA to the epithelia of several active developmental systems (heart, tooth bud, whisker follicle, submandibular gland) that overlie mesenchyme containing TGF-β1 peptide (Heine *et al.* 1987), thus suggesting a paracrine mode of action for this growth factor. This distribution of TGF-β2 gene expression is distinct from those of TGF-β1 and TGF-β3 in that it is absent from the epithelial cells of the medial edge. TGF-β2 RNA can first be detected in the hyperplastic nodules on the putative oral epithelia of the palatal shelves. These nodules will go on the form the palatal rugae (Fig. 12.2(*b*)) (Sakamoto *et al.* 1989). After shelf elevation, however, TGF-β2 expression is limited to the

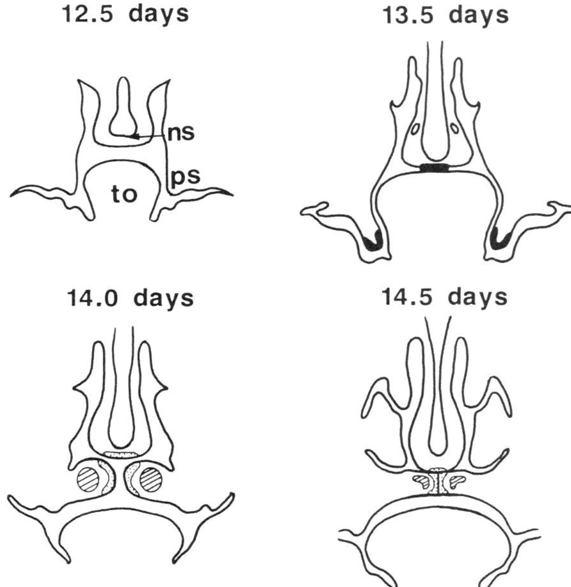



**Fig. 12.1** Diagramatic representation of stage- and site-specific distribution of TGF-$\beta$1, TGF-$\beta$2, and TGF-$\beta$3 RNA during palate development in the mouse embryo. The line drawings represent coronal sections through the mid-palate at the gestation (in days) indicated above the drawings. The diagrams show the palatal shelves (ps) growing down beside the tongue (to) and then elevating to fuse in the midline between the tongue and the nasal septum (ns). The filled black areas illustrates TGF-$\beta$3 expression, the diagonal striping TGF-$\beta$2 and the stippled areas combined expression of TGF-$\beta$1 and TGF-$\beta$3.

mesenchyme that underlies the MEE and epithelial seam (Fig. 12.3($a$,$c$)). With fusion of the MEE the TGF-$\beta$2 RNA becomes asymmetrically distributed in the palatal mesenchyme. A gradient of expression is apparent with highest concentrations adjacent to nasal epithelium and lowest concentrations underlying oral epithelium (Fig. 12.3($a$)). Recent immunohistochemical localization of TGF-$\beta$2 (Abbott and Birnbaum 1990) has suggested that the protein is epithelially-localized, thus implying that TGF-$\beta$2 may also have a paracrine mode of action in this system.

Pelton *et al.* (1990) coincidentally reported identical RNA localization data for TGF-$\beta$1, TGF-$\beta$2 and TGF-$\beta$3 during secondary palatogenesis, using gene-specific riboprobes complementary to different regions of the TGF-$\beta$ isoform transcripts.

## TGF-$\beta$s in retinoic acid treated embryos

The possibility that TGF-$\beta$2 may have a role in the *in vivo* actions of RA (Glick *et al.* 1989, 1991) has generated considerable interest. As a retinoid-susceptible

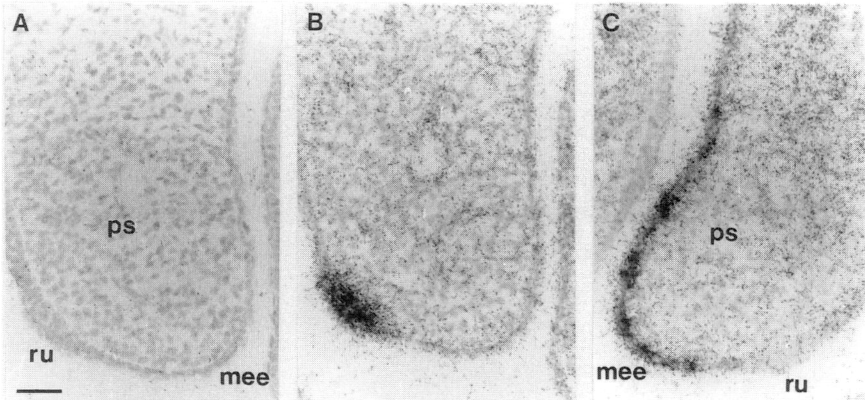


**Fig. 12.2** Epithelial expression of TGF-β2 and TGF-β3 in early palatogenesis. All sections are high-power, light-field images of coronal section through vertical palatal shelves of mouse embryos (13.5 GD). (*a*) The vertical palatal shelf (ps), showing an area of epithelial hyperplasia representing developing palatal ruga (ru) and the bilayered putative medial edge epithelium (mee). (*b*) TGF-β2 probe showing specific hybridization to the palatal ruga. (*c*) TGF-β3 probe hybridizing specifically to the MEE. Scale bar = 200 μm (*a*, *b*); 50 μm (*c*, *d*).

organ with a differential expression pattern of the TGF-β isoforms, the developing palate is of particular interest. We have, therefore, studied the effect of RA treatment on the expression patterns of TGF-β1–3 in the developing palates of C57Bl mouse embryos. Embryos were exposed to RA on 10 GD, as this is known to have a major effect on both the mesenchymal and epithelial components of the palatal shelves (Abbott *et al.* 1989). Pregnant mice were dosed with 100 mg/kg all-*trans*-RA in corn oil on 10 GD and the embryos collected on 12 GD and 14 GD; control dams were dosed with corn oil only.

The RA-exposed embryos studied had phenotypic abnormalities, including hypoplasia of the palatal shelves, that would be predicted from previous reports (Abbott *et al.* 1989). However, few alterations in TGF-β RNA distribution were seen in the palatal shelves of those embryos collected either on 12 GD (no specific hybridization in either control or RA-treated embryos) or 14 GD (Fig. 12.4). TGF-β1 and β3 RNAs had identical qualitative distribution patterns in the normal and RA-treated palates, as did the mesenchymally-expressed TGF-β2 RNA. *In situ* hybridization, however, does not give easily quantifiable result. Thus, we could not determine whether there were quantitative differences in TGF-β RNA expression between the experimental and control embryos.

One noticeable difference from the previous work, however, was the prescence of TGF-β2 RNA expression in the MEE of the RA-treated embryos

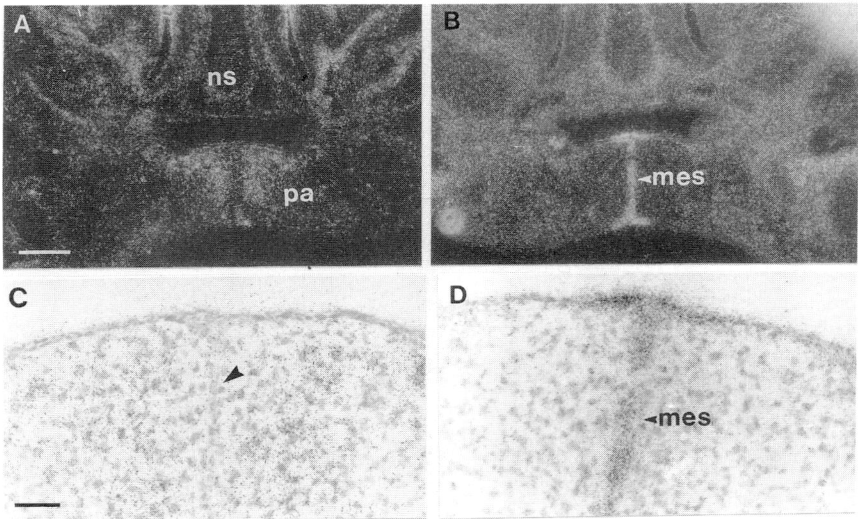


**Fig. 12.3** TGF-$\beta$2 and $\beta$3 in the fusing palate. All sections are coronal through the midpalate of 14.5-day mouse embryos. (*a* and *b*) are dark-field images, (*c* and *d*) are high-power light-field images. (*a*) TGF-$\beta$2 probe hybridizing to the paramedial mesenchyme of the newly-formed palate (pa). (*b*) TGF-$\beta$3 probe showing strong hybridization to the midline epithelial seam (mes). (*c*) High-power image of (*a*) showing TGF-$\beta$2 probe hybridization to the mesenchyme on either side of the mes (arrowed). (*d*) High-powered image of (*b*) showing hybridization of the TGF-$\beta$3 probe to the MES. Scale bar = 50 $\mu$m.

collected on 14 GD, despite the fact that mesenchymal expression remained predominant (Fig. 12.4(*c*)). This observation is particularly interesting with respect to the possible endogenous function of TGF-$\beta$2 in regulation of the state of growth and differentiation of epithelia. Our own studies (Fitzpatrick *et al.* 1990; Millan *et al.* 1991), and that of Pelton *et al.* (1989), have shown that TGF-$\beta$2 is widely expressed in differentiating epithelia, including olfactory, otic and bronchial epithelia, hypoplastic rugae, and the epidermis. Furthermore, TGF-$\beta$2 is induced in epithelia undergoing modified differentiation *in vitro* and *in vivo* following RA treatment, as discussed by Roberts *et al.* (Chapter 11). Glick *et al.* (1989) also demonstrated that, *in vitro*, the biological effects of RA on epithelial cells may, in part, be mediated by TGF-$\beta$2. On the basis of these observations, we previously suggested that TGF-$\beta$2 may be an endogenous regulator of epithelial homeostasis *in vivo*. It appears that in the palate, the altered differentiative pathway of the MEE cells after RA treatment could be mediated, in part, by TGF-$\beta$2. it is unlikely, however, that TGF-$\beta$2 RNA induction in the MEE is a primary response to RA, as it is seen at relatively late times following the treatment (i.e. at 14 GD but not at 12 GD).

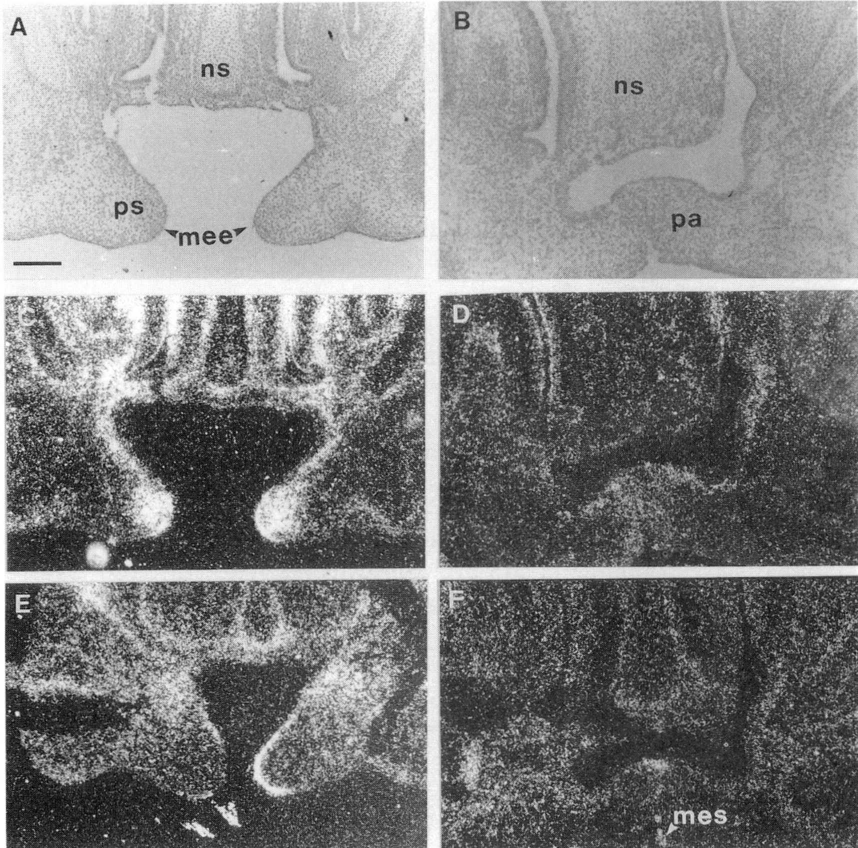


**Fig. 12.4** TGF-*β*2 and *β*3 in the developing palatal shelves of retinoic acid-exposed embryos. All sections are coronal sections through the midpalate of 14.5-day mouse embryos. (*a, c, e*) are sections from embryos collected from pregnant C57BL mice treated with 100 mg/kg of all-*trans*-retinoic acid on 10 GD. (*b, d, f*) are control sections from pregnant mice exposed to dosing vehicle only (corn oil) on 10 GD. (*a, b*) are light-field images, (*c, d, e, f*) are dark-field images. (*a*) RA-exposed embryo showing hypoplasia of the palatal shelves (ps). (*b*) Control embryo showing the newly formed palate (pa). (*c*) TGF-*β*2 probe hybridizing to the medial edge palatal mesenchyme, with some hybridization to the medial edge epithelium (mee). (*d*) TGF-*β*2 probe showing paramedial hybridization in the palate. (*e*) TGF-*β*3 probe showing hybridization to the medial edge epithlium (mee) (left side mee has sloughed off). (*f*) TGF-*β*3 probe hybridizing to the (disrupting) midline epithelial seam. Scale bar = 200 *μ*m.

Immunohistochemical localization of TGF-*β*1 and TGF-*β*2 proteins in palatal shelves of RA-treated mouse embryos has yielded some intriguing results (Abbott and Birnbaum 1990). An increase in the mesenchymal staining using an TGF-$\beta$1 antibody has been shown in the palatal shelves of embryos treated on 12 GD, but not in those treated on 10 GD. After exposure on either day, TFG-$\beta$2 was present in apparently increased amounts in the nasal epithelia of the shelves. This apparent upregulation of TGF-$\beta$2 may be the result of increased RNA concentrations that cannot be detected by the *in situ* technique. It is also possible that post-transcriptional or post-translational mechanisms may account for these differences (Glick *et al.* 1989).

## The role of TGF-$\beta$3 in palatogenesis

Assuming a paracrine mode of action for epithelially-transcribed TGF-$\beta$1 (and, perhaps, TGF-$\beta$3) in the developing palate (Lehnert and Akhurst 1988), putative functions for this peptide in the underlying mesenchyme may be suggested by its *in vitro* biological properties. One of the most interesting of these is the profound effects that TGF-$\beta$ has on extracellular matrix (ECM) deposition. TGF-$\beta$1 is known to induce the synthesis of collagens and fibronectin (Roberts *et al.* 1986; Ignotz *et al.* 1987), tenascin (Pearson *et al.* 1988) and chondroitin/dermatan proteoglycans (Hiraki *et al.* 1988; Sharpe and Ferguson 1988). Hydration of the proteoglycan network with the latter class of molecules as a major component, is thought to be important in palatal shelf elevation (Pratt *et al.* 1973; Brinkley and Morris Wiman 1987). In this respect, it is interesting that high levels of TGF-$\beta$3 RNA are observed 24 hours prior to palatal shelf elevation, and thus might contribute to this phenomenon.

The *in vivo* localization of tenascin to the medial edge palatal mesenchyme (MEPM), prior to and during palatal shelf fusion (Sharpe and Ferguson 1988), is also of considerable interest because the embryonic distribution of this molecule is almost completely correlated with the presence of epithelial TGF-$\beta$1 RNA (Lehnert and Akhurst 1988; Sharpe and Ferguson 1988; Chiquet-Ehrismann *et al.* 1989; Akhurst *et al.* 1990*b*). Tenascin has two properties that strongly suggest a functional role in palatal shelf mesenchyme. The first is the ability of tenascin to disrupt epithelial sheet continuity by breaking cell–cell and cell–substratum contacts (Chiquet-Ehrismann *et al.* 1989). The second is its ability to promote specifically the mobility of neural crest cells *in vitro* (Halfter *et al.* 1989). Both of these events would be necessary for fusion along the mid-line seam, a time when the epithelial sheet disrupts and there is much cell mixing.

The ability of TGF-$\beta$1 to stimulate chemotaxis, proliferation and differentiation of cells of mesenchymal origin (Moses *et al.* 1985; Postlethwaite *et al.* 1987; Robey *et al.* 1987) may suggest that similar effects may be seen in the MEPM. The *in vitro* studies of TGF-$\beta$1 action on MEPM have, however, given confusing results. Both Sharpe and Ferguson (1988) and Linask *et al.* (1991) have reported a growth inhibitory response of cultured cells from murine MEPM to TGF-$\beta$1. Linask *et al.* (1991) also studied cells cultured from human MEPM and found a mitogenic response to TGF-β1, with different TGF-β receptor subsets being expressed in the murine and human cultured cells. The response of palatal mesenchyme to TGF-β1 *in vivo* is, therefore, not clear and may differ with cell subsets, development stage and/or be species-specific.

It is our opinion that the epithelially-transcribed TGF-βs in the normal palate act on the underlying mesenchyme to produce a suitable microenvironment for cell differentiation and migration around the fusing ends of each palatal shelf. The fact that these expression patterns are not altered with RA exposure may not then be surprising as, if the shelves fail to come into contact or there is an epithelial barrier to fusion, then the proposed action of the epithelial TGF-βs will not be tested.

TGF-β2 RNA distribution during palatogenesis is, however, in marked contrast to that of TGFs β1 and β3. The predominantly mesenchymal localization of TGF-β2 RNA in the developing palate has been reported in other developing systems (Pelton *et al.* 1989) and would suggest an autocrine action on the growth and differentiation of these cells.

It was suggested by Pelton *et al.* (1989) that mesenchymal expression of TGF-β2 RNA might also be important in supporting growth of the overlying epithelium via secondary events, such as induction of TGF-α. In this context, it is interesting that the TGF-β2 RNA distribution is asymmetric with respect to the nasal and oral regions. Differential concentrations of growth factors within the mesenchyme could contribute to the generation of regional heterogeneity of the overlying epithelium.


CONCLUSIONS

The expression pattern of the genes encoding TGF-β1–3 during development of the secondary palate would argue an important role for these growth factors in the morphogenesis of this system. Following RA treament *in vivo*, there are few changes in the RNA distributions of these growth factors, suggesting that the gene products do not directly mediate the retinopathology of the palate. Further studies will, however, need to be performed, to examine the differential distributions of the proteins, as post-transcriptional controls may be operative (Glick *et al.* 1989).


ACKNOWLEDGEMENTS

We grateful to Dr R. Derynck (Genentech) for supplying the full-length TGF-β1 clone and Drs F. Denhez and P. Kondaiah (NIH) for supplying the TGF-β2 and TGF-β3 clones. Mrs Jean Hislop (RHSC) drew Fig. 12.1. The work in our laboratory is supported by grants from the CRC, MRC and

Wellcome Trust. DRF is in receipt of a Wellcome Trust Medical Graduate Research Training Fellowship.


REFERENCES

Abbott, B. D., Adamson, E. D., and Pratt, R. M. (1988). Retinoic acid alters EGF receptor expression during palatogenesis. *Development*, **102**, 853–67.

Abbott, B. D., Harris, M. W., and Birnbaum, L. S. (1989). Etiology of retinoic acid-induced cleft palate varies with the embryonic stage. *Teratology*, **40**, 533–53.

Abbott, B. D. and Birnbaum, L. S. (1990). Retinoic acid-induced alterations in the expression of growth factors in the embryonic mouse palatal shelves. *Teratology*, **42**, 597–610.

Akhurst, R. J., Lehnert, S. A., Gatherer, D., and Duffie, E. (1990a). The role of TGF beta in mouse development. *Annals of the New York Academy of Science*, **593**, 259–71.

Akhurst, R. J., Lehnert, S. A., Faissner, A. J. and Duffie, E. (1990b). TGF beta in murine morphogenetic processes: the early embryo and cardiogenesis. *Development*, **108**, 645–56.

Brinkley, L. L. and Morris Wiman, J. (1987). Computer-assisted analysis of hyaluronate distribution during morphogenesis of the mouse secondary palate. *Development*, **100**, 629–35.

Cate, R. L., Mattaliono, R. J., Hession, C., Tizard, R., Farber, N. M., Cheung, A. *et al.* (1986). Isolation of the bovine and human genes for Mullerian inhibiting substance and expression of the human gene in animal cells. *Cell*, **45**, 685–98.

Cheifetz, S., Weatherbee, J. A., Tsang, M. L. S., Anderson, J. K., Mole, J. E., Lucas, R. *et al.* (1987). The transforming growth factor-beta system, a complex pattern of cross-reactive ligands and receptors. *Cell*, **48**, 409–15.

Cheifetz, S., Herbandez, H., Laiho, M., ten Dijke, P., Iwata, K. K., and Massague, J. (1990). Distinct transforming growth factor-$\beta$ receptor subsets as determinants of cellular responsiveness to three TGF$\beta$ isoforms. *Journal of Biological Chemistry*, **265**, 20533–8.

Chiquet-Ehrismann, R., Kalla, P., and Pearson, C. A. (1989). Participation of tenascin and transforming growth factor beta in reciprocal epithelial-mesenchymal interactions of MCF7 cells and fibroblasts. *Cancer Research*, **49**, 4322–5.

Cohlan, S. Q. (1953). Excessive intake of Vitamin A as a cause of congenital abnormalities in the rat. *Science*, **117**, 535–6.

Davis, W. L., Crawford, L. A., Cooper, O. J., Farmer, G. R., Thomas, D., and Freeman, B. L. (1990). Generation of radical oxygen species by neural crest cells treated *in vitro* with isotretinion and 4-oxo-isotretinoin. *Journal Craniofacial Genetics and Developmental Biology*, **10**, 295–310.

Denhez, F., Lafayatis, R., Kondaiah, P., Roberts, A. B., and Sporn, M. B. (1990). Cloning by polymerase chaing reaction of a new mouse TGF beta, mTGF-beta3. *Growth Factors*, **3**, 139–46.

Derynck, R., Jarrett, J. A., Chen, E. Y., Eaton, D. H., Bell, J. R., Assoian, R. K. *et al.* (1985). Human transforming growth factor-beta cDNA sequence and expression in tumor cell lines. *Nature (Lond.)*, **316**, 701–5.

Fantel, A. G., Shepard, T. H., Newell-Morris, L. L., and Moffett, B. C. (1977).

Teratogenic effects of retinoic acid in pig-tail monkeys (*Macaca nemestrina*). I. General features. *Teratology*, **15**, 65–71.

Ferguson, M. W. J. and Honig, L. S. (1984). Epithelial-mesenchymal interactions during vertebrate palatogenesis. *Current Topics in Developmental Biology*, **19**, 138–64.

Ferguson, M. W. J. (1988). Palate development. *Development (Suppl.)*, **103**, 41–60.

FitzPatrick, D. R.. Denhez, F., Kondaiah, P., and Akhurst, R. J. (1990). Differential expression of TGF beta isoforms in murine palatogenesis. *Development*, **109**, 585–95.

Flanders, K. C., Roberts, A. B., Ling, N., Fleurdelys, B. S., and Sporn, M. B. (1988). Antibodies to peptide determinants in transforming growth factor-beta and their applications. *Biochemistry*, **27**, 739–46.

Flanders, K. C., Thompson, N. L., Cissel, D. S., Ellingsworth, L. R., Roberts, A. B., and Sporn, M. B. (1989). Epitope-dependent immunohistochemical localization of transforming growth factor-beta. *Journal Cell Biology*, **108**, 653–50.

Gatherer, D., ten Dijke, P., Baird, D. T., and Akhurst, R. J. (1990). Expression of TGFβ isoforms during first trimester human embryogenesis. *Development*, **110**, 445–60.

Glick, A. B., Flanders, K. C., Danielpour, D., Yuspa, S. H., and Sporn, M. B. (1989). Retinoic acid induces transforming growth factor-beta2 in cultured keratinocytes and mouse epidermis. *Cell Regulation*, **1**, 87–97.

Glick, A. B., McCune, B. K., Abdulkarem, N., Flanders, K. C., Lumadue, J. A., Smith, J. M. *et al.* (1991). Complex regulation of TGFβ expression by retinoic acid in the Vitamin A deficient rat. *Development*, **111**, 1081–6.

Greene, R. M. and Pratt, R. M. (1976). Developmental aspects of secondary palate development. *Journal of Embryology and Experimental Morphology*, **36**, 225–45.

Halfter, W., Chiquet-Ehrismann, R., and Tucker, R. P. (1989). The effect of tenascin and embryonic basal lamina on the behaviour and morphology of neural crest cells *in vitro*. *Development Biology*, **132**, 14–25.

Hanks, S. K., Armour, R., Baldwin, J. H., Maldonado, F., Spiess, J., and Holley, R. W. (1988). Amino acid sequence of BSC-1 cell growth inhibitor (polyergin) deduced from the nucleotide sequence of the cDNA. *Proceeding of the National Academy of Sciences USA*, **85**, 79–82.

Heine, U., Munoz, E. F., Flanders, K. C., Ellingsworth, L. R., Lam, H. Y., Thompson, N. L. *et al.* (1987). Role of transforming growth factor-beta in the development of the mouse embryo. *Journal Cell Biology*, **105**, 286–76.

Hiraki, Y., Inoue, H., Hiral, R., Kato, Y., and Suzuki, F. (1988). Effect of transforming growth factor beta on cell proliferation and glycosaminoglycan synthesis by rabbit growth-plate chondrocytes in culture. *Biochimica et Biophysica Acta*, **969**, 91–9.

Ignotz, R. A., Endo, T., and Massague, J. (1987). Regulation of fibronectin and type I collagen mRNA levels by transforming growth factor-beta. *Journal of Biological Chemistry*, **262**, 6443–6.

Kochhar, D. M. (1967). Teratogenic activity of retinoic acid. *Acta Pathologica, Microbiologica et Immunologica Scandinavica*, **A, 70**, 398–404.

Le Douarin, N. (1982) *The neural crest*. Cambridge University Press, Cambridge.

Lehnert, S. A. and Akhurst, R. J. (1988). Embryonic expression pattern of TGF beta type-1 RNA suggests both paracrine and autocrine mechanisms of action. *Development*, **104**, 263–73.

Linask, K. K., D'Angelo, M., Gehris, A. L., and Greene, R. M. (1991). Transforming growth factor-$\beta$ receptor profiles of human and murine embryonic palatal mesencymal cells. *Experimental Cell Research*, **192**, 1–9.

Massague, J. (1990). The Transforming Growth Factor-$\beta$ family. *Annunal Review of Cell Biology*, **6**, 597–641.

Millan, F. A., Kondaiah, P., Denhez, F., and Akhurst, R. J. (1991). Embryonic gene expression patterns of TGF betas 1, 2 and 3 suggest different developmental functions *in vivo*. *Development*, **111**, 131–44.

Miller, D. A., Lee, A., Matsui, Y., Chen, E. Y., Moses, H. L., and Derynck, R. (1989). Complementary DNA cloning of murine transforming growth factor beta3 (TGF beta3) precursor and the comparative expression of TGF beta1 and TGF beta3 and TGF beta1 in murine embryos and adult tissues. *Molecular Endocrinology*, **3**, 1926–34.

Morriss, G. (1973). The ultrastructural effects of excess maternal vitamin A on the primitive streak stage rat embryo. *Journal Embryology and Experimental Morphology*, **24**, 219–42.

Moses, H. L., Branum, E. L., Proper, J. A., and Robinson, R. A. (1981). Transforming growth factor production by chemically transformed cells. *Cancer Research*, **41**, 2842–8.

Moses, H. L., Tucker, R. F., Leof, E. B., Coffey, R. J., Halper, J., and Shipley, G. D. (1985) Type beta transforming growth factor is a growth stimulator and a growth inhibitor. In *Cancer cells*, (eds Feramisco, J., Ozanne, B., and Stiles, C.), pp. 65–71. Cold Spring Harbor Laboratory, Cold Spring Harbor.

Newall, D. R. and Edwards, J. R. G. (1981). The effect of Vitamin A on fusion of mouse palates. I. Retinyl palmitate and retinoic acid *in vivo*. *Teratology*, **23**, 115–24.

Nichols, D. H. (1986). Formation and distribution of neural crest mesenchyme to the first pharyngeal arch region of the mouse embryo. *American Journal of Anatomy*, **176**, 221–31.

Noden, D. M. (1986). Origins and patterning of craniofacial mesenchymal tissues. *Journal Craniofacial Genetics and Developmental Biology (Suppl.)* **2**, 15–32.

Ohta, M., Greenberger, J. S., Anklesaria, P., Bassols, A., and Massague, J. (1987). Two forms of transforming growth factor type beta distinguished by multipotential haemopoietic progenitor cells. *Nature*, **329**, 539–41.

Padgett, R. W., St Johnston, R. D., and Gelbart, W. M. (1987). A transcript from a Drosophila pattern gene predicts a protein homologous to the trasforming growth factor-beta family. *Nature (Lond)*, **325**, 81–4.

Pearson, C. A., Pearson, D., Shibahara, S. and Hofsteenge, J. (1988). Tenascin: cDNA cloning and induction by TGF-beta. *EMBO Journal*, **7**, 2977–82.

Pelton, R. W., Nomura, S., Moses, H. L., and Hogan, B. L. M. (1989). Expression of transforming growth factor beta-2 RNA during murine embryogenesis. *Development*, **106**, 759–67.

Pelton, R. W., Hogan, B. L. M., Miller, D. A., and Moses, H. L. (1990). Differential expression of genes encoding TGFs $\beta$1, $\beta$2, and $\beta$3 during murine palate formation. *Developmental Biology*, **141**, 456–60.

Postlethwaite, A. E., Keskioja, J., Moses, H. L., and Kang, A. H. (1987). Stimulation of the chemotactic migration of human fibroblasts by transforming growth factor beta. *Journal of Experimental Medicine*, **165**, 251–6.

Pratt, R. M., Goggins, J. F., Wilk, A. L., and King, C. T. (1973). Acid mucopolysaccharide synthesis in the secondary palate of the developing rat at the time of rotation and fusion. *Developmental Biology*, **32**, 230–7.

Pratt, R. M., Goulding, E. H., and Abbott, B. D. (1987). Retinoic acid inhibits migration of cranial neural crest cells in the cultured mouse embryo. *Journal Craniofacial Genetics and Developmental Biology*, **7**, 205–17.

Roberts, A. B., Anzano, M. A., Lamb, L. C., Smith, J. M., and Sporn, M. B. (1981). New class of transforming growth factors potentiated by epidermal growth factor. *Proceedings of the National Academy of Sciences USA*, **78**, 5339–43.

Roberts, A. B., Sporn, M. B., Assoian, R. K., Smith, J. M., Roche, N. S., and Wakefield, L. M. (1986). Transforming growth factor type beta: rapid induction of fibrosis and angiogenesis *in vivo* and stimulation of collagen formation *in vitro*. *Proceedings of the National Academy of Sciences USA*, **83**, 4167–71.

Roberts, A. B. and Sporn, M. B. (1990) The transforming growth factor betas. In *Peptide growth factors and their receptors—handbook of experimental pathology*, (eds M. B. Sporn and A. B. Roberts), pp. 419–72. Springer-Verlag, Heidelberg.

Robey, P. G., Young, M. F., Flanders, K. C., Roche, N. S., Kondaiah, P., Reddi, A. H. *et al.* (1987). Osteoblasts synthesize and respond to transforming growth factor-type beta (TGF-beta) *in vitro*. *Journal Cell Biology*, **105**, 457–63.

Rosa, F. W., Wilk, A. L., and Kelsey, F. O. (1986). Teratogen update: Vitamin A congeners. *Teratology*, **33**, 355–64.

Sakamoto, M. K., Nakamura, K., Handa, J., Kihara, T., and Tanomura, T. (1989). Morphogenesis of the secondary palate in mouse embryos with special reference to the development of rugae. *Anatomical Record*, **223**, 299–310.

Seyedin, S. M., Segarini, P. R., Rosen, D. M., Thompson, A. Y., and Bentz, H. (1987). Cartilage-inducing factor-B is a unique protein structurally and functionally related to transforming growth factor-beta. *Journal of Biological Chemistry*, **262**, 1946–9.

Sharpe, P. M. and Ferguson, M. W. (1988). Mesenchymal influences on epithelial differentiation in developing systems. *Journal of Cell Science (Suppl.)* **10**, 195–230.

Shenefelt, R. E. (1972). Morphogenesis of malformations in hamsters caused by retinoic acid: relation to dose and stage of treatment. Teratology, **5**, 103–18.

Slavkin, H. C. (1984). Morphogenesis of a complex organ. *Current Topics in Developmental Biology*, **19**, 1–16.

Smith, J. C. (1989). Mesoderm induction and mesoderm-inducing factors in early amphibian development. *Development*, **105**, 665–77.

Sulik, K. K., Johnston, M. C., Smiley, S. J., Speight, H. S., and Jarvis, B. E. (1987). Mandibulofacial dysostosis (Treacher Collins Syndrome): a new proposal for its pathogenesis. *American Journal Medical Genetics*, **27**, 359–72.

Sulik, K. K., Smiley, S. J., Turvey, T. A., Speight, H. S., and Johnston, M. C. (1989). Pathogenesis of cleft palate in Treacher Collins, Nager, and Miller syndromes. *Cleft Palate Journal*, **26**, 209–16.

ten Dijke, P., Hansen, P., Iwata, K. K., Pieler, C., and Foulkes, J. G. (1988). Identification of another member of the transforming growth factor type beta gene family. *Proceedings of the National Academy of Sciences USA*, **85**, 4715–19.

Thomsen, G. H. and Melton, D. A. (1990). Isolation of multiple new members of the transforming growth factor beta family from *Xenopus*. *Journal of Cellular Biochemistry*, **14E**, 65.

Thorogood, P. V., Smith, L., Nicol, A., McGinty, R., and Garrod, D. (1982). Effects of vitamin A on the behavior of migratory neural crest cells *in vitro*. *Journal of Cell Science*, **57**, 331–50.

Verwoerd, C. D. A. and Van Ooostrom, C. G. (1979). *Advances in anatomy, embryology and cell biology 58: Cephalic neural crest and placodes*. Springer-Verlag, Berlin.

Webster, W. S., Johnston, M. C., Lammer, E. J., and Sulik, K. K. (1986). Isotretinoin embryopathy and the cranial neural crest: an *in vivo* and *in vitro* study. *Journal Craniofacial Genetics and Development Biology*, **6**, 211–22.

Wilcox, J. N. and Derynck, R. (1988). Developmental expression of transforming growth factors alpha and beta in the mouse fetus. *Molecular and Cell Biology*, **8**, 3415–22.

Wiley, M. J., Cauwenbergs, P., and Taylor, I. M. (1983). Effects of retinoic acid on the development of the facial skeleton in hamsters: Early changes involving cranial neural crest cells. *Acta Anatomica*, **116**, 180–92.

Wrann, M., Bodmer, S., De-Martin, R., Siepl, C., Hofer-Warbinek, R., Frei, K. *et al.* (1987). T cell suppressor factor from human glioblastoma cells is a 12.5-kd protein closely related to transforming growth factor-beta. *EMBO Journal*, **6**, 1633–6.

# 13

# Morphogenesis-related changes in extracellular matrix induced by retinoic acid

Gillian Morriss-Kay and Radma Mahmood

Retinoic acid (RA) is a 'classic' teratogen in that it induces different patterns of malformations when mammalian embryos are exposed to it at different developmental stages. The developmental stage at which exposure to a teratogen leads to a specific malformation outcome is the critical period for that teratogen and malformation. For instance, the critical period for RA-induced craniofacial malformations is at or before the early neural plate stage of development; cleft palate and limb reduction defects have their own critical periods at later stages of embryogenesis.

Investigation of developmental events at the time of the critical period, in both normal and RA-treated embryos, can provide important insights into both the normal roles of RA and the early alterations at the onset of abnormal development. At the critical period for craniofacial malformation, embryonic structure is relatively simple (Fig. 13.1(a)), consisting of neural plate and primitive streak regions. In the primitive streak region, epithelial cells of the epiblast are converted to mesenchyme; these 'primary mesenchyme' cells migrate laterally and rostrally (anteriorly) within a glycosaminoglycan-rich extracellular matrix (Morriss and Solursh 1978a). Primary mesenchyme cells underlying the neural plate have a more elaborately structured matrix (see Fig. 13.4(a)), so that this part of the embryo has the characteristic appearance of interacting tissues. In this article we consider the possibility that early events in RA-induced craniofacial malformations involve alteration of the molecular composition and physical structure of the extracellular matrix. We begin by considering the evidence for identifying the early neural plate stage as the critical period.

THE CRITICAL PERIOD FOR INITIATION OF RETINOIC ACID-INDUCED CRANIOFACIAL ABNORMALITY

Craniofacial malformations are a characteristic response of early mammalian embryos to maternal retinoid excess during early pregnancy, and have been

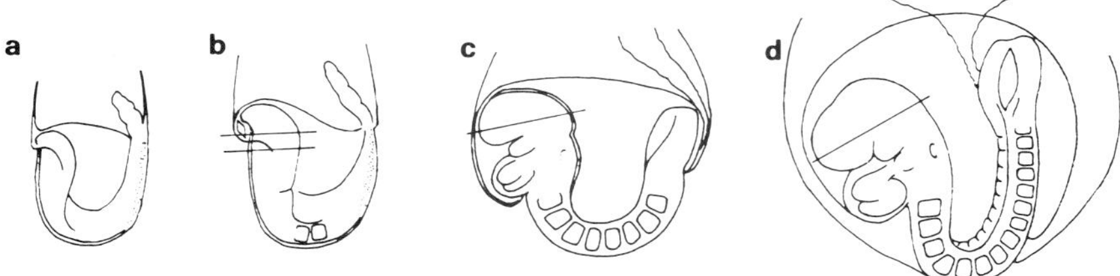

**Fig. 13.1** Diagrams illustrating (*a*) embryonic stage at the start of culture and stages of development of control embryos at (*b*) 3 hours, (*c*) 17 hours and (*d*) 24 hours of culture. Transverse lines indicate the planes of section illustrated in subsequent figures; the lower line on (*b*) corresponds to Fig. 13.2 and the upper line to Fig. 13.3. In all diagrams, the cranial region is on the left and the caudal region on the right. In (*a*), the cranial region consists only of neural plate, notochord (double line), primary mesenchyme and (on the outer surface) endoderm; the edge of the foregut is indicated by a curved line; the future heart tissue lies immediately rostral to it. The primitive streak is indicated by stippling. In (*b*), the preotic sulcus is present as a transverse line in the neural folds, rostral to the first two somites. The amnion (across the top of the embryo in (*a*) and (*b*), surrounding the embryo progressively in (*c*) and (*d*)) is present on some of the sections.

observed in a wide variety of mammalian species. The malformed structures include the jaws, secondary palate, middle and external ears, thymus, aorticopulmonary septum, and brain (see Lammer and Armstrong Chapter 21 for details and references). Most of these structures depend for their normal development on a major contribution from the cranial neural crest, and a strong case has been made for neural crest as the primary target of retinoid excess in this system (Wiley *et al.* 1983; Lammer 1985; Webster *et al.* 1986). However, this interpretation does not explain the observed central nervous sytem malformations.

Recent studies on mouse embryos exposed to all-*trans*-RA *in utero* (Murphy *et al.* Chapter 17) demonstrate that the critical period is before the first neural crest cells emigrate from the neural folds, suggesting that the neural epithelium (including its premigratory neural crest cells) may be the primary site of retinoid action. Exposure of rat or mouse embryos to RA excess after the onset of cranial neural crest cell migration has no effect on subsequent size, shape, or position of the pharyngeal arches (Ritchie and Webster 1991), except at very high concentrations (Lee *et al.* 1991). Neural crest cells exposed to RA *in vitro* show effects on cell adhesion and locomotion only after 24 hours (Smith-Thomas *et al.* 1987), a time period during which major translocations of neural crest cells occur *in vivo* (Tan and Morriss-Kay 1985, 1986).

Hence, although both experimental and clinical studies indicate that cranial neural crest cell migration is abnormal in RA-induced craniofacial dysmorphogenesis, this effect may be secondary to an effect on the hindbrain neuroepithelium (including the premigratory crest cells). The most severely affected part of the brain is the segmented region of the hindbrain, which lies between the midbrain/hindbrain boundary rostrally and the first occipital somite caudally (see Murphy *et al.* Chapter 17). Early neural crest cell migration from this region precedes rhombomere formation, but has a clear relationship to the development of the pharyngeal arches (Tan and Morriss-Kay 1985, 1986). In contrast, neural crest cells that populate the frontonasal and maxillary regions are derived from the midbrain neural folds; it is therefore interesting to note that the frontonasal region is not severely affected in retinoid-treated embryos, and that the abnormalities of the maxillary region may be attributable to ectopic migration of crest cells normally destined for the mandibular arch (Morriss and Thorogood 1978; Morriss-Kay 1991; Morriss-Kay *et al.* 1991).

A possible definition of 'RA excess' in the context of craniofacial malformations is an amount greater than that which saturates the RA binding capacity of cellular retinoic acid binding protein-I (CRABP I) in cells that contain it, allowing levels of unbound RA to rise in cells that normally use CRABP I to minimize the amount of RA reaching the nucleus. This concept of the function of CRABP I is supported by the observation that RA-induced gene expression in F9 teratocarcinoma cells requires higher levels of exogenous RA in a stably transfected line that overexpresses CRABP I than in the parent line (Boylan and Gudas 1991). The amount of RA that reaches the nucleus in these cells is therefore reduced when cytoplasmic CRABP I levels are high.

During early somite stages and during the formation of rhombomeres, the hindbrain neuroepithelium shows high levels of CRABP I protein (Dencker *et al.* 1990) and RNA transcripts (Ruberte *et al.* 1991). This correlation may be directly relevant to the vulnerability of the hindbrain to RA-induced dysmorphogenesis. However, at the late presomite stage, which is the critical period for RA-induced craniofacial morphogenesis, the future hindbrain region of the cranial neural plate has only just begun to differentiate, and there is no neuroepithelial CRABP I expression. At that stage CRABP I is expressed only in the primary mesenchyme, in all cells except those of the primitive streak itself (Ruberte *et al.* 1991). This distribution of CRABP I suggests that low nuclear RA levels are normal for migrating and postmigratory primary mesenchyme cells. Therefore, when free RA levels are raised, primary mesenchyme cell function may be specifically affected. This effect could mediate the observed effects on the hindbrain neural epithelium through epithelial–mesenchymal interactions during hindbrain differentiation.

We will now consider the possibility that primary mesenchyme, and its function in relation to tissue interaction with the overlying neuroepithelium, is the first tissue to be affected in the chain of events of RA-induced craniofacial malformations.

## TISSUE ORGANIZATION AT THE CRITICAL PERIOD

Observations from previous studies support the idea that primary mesenchyme is a specific target of RA excess. Reduced cell number in the cranial primary mesenchyme has been reported to occur in response to retinoid excess in rat embryos *in vivo* (Morriss 1972) and *in vitro* (Morriss and Steele 1974, 1977), while explanted primary mesenchyme *in vitro* showed reduced adhesion and poor locomotion (Morriss 1975). The neural epithelium, its basement membrane and its underlying primary mesenchyme have the typical appearance of interacting tissues, with a rich extracellular matrix whose constituents include hyaluronan, chrondroitin sulphate proteoglycans (CSPG), heparan sulphate proteoglycans (HSPG) (Solursh and Morriss 1977; Morriss and Solursh 1978*a,b*; Morriss-Kay and Tuckett 1989; Tuckett and Morriss-Kay 1989), fibronectin, laminin, and entactin (Tuckett and Morriss-Kay, 1986) and type IV collagen (see p. 170).

Tissue interactions between primary mesenchyme and neural (or preneural) hindbrain epithelium may be an essential component of the mechanism of segmentally organized gene expression: a number of genes that express in a segment-related pattern within the hindbrain are first expressed in the primary mesenchyme within and/or adjacent to the primitive streak, e.g. *int-2* (Wilkinson *et al.* 1988); *Hox-2.9* (Frohman *et al.* 1990); *Hox-1.6* (Murphy and Hill 1991). Of these genes, expression of *Hox-2.9* is known to be altered by RA both in embryonal carcinoma cells (see Boncinelli *et al.* Chapter 16) and in the mouse embryonic hindbrain (see Murphy *et al.* Chapter 17). *Hox-1.6*), which coexpresses with *Hox-2.9* in the early hindbrain and in the primary mesenchyme (Murphy and Hill 1991) is also RA-responsive, but so far this has only been demonstrated in teratocarcinoma cells (LaRosa and Gudas 1988, in which *Hox-1.6* is called ERA-1).

Extracellular matrix (ECM) plays active roles in both mesenchymal cell migration and epithelial–mesenchymal interactions. Neural crest cell migration and glandular morphogenesis are the best studied systems in this context. Dispersal of neural crest cells from an explanted neural tube requires a fibronectin and/or laminin-coated substratum (Perris *et al.* 1989). Antibodies to a receptor for fibronectin and laminin, or to a laminin–HSPG complex, perturb neural crest cell migration *in vivo* (Bronner-Fraser 1986; Bronner-Fraser and Lallier 1988). Adhesiveness of neural crest cells to a fibronectin-coated substratum can be modulated by CSPG (Tan *et al.* 1987). Adhesion of cells to extracellular fibronectin occurs both at the cell surface integrin (fibronectin) receptors and via cell surface HSPG (Woods *et al.* 1985, 1986). During epithelial morphogenesis to form glandular structures, the epithelial basement membrane contains fibronectin, laminin, type IV collagen, CSPG, and HSPG; a specific local pattern of degradation of some of these constituents is brought about by the adjacent mesenchyme at the sites of generation of epithelial curvature (Bernfield *et al.* 1984).

It is clear, therefore, that any agent affecting extracellular matrix composition during morphogenesis could have profound effects on morphogenesis-related cell behaviour.

## INTERACTIONS BETWEEN RETINOIDS AND EXTRACELLULAR MATRIX

Known effects of RA on ECM composition come mainly from studies on cartilage degradation, beginning with the work of Fell and Mellanby (1952). In fetal epiphyseal cartilage, RA induces release of proteoglycan from the cartilage matrix (Kistler 1986). Shapiro and Mott (1981) found that at $10^{-7}$ M RA (the concentration range used in our studies), glycosaminoglycan (GAG) synthesis by chondrocytes was reduced by 80 per cent; the released GAG consisted of heparan sulphate and chondroitin-4- and 6-sulphates. RA-induced cartilage matrix degradation requires the synthesis of RNA, protein and glycoprotein, and specifically alters the protein synthesis pattern (Kistler 1986). Retinoids also inhibit chondrogenesis in cultured limb bud cells (Lewis *et al.* 1978); the inhibitory activity of retinoids in this system correlates with their teratogenic activity (Kistler 1986). Conversely, RA can prevent degradation of mature cartilage matrix (see Pfahl Chapter 4, for details and references). In fibroblast cultures, RA has been reported to increase both the rate of synthesis and the degree of sulphation of heparan sulphate (Shapiro and Mott 1981).

RA also affects the synthesis and attachment of cell-surface-associated glycoproteins. Fibronectin was found to be reduced at the cell surface of avian neural crest cells in cultures containing $3.5 \times 10^{-5}$ M retinol; the rate of outgrowth of cells from the explant was half that of controls, and the cells were less adherent to the substratum (Thorogood *et al.* 1982). However, in a similar study using retinol ($7 \times 10^{-5}$ M) and 13-*cis*-RA ($6.7 \times 10^{-6}$ M and $6.7 \times 10^{-5}$ M), neural crest cell outgrowth and adhesiveness were reduced but cell surface fibronectin and fibronectin receptor were unaffected (Smith-Thomas *et al.* 1987). In F9 embryonic teratocarcinoma stem cells, the gene encoding the B1 subunit of laminin is transcriptionally activated by RA (1 $\mu$M) during differentiation into extraembryonic parietal endoderm cells (Vasios *et al.* 1989); collagen type IV is also activated (Wang and Gudas 1983).

## EFFECTS OF RETINOIC ACID ON EXTRACELLULAR MATRIX COMPOSITION IN EARLY EMBRYOS

To investigate the effects of RA on ECM composition during the critical period for RA-induced craniofacial morphogenesis, we cultured late presomite, early neural plate stage rat embryos (day 9.5) in medium containing 0.25 $\mu$g ml$^{-1}$ ($8.3 \times 10^{-7}$ M) RA for 3, 17, or 24 hours. RA was dissolved in ethanol (0.25 mg ml$^{-1}$); 1 $\mu$l of this solution or 1 $\mu$l ethanol alone was added for each ml of culture medium (1:1 Tyrode's saline:rat serum). The culture technique was as described previously (Morriss-Kay and Tuckett 1989). The bottles were wrapped in aluminium foil to prevent photo-oxidation of RA. Embryonic appearance at the start of culture and at each of the three termination times is shown in Fig. 13.1 (see p. 166). After culture, the embryos were fixed in St Marie's fixative, dehydrated and embedded in paraplast. Sections were cut at 8 $\mu$m and mounted 12 to a slide on gelatin-coated slides for immunocytochemistry to localize the following ECM components: fibronectin, laminin, type IV collagen, CSPG, and HSPG. The primary antibodies were: goat anti-rat fibronectin (CP laboratories); rabbit anti-mouse laminin (Bethesda Research Laboratories); a rabbit monoclonal antibody to type IV collagen (ICN Biomedicals); CS-56, a monoclonal antibody to the GAG moiety of CSPG (Avnur and Geiger 1984) (ICN Biomedicals); antibody to the core protein of HSPG (Hassell *et al.* 1980; gift from John Hassell). Specificity of the commercially obtained antibodies was ascertained by preabsorption with purified antigens and by immunoblotting. The sections were incubated with primary antibody at a dilution of 1:50 (anti-laminin, anti-type IV collagen, and CS-56), 1:25 (anti-fibronectin), or 1:10 (anti-HSPG). The antigens were localized in permeabilized tissue sections by means of the avidin–biotin–peroxidase technique (kits from Vector Laboratories or Sigma). Control sections incubated without primary antibody showed no staining with any of the secondary antibodies, and are not illustrated. Sections showing features mentioned in the text are illustrated in Figs 13.2–13.5.

Morphological and immunohistochemical differences between control and RA-exposed embryos were already apparent in embryos cultured for 3 hours. Neural folds were convex at this stage, in both normal and RA-treated embryos. Mesenchymal cell contact with the overlying neuroepithelial basement membrane was extensive in control embryos; direct cell–basement membrane contact was supplemented with a rich ECM, which was strongly positive for CSPG (Fig. 13.2(c)), HSPG (Fig. 13.4(a)) and fibronectin, faintly positive for type IV collagen, and negative for laminin. In RA-treated embryos, this complex ECM was very greatly reduced (Figs 13.2(d) and 13.4(b)); the neuroepithelial basement membrane was thinner than in control embryos, and the ECM meshwork beneath it was sparse or absent. The composition of the basement membrane was altered, showing reduced staining for HSPG, while CSPG staining here was undetectable (Figs 13.2(c, d), 13.4(a, b)). No differences were observed for laminin, fibronectin, or type IV collagen. (In fact, type IV collagen staining of the neuroepithelial basement membrane was faint and incomplete in both control and RA-treated embryos at this stage.)

In some RA-treated embryos the neural and surface epithelia projected laterally without any mesenchyme, so that the two basement membranes were

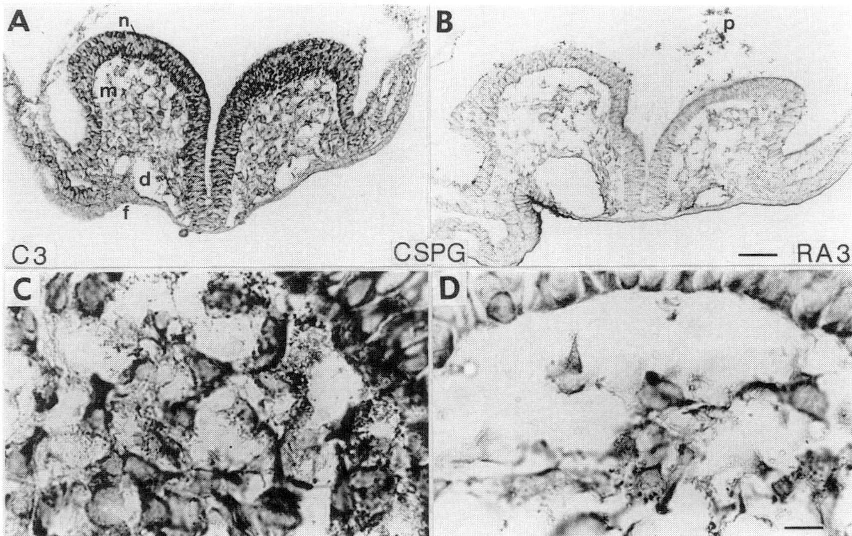



**Fig. 13.2**  Low (*a*, *b*) and high (*c*, *d*) power views of sections through the cranial neural folds of control (*a*, *b*) and RA-treated (*c*, *d*) embryos cultured for 3 hours, stained immunohistochemically for CSPG. (*c*) and (*d*) show part of the neural epithelium and underlying mesenchyme. d, dorsal aorta; f, endoderm of the future foregut; m, primary mesenchyme; n, neural epithelium; p, precipitate of CSPG-positive material. Scale bars = 50 μm (*a*, *b*); 10 μm (*c*, *d*).

actually apposed (Fig. 13.3(b, c)). The apposed basement membranes were positive for fibronectin (Fig. 13.3(c)) and laminin but negative or only sparsely reactive for HSPG (Fig. 13.3(c)) and negative for CSPG. CSPG reactivity was greatly reduced throughout the embryo, both intracellularly and in the ECM (Fig. 13.2(b, d)).

In the primitive streak region, CSPG staining was much reduced in RA-treated embryos compared with controls; none of the other antibodies showed any differences in this region.

These differences can be summarized as follows. In the early cranial neural folds of control embryos, the neuroepithelial basement membrane contains fibronectin, laminin, CSPG, and HSPG, with small amounts of type IV collagen. Except for laminin, these ECM components extend below the basement membrane as a meshwork of strands and granules, making extensive contact with the mesenchymal cells. The same ECM components are present in the mesenchymal cell surface-associated material. After 3 hours of culture in RA-containing medium, most of the basement membrane-associated ECM meshwork is lost. CSPG levels are greatly reduced throughout the embryo. HSPG is reduced in the neuroepithelial basement membrane, and to a lesser extent in the mesenchymal cell surface material. In

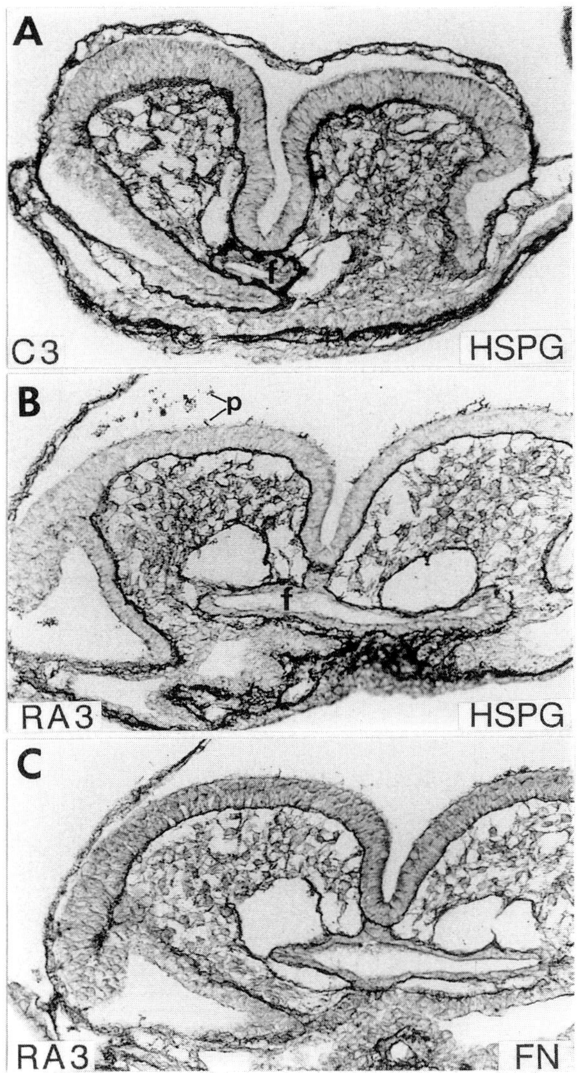

**Fig. 13.3** Transverse sections through the cranial folds of control (*a*) and RA-treated (*b, c*) embryos cultured for 3 hours. Immunostaining for HSPG or fibronectin as indicated. (*b*) and (*c*) are parallel sections from the same embryo; at the left-hand edge of the neural fold there is a projection of apposed neural and surface epithelia. f, foregut (cardiogenic tissue lies beneath it); p, precipitate of HSPG-positive material; other structures as Fig. 13.2. Scale bar = 50 μm.

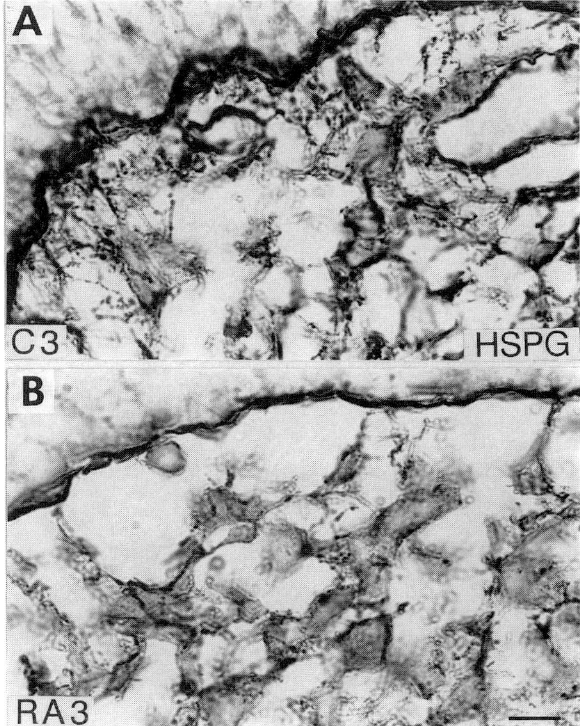



**Fig. 13.4** High power views of the neuroepithelial–mesenchymal interface immunostained for HSPG. (*a*) The control section shows a thick HSPG-containing basement membrane with associated strands and granules of HSPG-positive material, which extend into the mesenchyme and are continuous with cell surface-associated material. (*b*) The HSPG-positive ECM is very much reduced in the RA-treated embryo section. Scale bar = 10 $\mu$m.

the primitive streak region, CSPG levels are reduced, but all of the other components are present at the same levels as in control embryos. These observations suggest that exposure to RA for 3 hours causes considerable loss of CSPG and some loss of HSPG, but that other components are only lost where there is structural breakdown of a complex ECM.

After 17 and 24 hours of culture, the neuroepithelial–mesenchymal interface of control embryos no longer showed the ECM-rich structure seen in embryos cultured for 3 hours. In fact, a large blood vessel subjacent to up to 50 per cent of the width of the hindbrain neuroepithelial basement membrane separated the two tissues during the later stages of neurulation (Fig. 13.5(a, b)). Mesenchymal–neuroepithelial contact appeared to be similar in RA-treated and control embryos, but contacts between the mesenchyme and surface ectoderm were much reduced in RA-treated embryos (Fig. 13.5(b, d)).

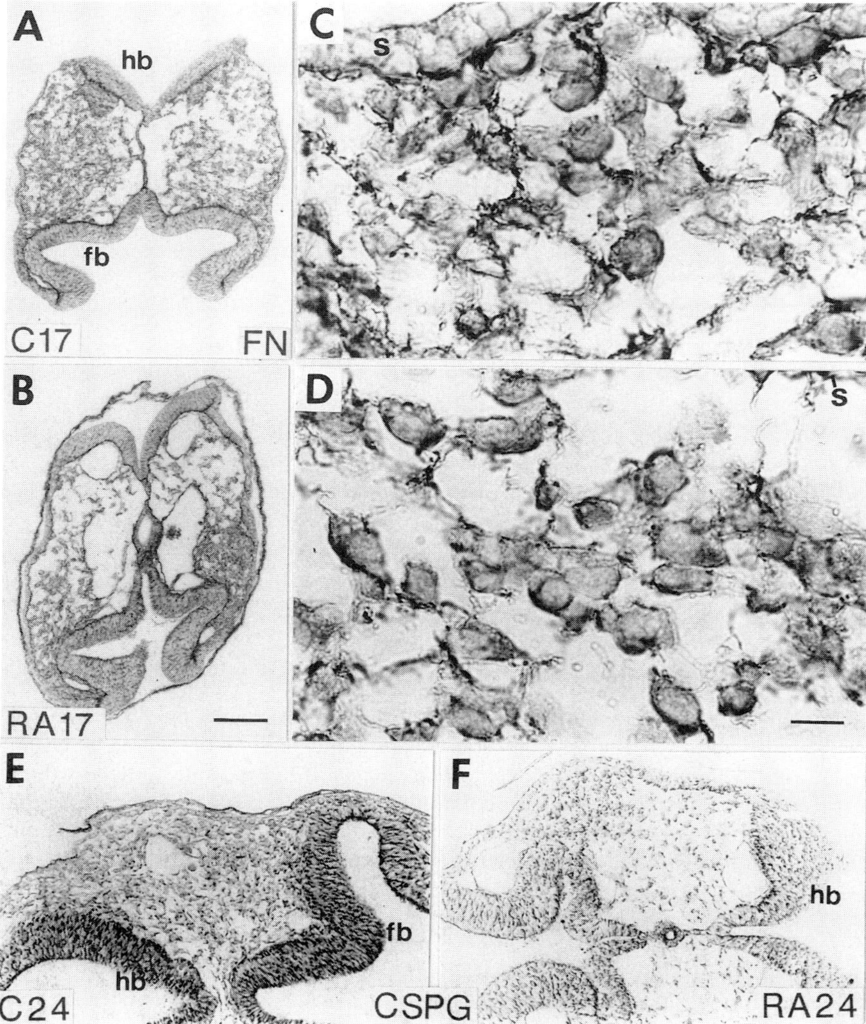



**Fig. 13.5** (*a, b*) After 17 hours of culture, there was no difference in the intensity of immunostaining for fibronectin in basement membranes. Apparent differences in mesenchymal cell surface staining were due to differences in cell density, as revealed at high power (*c, d*). The hindbrain (hb) neural folds are still convex in the RA-treated embryo (*b*), even though the forebrain (fb) neural folds are at a more advanced stage of closure than those of the control. Mesenchyme is closely apposed to surface ectoderm (s) in controls but separated from it in RA-treated embryos. (*e, f*) sections through the heads of control (*e*) and RA-treated (*f*) embryos cultured for 24 hours. CSPG immunoreactivity is generally lower in RA-treated embryos. Bar lines (*a, b, e, f*) 100 μm; (*c, d*) 10 μm.

Immunoreactivity for basement membrane and cell surface-associated fibronectin (Fig. 13.5), laminin and HSPG were equivalent in RA-treated and control embryos, while CSPG immunoreactivity was much reduced throughout the embryo (Fig. 13.5(e, f)), as observed at 3 hours. Type IV collagen staining was now strong in all basement membranes of both control and treated embryos. Conversion of hindbrain neural fold shape from convex to concave was delayed in RA-treated embryos (Fig. 13.5(b)); at 24 hours most control embryos had completed cranial neurulation whereas RA-treated embryos had not (Fig. 13.5(e, f)). Although these stages are later than the critical period, they provide useful information in indicating that of the ECM components studied, only CSPG levels were affected by long term (24 hours) exposure to a teratogenic concentration of RA.

RETINOIC ACID-INDUCED EXTRACELLULAR MATRIX CHANGES IN EMBRYOS: COMPARISONS WITH OTHER SYSTEMS

It is clear that during the critical period for RA-induced craniofacial malformations, when mesenchymal–epithelial interactions may be essential for the establishment of segmentally organized gene expression in the neuroepithelium, the structure and molecular composition of the ECM at the interface of the interacting tissues is altered by RA. These alterations were clear after only 3 hours of exposure to RA, a time period that, at this stage, is sufficient to cause malformations of the hindbrain and pharyngeal arches in embryos subsequently cultured in addition-free medium (our unpublished observations for RA; Ritchie and Webster 1991, for 13-*cis*-RA). Changes in basement membrane structure following exposure of embryos to RA for only 30 minutes have been observed by electron microscopy (Morriss and Steele 1977).

Other documented effects of RA on ECM composition require longer exposure times. In teratocarcinoma cell cultures, both collagen IV ($\alpha$2) and laminin B1 show only a slight increase at 12 hours, with a dramatic increase between 12 and 24 hours after addition of RA. These molecular changes reflect the process of epithelialization, which occurs in response to RA, including *de novo* synthesis of a basement membrane. No changes in ECM gene expression in RA-treated teratocarcinoma cell lines that resemble the changes reported here in early mammalian embryos have been reported. On the other hand, our observations show significant similarities to the RA-induced loss of proteoglycan from cartilage matrix observed by Kistler (1986 and references therein) and Shapiro and Mott (1981). That system therefore has more in common with the RA-induced ECM changes in embryos than has the F9 teratocarcinoma cell system. Embryonal carcinoma cell lines with the capacity to undergo neuronal differentiation in response to RA (Andrews 1984) and that show RA-inducible expression of homeobox genes characteristic of early embryos (Boncinelli, Chapter 16) have not been explored with respect to proteoglycan-related gene expression. It would be useful to know which cell line most resembles the response of intact embryos to RA.

## ARE EXTRACELLULAR MATRIX ALTERATIONS SUFFICIENT TO EXPLAIN RETINOIC ACID-INDUCED CRANIOFACIAL DEFECTS?

If loss of CSPG and HSPG is directly responsible for RA-induced alterations in the normal pattern of morphogenesis, we would expect to see some similarities between RA-treated embryos and those in which these two ECM components have been altered by exposure of embryos of the same stages to enzymes (chondroitinase ABC and heparitinase) that specifically degrade them, or to $\beta$-D-xyloside, which inhibits proteoglycan but not GAG synthesis (Morriss-Kay and Tuckett 1989; Tuckett and Morriss-Kay 1989; Morriss-Kay and Crutch 1982, and references therein). There are some similarities in that both $\beta$-D-xyloside and heparitinase treatments were associated with delay or inhibition of cranial neural fold morphogenesis and poor differentiation of rhombomeres; also, degradation of either CSPG or HSPG was associated with reduced pharyngeal arches and altered otocyst position. These comparisons suggest that ECM-related effects of RA play an important role in RA-induced alterations of early craniofacial morphogenesis, probably through loss of the normal interactive contact between the neural epithelium and its underlying mesenchyme.

The observed effects on morphogenesis are not identical in RA-treated and enzyme- or $\beta$-D-xyloside-treated embryos, suggesting that non-ECM-related factors must also be important. For instance, RA-induced activation of *Hox-2.9* in mouse embryos (Murphy *et al.* Chapter 17) is unlikely to be mediated by loss of proteoglycan, as it also occurs rapidly in embryonal carcinoma cells. However, it is possible that ECM alterations may play a role in altering the domain of expression of *Hox-2.9*, the establishment of which may involve epithelial–mesenchymal interactions (Frohman *et al.* 1990). Formation of clear boundaries between domains of segmental gene expression is normally correlated with morphological development of rhombomeres; both of these processes fail in RA-treated embryos (Murphy *et al.* Chapter 17). This failure could be ECM-related, as rhombomere formation involves specific organization of cytoskeletal structures (Tuckett and Morriss-Kay 1985), and basement membrane composition is known to affect epithelial cytoskeletal organization (Bernfield *et al.* 1984).

## CONCLUSIONS

Late presomite, early neural plate stage rat embryos have a highly structured extracellular matrix at the interface of the neuroepithelium and its underlying mesenchyme, consisting of a basement membrane and a web of strands and granules that greatly increase the amount of physical contact between the mesenchyme cells and the basement membrane. The molecular components of this matrix include CSPG, HSPG, laminin, fibronectin, and type IV collagen. Exposure to RA is associated with loss of HSPG from this part of the ECM, and loss of CSPG from the embryo as a whole. ECM composition is known to have important effects on the epithelial response to mesenchyme during glandular morphogenesis. If it is similarly involved in tissue interaction between primary mesenchyme and the early hindbrain neuroepithelium, disturbance of this interaction may be one of the primary events in retinoid-induced teratogenesis. It is of particular interest in this respect that the teratogenic activity of different retinoids correlates well with their chondrogenesis-inhibiting activity.

Except for the observation that CSPG was lost from the primitive streak region and migrating primary mesenchyme, this study did not shed any light on the possibility that decreased primary mesenchyme cell migration might play a role in mediating RA-induced craniofacial malformations.


## ACKNOWLEDGEMENTS

We thank Martin Barker and Colin Beesley for technical and photographic assistance, Professor Ray Guillery for helpful comments on the manuscript, and Hoffman-La Roche for financial support. R.M. is an MRC scholar. Preliminary work related to the new data presented here was carried out using immunofluorescence as part of undergraduate research projects. We thank Emma Hollick (CSPG), Alec McEwan (fibronectin), and Judith James (laminin) for their contributions.


## REFERENCES

Andrews, P. W. (1984). Retinoic acid induces neuronal differentiation of a cloned human embryonal carcinoma cell line *in vitro*. *Developmental Biology*, **103**, 285–93.

Avnur, Z. and Geiger, B. (1984). Immunocytochemical localization of native chondroitin-sulfate in tissues and cultured cells using specific monoclonal antibody. *Cell*, **38**, 811–22.

Bernfield, M., Banerjee, S. D., Koda, J. E., and Rapraeger, A. C. (1984). Remodelling of the basement membrane as a mechanism of morphogenetic tissue interaction. In *The role of extracellular matrix in development, 42nd symposium of the Society for Developmental Biology*, (ed. R. L. Trelstad), pp. 545–72. Alan R. Liss Inc., New York.

Boylan, J. F. and Gudas, L. (1991). Overexpression of the cellular retinoic acid binding protein-I (CRABP-I) results in a reduction in differentiation-specific gene expression in F9 teratocarcinoma cells. *Journal of Cell Biology*, **112**, 965–79.

Bronner-Fraser, M. (1986). An antibody to a receptor for fibronectin and laminin perturbs cranial neural crest cell development *in vivo. Developmental Biology*, **117**, 528–36.

Bronner-Fraser, M. and Lallier, T. (1988). A monoclonal antibody against a laminin–heparan sulfate proteoglycan complex perturbs cranial neural crest migration *in vivo. Journal of Cell Biology*, **106**, 1321–9.

Dencker, L. Annerwall, E., Busch, C., and Eriksson, U. (1990). Localization of specific retinoid binding sites and expression of cellular retinoic acid binding protein (CRABP) in the early mouse embryo. *Development*, **110**, 343–52.

Fell, H. B. and Mellanby, E. (1952). The effect of hypervitaminosis A on embryonic limb-bones cultivated *in vitro. Journal of Physiology*, **116**, 320–49.

Frohman, M. A., Boyle, M., and Martin, G. R. (1990). Isolation of the mouse *Hox-2.9* gene; analysis of embryonic expression suggests that positional information along the anterior-posterior axis is specified by mesoderm. *Development*, **110**, 589–608.

Hassell, J. R., Robey, P. G., Barrach, H-J., Wilczek, J., Rennard, S. I., and Martin, G. R. (1980). Isolation of a heparan sulfate-containing proteoglycan from basement membrane. *Proceedings of the National Academy of Sciences USA*, **77**, 4494–8.

Kistler, A. (1986). Hypervitaminosis A: side-effects of retinoids. *Biochemical Society Transactions*, **14**, 936–9.

Lammer, E. J. (1985). On the plausibility of retinoids adversely influencing neural crest cell activity. *Proceedings of the Greenwood Genetics Center.* **4**, 29–32.

Lammer, E. J., Chen, M., Hoar, R., Agnish, N. D., Benke, P., Braun, J., Curry, C., Fernhoff. P., Grix, A., Lott, I., Richard, J., and Sun, S. (1985). Retinoic acid embryopathy. *New England Journal of Medicine*, **313**, 837–41.

Lee, Q. P., Juchau, M. R., and Creech Kraft, J. M. (1991). Microinjection of cultured rat embryos: studies with retinoids. *Teratology*, **44**, 313–23.

Lewis, C. A., Pratt, R. M., Pennypacker, J. P., and Hassell, J. R. (1978). Inhibition of limb chondrogenesis *in vitro* by vitamin A: alterations in cell surface characteristics. *Developmental Biology*, **64**, 31–47.

Morriss, G. M. (1972). Morphogenesis of the malformations induced in rat embryos by hypervitaminosis A. *Journal of Anatomy*, **113**, 241–50.

Morriss, G. M. (1975) Abnormal cell migration as a possible factor in the genesis of vitamin A-induced craniofacial anomalies. In *New approaches to the evaluation of abnormal embryonic development* (ed. D. Neubert and H-J. Merker), pp. 678–87. Georg Thieme Verlag, Stuttgart.

Morriss, G. M. and Solursh, M. (1978*a*). Regional differences in mesenchymal cell morphology and glycosaminoglycans in early neural-fold stage rat embryos. *Journal of Embryology and Experimental Morphology*, **46**, 37–52.

Morriss, G. M. and Solursh, M. (1978*b*). The role of primary mesenchyme in normal and abnormal morphogenesis of mammalian neural folds. *Zoon*, **6**, 33–8.

Morriss, G. M. and Steele, C. E. (1974). The effect of excess vitamin A on the development of rat embryos in culture. *Journal of Embryology and Experimental Morphology*, **32**, 505–14.

Morriss, G. M. and Steele, C. E. (1977). Comparison of the effects of retinol and retinoic acid on postimplantation rat embryos *in vitro. Teratology*, **15**, 109–19.

Morriss, G. M. and Thorogood, P. V. (1978). An approach to cranial neural crest cell migration and differentiation in mammalian embryos. In *Development in mammals*, vol. 3, (ed. M. H. Johnson), pp. 363–412. North-Holland, Amsterdam.

Morriss-Kay, G. M. (1991). Retinoic acid, neural crest, and craniofacial development. *Seminars in Development* **2**, 211–18.

Morriss-Kay, G. M. and Crutch, B. (1982). Culture of rat embryos with *β-D*-xyloside: evidence of a role for proteoglycans in neurulation. *Journal of Anatomy*, **134**, 491–50.

Morriss-Kay, G., Murphy, P., Hill, R. and Davidson, D. (1991). Effects of retinoic acid excess on expression of *Hox-2.9* and *Krox-20* and on morphological segmentation of the hindbrain of mouse embryos. *EMBO Journal*, **10**, 2985–95.

Morriss-Kay, G. M. and Tuckett, F. (1989). Immunohistochemical localization of chondroitin sulphate proteoglycans and the effects of chondroitinase ABC in 9- to 11-day rat embryos. *Development*, **106**, 787–98.

Murphy, P. and Hill, R. E. (1991). Expression of the mouse *labial*-like homeobox-containing genes, *Hox-2.9* and *Hox-1.6,* during segmentation of the hindbrain. *Development*, **111**, 61–74.

Perris, R., Paulsson, M., and Bronner-Fraser, M. (1989). Molecular mechanisms of avian neural crest cell migration on fibronectin and laminin. *Developmental Biology*, **136**, 222–38.

Ritchie, H. and Webster, W. (1991). Parameters determining isotretinoin teratogenicity in rat embryo culture. *Teratology*, **43**, 71–81.

Ruberte, E., Dollé, P., Chambon, P., and Morriss-Kay, G. (1991). Retinoic acid receptors and cellular binding proteins: II. Their differential pattern of transcription during early morphogenesis in mouse embryos. *Development*, **111**, 45–60.

Shapiro, S. S. and Mott, D. J. (1981). Modulation of glycosaminoglycan biosynthesis by retinoids. *Annals of the New York Academy of Sciences*, **359**, 306–21.

Smith-Thomas, L., Lott, I., and Bronner-Fraser, M. (1987). Effects of isotretinoin on the behavior of neural crest cells *in vitro*. *Developmental Biology*, **123**, 276–81.

Solursh, M. and Morriss, G. M. (1977). Glycosaminoglycan synthesis in rat embryos during formation of the primary mesenchyme and neural folds. *Developmental Biology*, **57**, 75–86.

Tan, S. S., Crossin, K. L., Hoffman, S., and Edelman, G. (1987). Asymmetric expression in somites of cytotactin and its proteoglycan ligand is correlated with neural crest cell distribution. *Proceedings of the National Academy of Sciences USA*, **84**, 7977–81.

Tan, S. S. and Morriss-Kay, G. M. (1985). The development and distribution of the cranial neural crest in the rat embryo. *Cell and Tissue Research*, **240**, 403–16.

Tan, S. S. and Morriss-Kay, G. M. (1986). Analysis of cranial neural crest cell migration and early fates in postimplantation rat chimaeras. *Journal of Embryology and Experimental Morphology*, **98**, 21–58.

Thorogood, P., Smith, L., Nicol, A., McGinty, R., and Garrod, D. (1982). Effects of vitamin A on the behaviour of migratory neural crest cells *in vitro*. *Journal of Cell Science*, **57**, 331–50.

Tuckett, F. and Morriss-Kay, G. M. (1985). The ontogenesis of cranial neuromeres in the rat embryo. II. A transmission electron microscope study. *Journal of Embryology and Experimental Morphology*, **88**, 231–47.

Tuckett, F. and Morriss-Kay, G. M. (1986). The distribution of fibronectin, laminin and entactin in the neurulating rat embryo studied by indirect immunofluorescence. *Journal of Embryology and Experimental Morphology*, **94**, 95–112.

Tuckett, F. and Morriss-Kay, G. M. (1989). A role for heparan sulphate proteoglycan in the rat embryo: effects of heparitinase treatment during early organogenesis. *Anatomy and Embryology*, **180**, 393–400.

Vasios, G. W., Gold, J. D., Petkovich, M., Chambon, P., and Gudas, L. J. (1989). A retinoic acid-responsive element is present in the 5′ flanking region of the laminin B1 gene. *Proceedings of the National Academy of Sciences USA*, **86**, 9099–103.

Wang, S-Y. and Gudas, L. J. (1983). Isolation of cDNA clones specific for collagen IV and laminin from mouse teratocarcinoma cells. *Proceedings of the National Academy of Sciences USA*, **80**, 5880–4.

Webster, W. S., Johnston, M. C., Lammer, E. J., and Sulik, K. K. (1986). Isotretinoin embryopathy and the cranial neural crest: an *in vivo* and *in vitro* study. *Journal of Craniofacial Genetics and Developmental Biology*, **6**, 211–22.

Wiley, M. J., Cauwenbergs, P., and Taylor, I. M. (1983). Effects of retinoic acid on the development of the facial skeleton in hamsters: early changes involving cranial neural crest cells. *Acta Anatomica ( Basel )*, **116**, 180–92.

Wilkinson, D. G., Peters, G., Dickson, C., and McMahon, A. P. (1988). Expression of the FGF-related proto-oncogene *int-2* during gastrulation and neurulation in the mouse. *EMBO Journal*, **7**, 691–5.

Woods, A., Couchman, J. R., and Höök, M. (1985). Heparan sulfate proteoglycans of rat embryo fibroblasts. *Journal of Biological Chemistry*, **260**, 10872–9.

Woods, A., Couchman, J. R., Johansson, S., and Höök, M. (1986). Adhesion and cytoskeletal organization of fibroblasts in response to fibronectin fragments. *EMBO Journal*, **5**, 665–70.

# 14

# Diverse roles for retinoids in modifying skin, skin appendages, and oral mucosa in mammals and birds

Margaret H. Hardy

A number of changes brought about in mammalian or avian skin by exposure to excess retinoids will be discussed in this chapter. Their investigation has revealed three distinct roles played by varying retinoid levels with respect to the differentiation and functions of skin and some related tissues; they will be presented in the order of their discovery. The first role is to modulate the differentiation of epidermis and other keratinizing epithelia towards either keratin synthesis or mucus production. The second role is to initiate totally new, functioning mucous glands—structures unknown in the skin of birds or mammals. The third role is to divert skin on the avian foot from one recognizable morphogenetic programme (scale formation) to a different, but equally recognizable one (feather production).

MODULATION OF EPITHELIA PROGRAMMED FOR KERATINIZATION

The skin of adult mammals is not especially vulnerable to damage by overdosing with retinoids, although massive doses can greatly increase epidermal thickness (Moore 1957). Paradoxically, it was a study of the damage by excess vitamin A to skeletal development in chick limbs in organ culture that led to the accidental discovery by Fell and Mellanby of a more profound effect of the vitamin on developing skin (Fell and Mellanby 1952, 1953). In the high vitamin A medium, the primitive epidermal layer on a fragment of 7-day chick embryo skin became a cuboidal, mucus-secreting epithelium instead of a stratified, squamous keratinizing one. Further studies in organ culture by Fell and collaborators (Fell and Mellanby 1953; Fell 1957; Fitton Jackson and Fell 1963) showed that ciliated and mucous cells differentiated on the surface of the stratifying epithelium, which then resembled the epithelial lining of mammalian nasal passages or trachea. The change, called a *mucous metaplasia*, persisted for more than 2 weeks in culture in the presence of excess retinoid. When the explant was then transferred to the more physiological retinoid level of the control medium, containing only fowl plasma and chicken embryo extract, the superficial mucus-secreting and ciliated cells were shed and a normally keratinizing epidermis differentiated from the underlying epithelial cells. If the chick embryos were slightly older when the skin was explanted to vitamin A medium, the metaplasia was reduced—there was no mucus secretion by the surface cells, but no keratinization of these cells either. In skin explants from still older embryos, in which the first keratinized cells were already appearing, excess vitamin A had no effect on the epidermis.

The experiments of Fell and others in the 1950s created much interest in medical and pharmaceutical circles, which led eventually to the topical use of retinoids to reduce hyperkeratinization in acne, psoriasis and several other skin disorders, and to encouraging advances in the treatment of some epithelial tumours. Many laboratory animal models or *in vitro* skin cell culture models have been useful in these endeavours, but none have produced in mammalian epidermis a full mucous metaplasia resembling that shown in chick skin.

Since the late 1960s, our laboratory has followed the traditional organ culture methods, first with embryonic mouse skin on the surface of plasma clot cultures, and then in organ culture dishes on lens paper above serum-supplemented synthetic media. We have found that full mucous metaplasia is produced in mammalian skin at a very early age, when little differentiation has occurred (Fig 14.1). Figure 14.2 highlights some results of our histological, histochemical, and ultrastructural analyses of mouse embryo skin in organ culture during the 1970s (Hardy 1967; Sweeny and Hardy 1976; Bellows and Hardy 1977; Hardy *et al.* 1978). It shows:

1. That if retinoid treatment begins early enough, ciliated and mucus-secreting epithelium is produced.
2. There is a step-wise reduction in the extent of metaplasia as the retinoid is applied at later ages, and new products (Alcian blue-positive bodies) are produced.
3. After the age of 16 days postcoitum, retinoids do not prevent the normal differentiation into a keratinized epidermis.

Another keratinizing epithelium that responds to excess retinoids is the lining of the hamster cheek pouch. It arises from the ectodermally derived epithelium inside the corner of the mouth before birth as a solid tongue of undifferentiated epithelial cells within the cheek. By 12 days after birth the tongue of epithelium has thickened and keratinized inwardly to form a small cavity, which becomes the lumen of the pouch (Hardy *et al.* 1986). Lawrence and Bern (1960) were the first to demonstrate that the continuous, long-term local application of a retinoid inside the pouch of the adult hamster could cause a complete mucous metaplasia of pouch epithelium. Mock and Main (1978) reported that the epithelial pouch rudiment of the newborn hamster with its surrounding mesenchymal stroma could develop normally into a

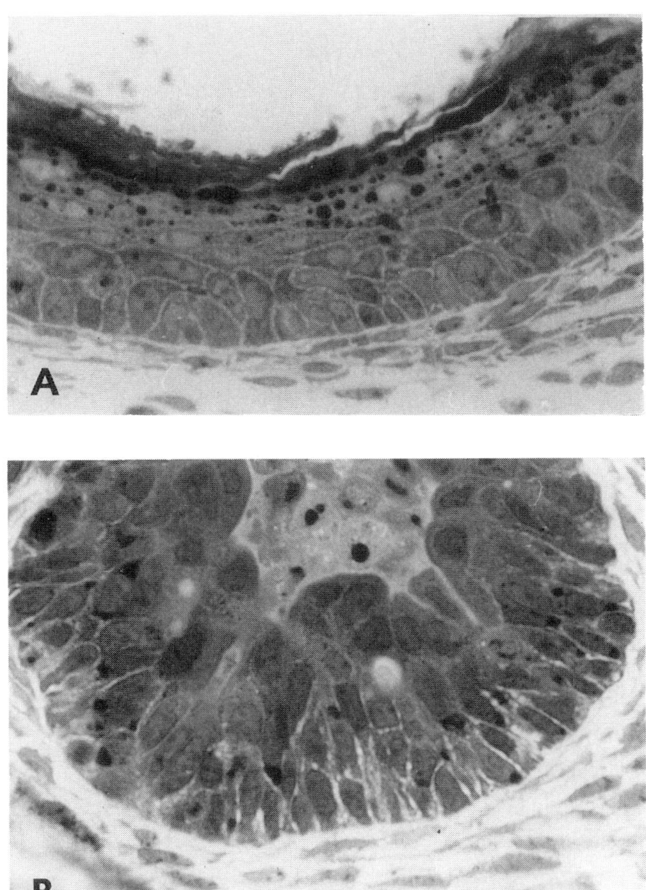

**Fig. 14.1** (*a*) One micron Epon section of skin from a mouse embryo 12 days postcoitum after 10 days in standard medium, showing the stratified, squamous, keratinized epidermis that had differentiated *in vitro*. Note the dark, flattened, keratinized cells on the surface layers and the dark keratohyalin granules in the layer below. (*b*) Section of skin from the same embryo after 10 days in medium supplemented with 6 µg retinol per ml. The uppermost cells retained a polyhedral form, and above them is a central cavity containing secreted material. At higher magnifications the ciliated borders of some cells are visible. × 865.

keratinized pouch lining in an organotypic culture. When excess retinoid was added to the medium, the pouch developed a mucus-secreting epithelium instead of a keratinizing one. Covant and Hardy (1988) showed that, while a newborn pouch in control medium produced a keratinized layer by 7 days, a pouch cultured for the first 3 days in $2.0 \times 10^{-5}$M retinoic acid (RA) showed

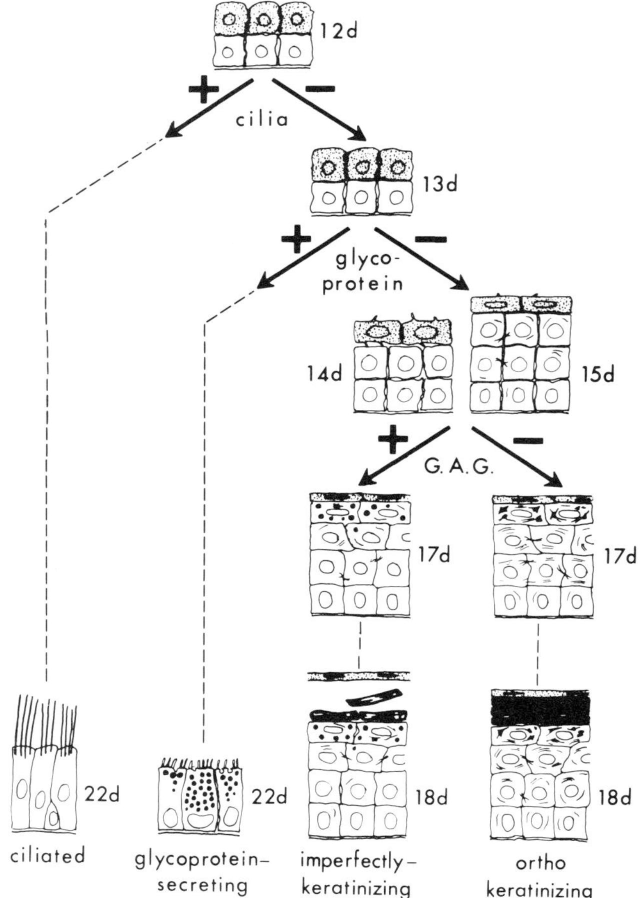



**Fig. 14.2** Schematic representation showing the fate of embryonic ectoderm of the mouse in organ cultures of skin treated with excess vitamin A at different ages (+treated; −untreated). Ectoderm treated at 12 days can develop into ciliated epithelium (+/−glycoprotein-secreting cells). When treated at 13 days cilia are no longer formed, but glycoprotein synthesis is still possible. When treated at 14 days, the stratified, squamous differentiation had already been determined, but glycosamino-glycan (GAG) synthesis (Alcian-positive bodies) and imperfect keratinization are possible.

some early mucous changes, and then, after transfer to normal medium, showed a delay in the keratinization process for a further 8 days. Seven days of exposure of newborn pouches to RA delayed keratinization for up to 21 days, and some explants showed goblet cells secreting mucus on the surface of the stratified cuboidal epithelium. Thus, there is abundant evidence of modulation between mucus-secreting and keratinizing types of epithelium in response to varying retinoid levels, in both the newborn and the adult hamster.

Most of the metaplastic changes in mouse and hamster skin *in vitro* were brought about by addition to the plasma- or serum-containing medium of 6 $\mu$g/ml of retinol, retinyl acetate or RA ($1.8$–$2.1 \times 10^{-5}$M). The medium was renewed every 2–3 days. Even at the end of 2 days of incubation, the total retinoid concentration of the medium for the newborn hamster cheek pouch did not fall by more than 50 per cent. Thus, there was continuous exposure to levels that at their lowest were more than 10 times higher than that of the untreated cheek pouch *in vivo* (Robinson *et al.* 1988, 1990).

## MUCOUS GLAND MORPHOGENESIS IN MAMMALS

To explain the second role of retinoids in modifying development in skin, it will be necessary to review some features of skin appendage morphogenesis in mammals (Fig. 14.3).

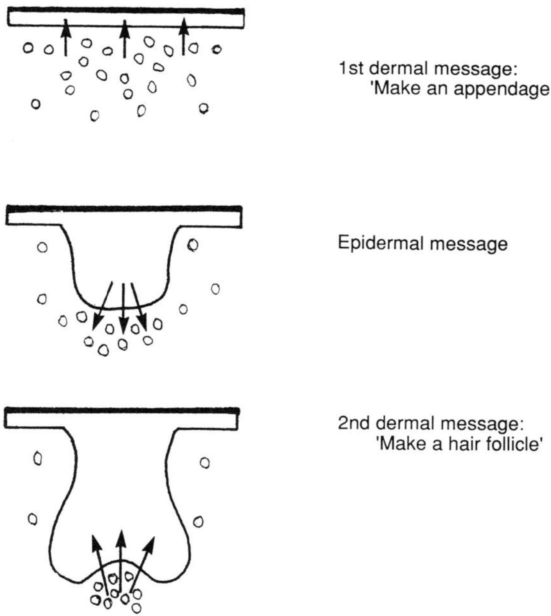


**Fig. 14.3** Epithelial–mesenchymal interactions during hair follicle development in mammals. The thick line represents the upper margin of the epidermis. The thin line represents the lower margin of the epidermis and the margins of the 'hair bud'. Small circles represent the mesenchyme cells of the dermis.

A hair follicle begins as a thickening (placode) of the epidermis (stage 1), which may be adjacent to an aggregation of mesenchymal cells. The epidermal cells of the placode move into the dermis to form an elongated epithelial plug (stage 2) and then surround a mesenchymal cell aggregation, which thus becomes the dermal papilla (stage 3). The epithelial cells surrounding the dermal papilla then undergo rapid division, giving rise to the differentiating inner root sheath and hair. From a series of studies in which the epithelial and mesenchymal layers of chick, mouse, and lizard skin were separated and recombined with other layers, it has been established that at least three tissue interactions are essential for the formation of a hair follicle or other skin appendage (Sengel 1976*a,b*; Dhouailly 1977; Viallet and Dhouailly Chapter 15), as shown in Fig. 14.3. The only glands in the hairy skin of mice are the small sebaceous glands, each of which arises as a bud from one side of the upper part of a hair follicle, and whose cells break down to form an oily secretion. There are no sweat glands on the hairy skin of mice, and no mucus-secreting glands on the skin of any mammal. The large whisker follicles of mice also have small sebaceous glands but no sweat glands.

The initiation and complete development of mouse pelage hair follicles and whisker follicles can be observed in the living state in organ cultures in biological medium, using Maximow slides (Hardy 1949, 1969). The addition to the medium of a naturally occurring retinoid (retinol or retinyl acetate) in high concentrations (2–8 $\mu$g/ml) was without toxic effects (Hardy 1967), but interfered with the development of both types of follicles in a dose-dependent and stage-dependent manner (Hardy 1968). In the case of pelage follicles, retinoid added to skin explanted at 13.5 days of gestation, half a day before the first wave of follicle initiation *in vivo*, prevented their initiation. If the retinoid was added to skin explanted at 15 days, one day after initiation, the initiated follicles developed normally for 4 days, then regressed, losing their dermal papillae. The timing of the observed changes in these and other experiments suggest that the sensitive periods for interference by the retinoids in follicle development might be: (i) just before stage 1 follicles formed (i.e. the period of the first dermal message); and (ii) just before stage 4 began (the period of the second dermal message).

The first vibrissa follicles are initiated 2 days earlier than pelage follicles and differentiate a little more rapidly (Fig. 14.4. column 1). They form in sequence along five rows on the upper lip, so that many stages can be observed in one explant. In explants from 14 day embryos, most of those follicles that had reached the end of stage 3 after 3 days in culture with excess retinoids underwent regression, as in the case of the pelage follicles at a comparable stage (Fig. 14.4. column 4). Dermal papilla cells were dispersed, and follicle bases became rounded. However, at the same time there was abnormal budding from the lateral follicle walls, frequently followed by elongation and branching of the buds, lumen formation, and mucus secretion (Fig. 14.5). A similar morphogenesis began from the lateral follicle wall with follicles that

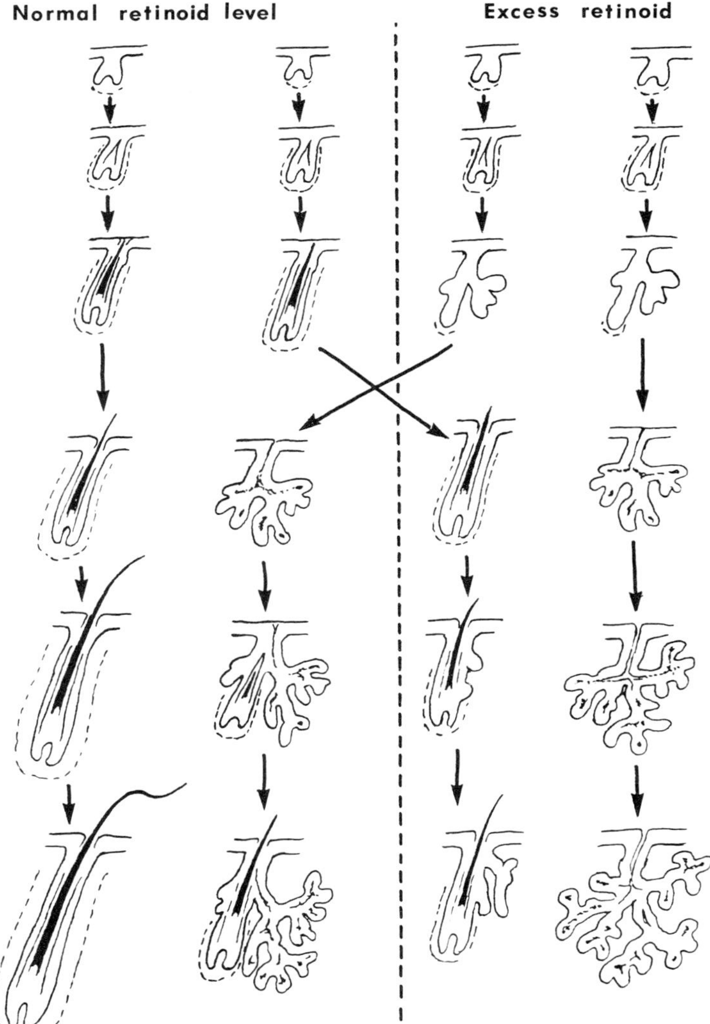



**Fig. 14.4** Diagram showing results of treatment reversal experiments. Column 1 shows the typical appearance of a stage 3b vibrissa follicle during 14 days in standard medium with a physiologically 'normal' quantity of vitamin A. Column 2—the upper three sketches indicate development during 7 days in standard medium. These explants were transferred to excess vitamin A medium, (arrow to column 3) where hair growth was maintained for a further 7 days, but a few small buds developed, suggesting early gland morphogenesis. Column 3—the upper three sketches indicate glandular morphogenesis and loss of dermal papilla in excess vitamin A. After transfer to standard medium (arrow to colunm 2), some follicles recovered hair growth but continued glandular morphogenesis and differentiation. Column 4 shows the effect of 14 days in excess vitamin A.

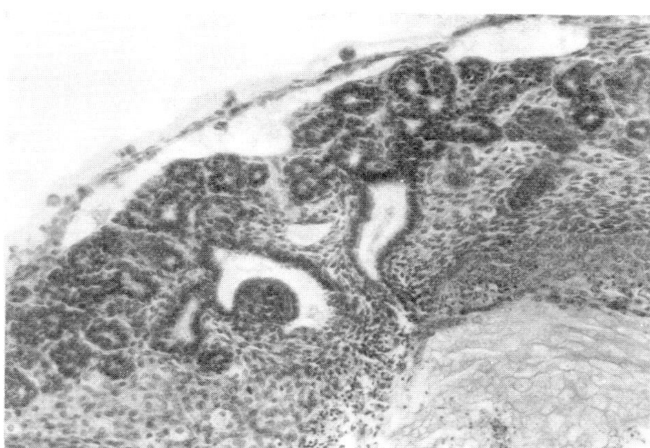

**Fig. 14.5** Histological section through an explant of 14-day embryo skin after 14 days with 4 $\mu$g/ml retinol added to the control medium of 75 per cent fowl plasma and 25 per cent chicken embryo extract. Two vibrissa follicles have been completely transformed into branching, mucus-secreting glands. Haematoxylin, Alcian blue, × 100.

reached stage 4 after 3 days *in vitro*, but instead of regression at the base of the follicle there was merely delay, then resumption of normal cell division and whisker formation. In explants from 15-day embryos, only the latest-forming follicles (stage 3c or less by 3 days *in vitro*) showed glandular morphogenesis. It appears that gland induction by retinoid may be possible only in a relatively short period of follicle development, which is around the time of the second dermal message.

An ultrastructural study of the retinoid-treated and control explants of upper lip skin (Hardy *et al.* 1983) showed that after 1 day in retinoid medium, large gaps appeared in the previously intact basal lamina surrounding the epithelial follicle wall (Fig. 14.6). After 2 days, processes of epithelial cells then made direct contacts with the plasma membranes of mesenchymal cells through these gaps. A day later, groups of epithelial cells had moved through one or more gaps as solid buds, and subsequently gave rise to a mucus-secreting gland. In the same study, in both control and retinoid-treated explants, basal lamina gaps and heterotypic cell contacts were also found in the dermal papilla area at a specific period of follicle development. Further research (Goldberg and Hardy 1983; Hardy and Goldberg 1983) confirmed that gaps and contacts at the dermal papilla were also consistent features of vibrissa follicles *in vivo* between stages 3b and 5, i.e. around the stage of the normal second dermal interaction and the beginning of epithelial cell response. It was proposed that the normal second dermal interaction in hair follicles was one of a growing number of morphogenetic processes in which tissue interaction through heterotypic cell contact was considered to be required. These included the branching morphogenesis of salivary glands

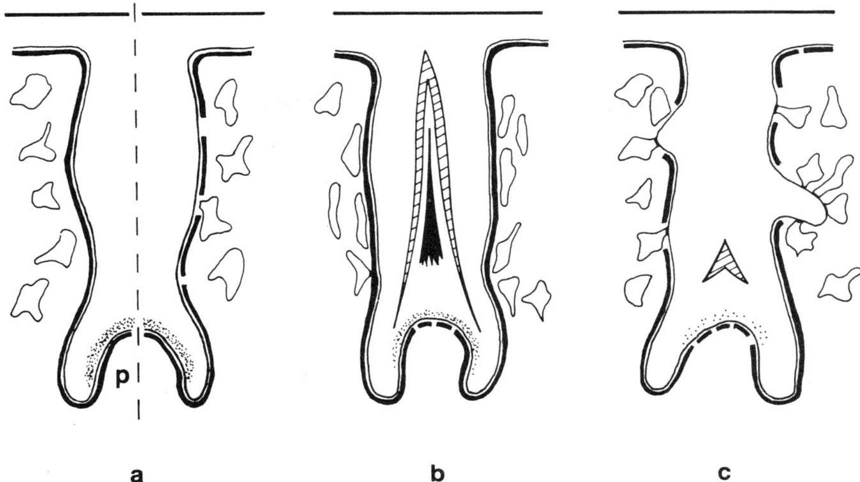



**Fig. 14.6** Schematic diagram of developing vibrissa follicles, highlighting some events revealed by electron microscopy. On the left side of follicle (*a*) the basal lamina is represented by a thickened line surrounding the epithelial part of the follicle, which is at stage 3c, after 1 day in a standard medium. The right side of follicle (*a*) shows the differences observed when a follicle has had 1 day in medium with 4–6 $\mu$g/ml of retinyl acetate. Gaps have appeared at random along the lateral follicle walls, with mesenchymal cells entering the gaps. After 3 days in standard medium, follicle (*b*) has progressed to stage 6, with an intact basal lamina. After 3 days in retinyl acetate supplemented medium, follicle (*c*) shows direct contacts between plasma membranes of mesenchymal and epithelial cells, and buds of epithelium extending through the gaps.

(Cutler 1977), and the development of duodenal mucosa (Matham *et al.* 1972) and tooth germs (Slavkin and Bringas 1976). The abnormal induction of mucous glands from the vibrissa follicle wall by retinoids may also require heterotypic cell contact.

Mucous gland morphogenesis caused by excess retinoid is not limited to the developing whisker follicles of mice. Similar glands occasionally developed directly from the epidermis in the upper lip skin in organ culture in the experiments reported above (Hardy 1968). The paper by Lawrence and Bern (1960) on mucous metaplasia of epithelium in the adult cheek pouch treated with a retinoid implant also illustrated secretory glands arising directly from the pouch epithelium. Mock and Main (1978) referred to similar glands in the newborn pouch treated with retinoid *in vitro*. This common response from a variety of epithelial tissues to retinoid in the presence of mesenchymal tissue suggests some type of instructive interaction, as defined by Wessells (1977). A distinguishing feature of an instructive interaction, as opposed to a permissive one, is that it changes the direction of a tissue to one it would not otherwise have followed. This is certainly true in these examples of gland morphogenesis.

Unlike modulations, the results of instructive interactions are usually stable, as shown in the central columns of Fig. 14.4. The glands induced in 13-day whisker pads after 3 days *in vitro* with retinoid continued to grow and differentiate, after retinoid removal, for the duration of culture, a further 11 days (Hardy and Bellows 1978). Likewise, we found that the glands induced in hamster cheek pouch by 7 days of retinoid treatment continued to develop for the remaining 14 days *in vitro* (Covant and Hardy 1988).

To see whether retinoids needed to act through the mesenchyme or could act directly on the epithelium to bring about the gland formation, Covant and Hardy (1990) combined retinoid-treated or untreated mesenchyme with treated or untreated epithelium in every possible combination, and then cultured the recombinants. Glands were initiated in a proportion of all types of recombinants containing retinoid-treated mesenchyme but none in any of the recombinants containing untreated mesenchyme. It was concluded that the message for gland morphogenesis had to come through the mesenchyme in order to be effective. Hardy *et al.* (1990) came to the same conclusion when they tested recombinants of mouse upper lip dermis and epidermis in chick chorioallantoic grafts instead of cultures. None of the recombinants with untreated dermis produced glands, but 26 per cent of those with retinoid-treated dermis were successful in this respect. When chick embryonic dermis from the tarsometatarsal region was used instead of mouse dermis, 75 per cent of the recombinants with retinoid treated dermis produced glands (Fig. 14.7). It was surprising to find that 29 per cent of recombinants without retinoid treatment of dermis also had glands. This was attributed to the high concentrations of retinoids found in the untreated chick skin, four times higher than in untreated mouse skin. The very high level of cellular RA binding protein (CRABP) in chick embryonic skin (Gates and King 1985) may also have an effect on the retinoid dynamics.


SWITCHING OF A MORPHOGENETIC PROGRAMME IN BIRDS


There are two main categories of skin appendages in birds—the feathers, which cover most of the body, and the scales which, in the majority of domestic breeds of *Gallus gallus,* are the only appendages on the lower legs and feet. The first scales to appear on the foot are the large, rectangular scuta, followed, in other regions, by the smaller scutella and finally by the rounded reticular scales that occupy the plantar surface. Numerous experiments with chick embryo skin, using recombinants of epidermis and dermis in organ cultures or in grafts to the chick chorioallantoic membrane, have shown that there must be a series of three tissue interactions (Sengel 1976a; Cadi *et al.* 1983; Dhouailly and Sawyer 1984), which are similar to those we have summarized for mammals (see Fig. 14.3). The first dermal message directs the location, size, and shape of the placode but does not specify the type of appendage. The

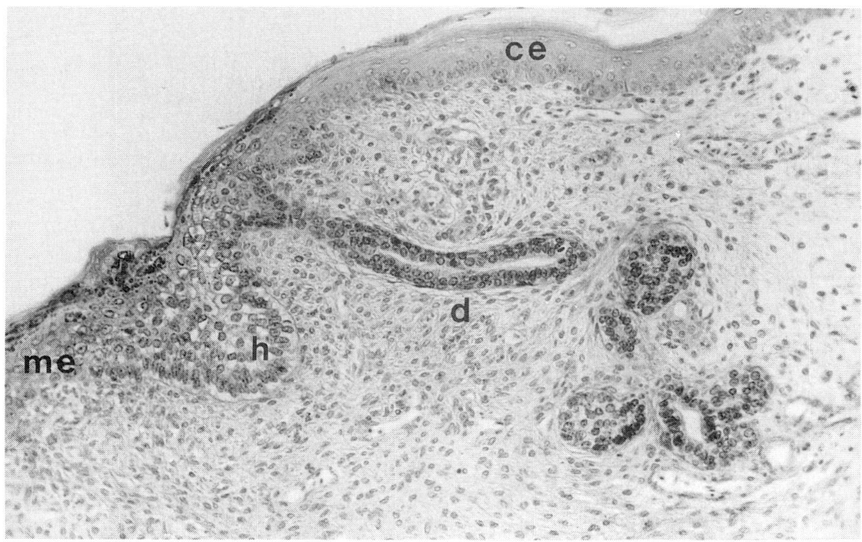


**Fig. 14.7** Mouse epidermis (me) can be distinguished from chick chorioallantoic epithelium (ce) by its darkly staining nuclei. An arrested hair bud is shown (h). The duct (d) of a mucous gland is growing directly from the mouse epidermis, and branches into smaller ducts and paler-staining terminal buds. × 350.

tarsometatarsal region of skin, like all the other regions, has the potential to produce feathers from an early age, but from as early as 8.5 days of incubation the tarsometatarsal epidermis has a bias towards forming scales. From about 9.5 days of incubation, the first dermal message reaching the anterior surface of the tarsometatarsal region is an instruction for placode formation to begin in the epidermis. The pattern and the somewhat rectangular (scutate) shape of these scale placodes are specified by the bias of the epidermis. Next comes the epidermal message, to organize the dermis in a pattern to correspond to the scale placodes, including the formation of a single dermal aggregation under each placode. The second dermal message, following at about 13 days, is to proceed with histogenesis of the appendages already specified, including the synthesis of scale-type $\beta$-keratins.

When incubating chicks were treated with RA by injection into the amniotic sac at 10, 11, and 12 days, and the embryos were recovered at 17 days, Dhouailly and Hardy (1978) found that elongated feather filaments had formed on the normally scale-forming areas of the feet (Fig. 14.8). Hatched chickens had fluffy down feathers on their feet. A more extensive analysis (Dhouailly *et al.* 1980) showed that a single injection of 125 $\mu$g RA caused feathers to appear on the developing foot scales in different locations, depending on the age of the embryos at the time of treatment. Injection at

**Fig. 14.8** Anterior view of the foot of a 17-day chicken embryo, which had received injection of 125 μg retinoic acid in ethanol on days 10, 11, and 12. Developing feathers, known as feather filaments, have formed at the margins of the scales as a result of the retinoic acid injection. Chicks injected with ethanol only showed smooth scales with no feather filaments.

10 days resulted in feathers most frequently on the scuta and scutella. After injection at 11 days, the feather filaments were usually found on the tarsometatarsal scutella and the digital reticula. After injection at 12 days the incidence of feathered feet was reduced, and feather filaments were confined to the reticular scales. In each case, the feathers appeared on those scale rudiments that were just starting to form at the time of injection, being at the placode or asymmetrical placode stage (Sawyer 1972). The RA injection did

not prevent the eventual completion of scale development, but did permit the initiation in less than 24 hours of small feather-type placodes within the area of a scale placode (Dhouailly and Sengel 1983). Three days after the injection, RA was no longer detectable in the tissues by high pressure liquid chromatography, and the scales had resumed normal development, but bore feather filaments on their margins.

Whatever the mechanism by which RA affects the development of scales and feathers in the foot region, it is clearly altering a delicate balance between two normal morphogenetic programmes, each already 'authorized' in nature, rather than creating a new one.


COMMENTS ON POSSIBLE MECHANISMS OF RETINOID ACTIONS IN SKIN


The modulating action of retinoids on epidermis and other epithelia has, not surprisingly, received more attention from molecular biologists than the other two actions discussed in this paper. Analysis has been facilitated by the successful growth of isolated, dissociated keratinocytes in cell culture and by recent modification of the substrates with a view to achieving complete differentiation *in vitro*. The present state of knowledge of the control of epidermal differentiation, based largely on these *in vitro* models, is the subject of a recent review by Fuchs (1990), from which I will summarize an interpretation of the role of retinoids in epidermis. Stratifying epithelia such as the epidermis require a balance between cell proliferation in the basal layer and progressive cell differentiation in the upper layers. Both epidermal growth factor and transforming growth factor-$\alpha$ (TGF-$\alpha$) have been shown to stimulate proliferation in the undifferentiated basal keratinocyte population in mammalian cell cultures, while members of the TGF-$\beta$ group inhibit proliferation and promote the early stages of terminal differentiation in the suprabasal cells. Physiological levels of retinoids, such as retinol, *in vivo* and *in vitro*, block the differentiation of basal epidermal cells. This is probably achieved via intracellular retinoid conversion to RA, and subsequent interaction of RA with the retinoic acid receptors (RARs) present in those cells. Excess RA added to cultured keratinocytes has been shown to act in this way to suppress the expression of keratins K1/K10, K6/K16, cornified envelope production and filaggrin expression (the latter responsible for keratohyalin granules). In other experiments excess RA at $10^{-6}$ M inhibited cell proliferation and concomitantly induced the active form of TGF-$\beta$2 in mouse keratinocytes, suggesting the existence of a regulatory loop. I think that these findings from the cell culture model of keratinocytes give us a satisfying explanation of most of the loss of keratinized cell differentiation observed in organ culture of full-thickness skin. However, the alternative pathway of differentiation to ciliated and mucus-secreting epithelium in organ cultures of full-thickness skins remains unexplained. Two possibilities are suggested, either: (i) the primitive ectoderm, in organ culture with excess retinoid, is merely following a differentiation pathway common to such remote ancestors as amphibians or fishes; or (ii) it is responding to positive signals, probably from the dermis, that are there because of the retinoid treatment. The abundance of CRABP mRNA transcripts in mouse limb dermis as early as 12.5 days postcoitum, and the presence of some RAR-$\alpha$ and RAR-$\gamma$ transcripts in the epidermis as well as the dermis by 14.5 days (Dollé *et al.* 1989, 1990) makes the latter theory plausible. The fact that epidermal cell cultures, even on a cell-free substrate of extracellular matrix constituents, do not undergo mucous metaplasia in response to excess retinoids, also supports this second alternative. Further evidence of retinoic acid receptor expression in relation to the second dermal message in normal development and in retinoid-induced glandular morphogenesis of whisker follicles has just become available (Viallet and Dhouailly, Chapter 15).

*In situ* hybridization has revealed very interesting distribution patterns of the TGF-$\beta$ proteins in developing mouse embryo skin (Lehnert and Akhurst 1988; Pelton *et al.* 1989; Millan *et al.* 1991). The findings *in vivo* support the opinion based on cell culture behaviour that TGF-$\beta$1 and TGF-$\beta$2 are involved with coordinating the regulation of epidermal differentiation. They may also be directly involved in the mesenchymal—epithelial interactions of the skin *per se*, the existence of which has been known for many years (Billingham and Silvers 1968).

In developing hair follicles, especially those of vibrissae, the activity of retinoid-related molecules and growth factors show very specific patterns, principally revealed by *in situ* hybridization. Retinoids were investigated by Dollé *et al.* (1990) and Ruberte *et al.* (1990, 1991). The RA receptor RAR-$\gamma$ transcripts were associated with whisker follicle development from the early placode stage, both in the epithelial cells of the placode and the underlying mesenchyme. Cellular RA binding protein (CRABP) transcripts were confined to the dermis, and were particularly abundant next to the developing epidermis and lateral follicle walls. A suggested function is to sequester free RA and provide a low level of active RA to the adjacent epithelial cells. If this CRABP is unable to handle a large retinoid excess, it might explain why gland morphogenesis can start from the lateral walls.

A possibly significant similarity between CRABP mRNA and TGF-$\beta$1 protein in early vibrissa mesenchyme was mentioned by Dollé *et al.* (1990). Many of the transforming growth factor beta mRNAs also show localization in different parts of the hair follicle during its development (Lehnert and Akhurst 1988; Lyons *et al.* 1989, 1990; Pelton *et al.* 1989; Schmid *et al.* 1991; Millan *et al.* 1991). TGF-$\beta$1 and TGF-$\beta$2 appear to play complementary roles, and are expressed in epithelial and mesenchymal parts of the follicles at different stages. Autocrine and paracrine modes of activity have been proposed. Another member of the TGF-$\beta$-like family of genes is bone morphogenetic protein (BMP)-2a, which Lyons *et al.* (1990) found was restricted to the epithelial placode in early vibrissa development. This particular population of cells retained its unique BMP-2a expression pattern as it became the tip of the follicle plug extending into the dermis, then the population of rapidly dividing cells which give rise to the inner root sheath and hair. The cells that were BMP-2a positive are the ones that are believed to participate in all three of the epithelial–mesenchymal interactions defined at the beginning of this paper. How all these molecules work together, along with extracellular matrix molecules and cell adhesion molecules, to construct a normal hair follicle, or an abnormal one in the presence of excess retinoids, is still an intriguing question.

ACKNOWLEDGEMENT

The author's studies of retinoid effects on development of skin and appendages have been interspersed with many others over the past 20 years. They owe much to the encouragement of fellow scientists such as Dr Richard Bunge at the Columbia University College of Physicians and Surgeons and Dr J. P. W. Gilman at the Ontario Veterinary College, University of Guelph, as well as to the many colleagues and graduate students at University of Guelph whose names appear as co-authors in the reference list. Financial support was provided by the Natural Sciences and Engineering Research Council of Canada.

Figures 1–6 are original illustrations that are reprinted with the permission of the publishers as follows: Figure 1 appeared first in a paper by P. R. Sweeny and M. H. Hardy (1976) Ciliated and secretory epidermis produced from embryonic mammalian skin in organ culture by vitamin A, *Anatomical Record* **185**, 93–100. Figures 2 and 4 were in the chapter by Hardy (1983) in *Epithelial–mesenchymal interactions in development*, (ed. R.H. Sawyer and J. F. Fallon), Praeger Press. Figure 3 is from Hardy (1989), published by the Tissue Culture Association. Figure 5 is from Hardy (1968), published by the Company of Biologists, and Figure 6 from Hardy *et al.* (1983), published by Williams and Wilkins.

REFERENCES

Bellows, C. G. and Hardy, M. H. (1977). Histochemical evidence of mucosubstances in the metaplastic epidermis and hair follicles produced in vitro in the presence of excess vitamin A. *Anatomical Record*, **187**, 257–72.
Billingham, R. E. and Silvers, W. K. (1968). Dermoepidermal interactions and epithelial specificity. In *Epithelial–mesenchymal interactions*, (ed. R. Fleischmajer and R. E. Billingham), pp. 252–66. Williams and Wilkins, Baltimore.

Cadi, R., Dhouailly, D., and Sengel, P. (1983). Use of retinoic acid for the analysis of dermal-epidermal interactions in the tarso-metatarsal skin of the chick embryo. *Developmental Biology*, **100**, 489–95.

Covant, H. A. and Hardy, M. H. (1988). Stability of the glandular morphogenesis produced by retinoids in the newborn hamster cheek pouch in vitro. *Journal of Experimental Zoology*, **246**, 139–49.

Covant, H. A. and Hardy, M. H. (1990). Excess retinoid acts through the stroma to produce mucous glands from newborn hamster cheek pouch in vitro. *Journal of Experimental Zoology*, **253**, 271–9.

Cutler, L. S. (1977). Intercellular contacts of the epithelial-mesenchymal interface of developing rat submandibular gland *in vivo*. *Journal of Embryology and Experimental Morphology*, **39**, 71–7.

Dhouailly, D. (1977). Dermis-epidermal interactions during morphogenesis of cutaneous appendages in mammals. *Frontiers in Matrix Biology*, **4**, 86–121.

Dhouailly, D. and Hardy, M. H. (1978). Retinoic acid causes the development of feathers in the scale-forming integument of the chick embryo. *Wilhelm Roux's Archives*, **185**, 195–200.

Dhouailly, D. and Sawyer, R. H. (1984). Avian scale development XI. Initial appearance of the dermal defect in scaleless skin. *Developmental Biology*, **105**, 343–50.

Dhouailly, D. and Sengel, P. (1983). Feather-forming properties of the foot integument in avian embryos. In *Epithelial–mesenchymal interactions in development*, (ed. R. H. Sawyer and J. F. Fallon), pp. 147–61. Praeger Publishers, New York.

Dhouailly, D., Hardy, M. H., and Sengel, P. (1980). Formation of feathers on chick foot scales: a stage-dependent morphogenetic response to retinoic acid. *Journal of Embryology and Experimental Morphology*, **58**, 63–78.

Dollé, P., Ruberte, E., Kastner, P., Petkovich, M., Stoner, C. M., Gudas, L. J., and Chambon, P. (1989). Differential expression of genes encoding $\alpha$, $\beta$ and $\gamma$ retinoic acid receptors and CRABP in the developing limbs of the mouse. *Nature*, **342**, 702–5.

Dollé, P., Ruberte, E., Leroy, P., Morriss-Kay, G., and Chambon, P. (1990). Retinoic acid receptors and cellular retinoid binding proteins. I. A systematic study of their differential pattern of transcription during mouse organogenesis. *Development*, **110**, 1133–51.

Fell, H. B. (1957). The effect of excess vitamin A on cultures of embryonic chicken skin explanted at different stages of differentiation. *Proceedings of the Royal Society*, **B146**, 242–56.

Fell, H. B. and Mellanby, E. (1952). The effect of hypervitaminosis A on embryonic limb-bones cultivated *in vitro*. *Journal of Physiology*, **116**, 320–49.

Fell, H. B. and Mellanby, E. (1953). Metaplasia produced in cultures of chick ectoderm by high vitamin A. *Journal of Physiology*, **119**, 470–88.

Fitton Jackson, S. and Fell, H. B. (1963). Epidermal fine structure in embryonic chicken skin during atypical differentiation induced by vitamin A in culture. *Developmental Biology*, **7**, 394–419.

Fuchs, E. (1990). Epidermal differentiation: the bare essentials. *Journal of Cell Biology*, **111**, 2807–14.

Gates, R. E. and King, L. E. (1985). Cytoplasmic vitamin A binding proteins in chick embryo dermis and epidermis. *Journal of Investigative Dermatology*, **85**, 279–83.

Goldberg, E. A. and Hardy, M. H. (1983). Heterotypic cell contacts and basal lamina morphology during hair follicle development in the mouse. *Canadian Journal of Zoology*, **61**, 2703–19.

Hardy, M. H. (1949). The development of mouse hair *in vitro* with some observations on pigmentation. *Journal of Anatomy*, **83**, 364–84.

Hardy, M. H. (1967). Responses in embryonic mouse skin to excess vitamin A in organotypic cultures from the trunk, upper lip and lower jaw. *Experimental Cell Research*, **46**, 367–84.

Hardy, M. H. (1968). Glandular metaplasia of hair follicles and other responses to vitamin A excess in cultures of rodent skin. *Journal of Embryology and Experimental Morphology*, **19**, 157–80.

Hardy, M. H. (1969). The differentiation of hair follicles and hairs in organ culture. In *Advances in biology of skin*, vol. 9, (eds W. Montagna and R. L. Dobson), pp. 35–60. Pergamon Press, Oxford.

Hardy, M. H. (1989). The use of retinoids as probes for analysing morphogenesis of glands from epithelial tissues. *In Vitro, Cell and Developmental Biology*, **25**, 454–9.

Hardy, M. H. and Bellows, C. G. (1978). The stability of vitamin A-induced metaplasia of mouse vibrissa follicles in vitro. *Journal of Investigative Dermatology*, **71**, 236–41.

Hardy, M. H. and Goldberg, E. A. (1983). Morphological changes at the basement membrane during some tissue interactions in the integument. *Canadian Journal of Biochemistry and Cell Biology*, **61**, 957–66.

Hardy, M. H., Sweeny, P. R., and Bellows, C. G. (1978). The effects of vitamin A on the epidermis of the foetal mouse in organ culture—an ultrastructual study. *Journal of Ultrastructure Research*, **64**, 246–60.

Hardy, M. H., Van Exan, R. J., Sonstegard, K. S., and Sweeny, P. R. (1983). Basal lamina changes during tissue interactions in hair follicles—an in vitro study of normal dermal papillae and vitamin A-induced glandular morphogenesis. *Journal of Investigative Dermatology*, **80**, 27–34.

Hardy, M. H., Vrablic, O. E., Covant, H. A., and Kandarkar, S. V. (1986). The development of the Syrian hamster cheek pouch. *Anatomical Record*, **214**, 273–82.

Hardy, M. H., Dhouailly, D., Törma, H., and Vahlquist, A. (1990). Either chick embryo dermis or retinoid-treated mouse dermis can initiate glandular morphogenesis from mammalian epidermis. *Journal of Experimental Zoology*, **256**, 279–89.

Lawrence, D. J. and Bern, H. A. (1960). Mucous metaplasia and mucous gland formation in keratinized adult epithelium *in situ* treated with vitamin A. *Experimental Cell Research*, **21**, 443–6.

Lehnert, S. A. and Akhurst, R. J. (1988). Embryonic expression pattern of TGF beta type-1 RNA suggests both paracrine and autocrine mechanisms of action. *Development*, **104**, 263–73.

Lyons, K. M., Pelton, R. W., and Hogan, B. L. M. (1989). Patterns of expression of murine Vgr-1 and BMP-2a RNA suggest that transforming growth factor-$\beta$-like genes coordinately regulate aspects of embryonic development. *Genes and Development*, **3**, 1657–88.

Lyons, K. M., Pelton, R. W., and Hogan, B. L. M. (1990). Organogenesis and pattern formation in the mouse: RNA distribution patterns suggest a role for Bone Morphogenetic Protein-2A (BMP-2A). *Development*, **109**, 833–44.

Mathan, M., Hermos J., and Trier, J. (1972). Structural features of the epithelio-mesenchymal interface of rat duodenal mucosa during development. *Journal of Cell Biology*, **52**, 577–88.

Millan, F. A., Denhez, F., Kondaiah, P., and Akhurst, R. J. (1991). Embryonic gene expression patterns of TGF$\beta$1, $\beta$2 and $\beta$3 suggest different developmental functions *in vivo*. *Development*, **111**, 131–44.

Mock, D. and Main, J. H. P. (1978). The effect of vitamin A on hamster cheek pouch in organ culture. *Journal of Dental Research*, **58**, 635–7.

Moore, T. (1957), *Vitamin A*, p. 11. Elsevier, Amsterdam.

Pelton, R. W., Nomura, S., Moses, H. L., and Hogan, B. L. M. (1989). Expression of transforming growth factor $\beta$2 RNA during murine embryogenesis. *Development*, **106**, 759–67.

Robinson, M. E., Verrinder Gibbins, A. M., and Hardy, M. H. (1988). Normal levels of vitamin A in tissues of the Syrian hamster (*Mesocricetus auratus*) determined colorimetrically. *Laboratory Animals*, **22**, 107–11.

Robinson, M. E., Verrinder Gibbins, A. M., and Hardy, M. H. (1990). Persistence of added retinoids in organ culture during induction of mucous metaplasia and glandular morphogenesis in hamster cheek pouches. *Experientia*, **46**, 513–17.

Ruberte, E., Dollé, P., Krust, A., Zelent, A., Morriss-Kay, G., and Chambon, P. (1990). Specific spatial and temporal distribution of retinoic acid receptor gamma transcripts during mouse embryogenesis. *Development*, **108**, 213–22.

Ruberte, E., Dollé, P., Chambon, P., and Morriss-Kay, G. (1991). Retinoic acid receptors and cellular retinoid binding proteins II. *Development*, **111**, 45–60.

Sawyer, R. H. (1972). Avian scale development. I. Histogenesis and morphogenesis of the epidermis and dermis during morphogenesis of the scale ridge. *Journal of Experimental Zoology*, **181**, 365–84.

Schmid, P., Cox, D., Bilbe, G., Maier, R., and McMaster, G. K. (1991). Differential expression of TGF $\beta$1, $\beta$2 and $\beta$3 genes during mouse embryogensis. *Development*, **111**, 117–30.

Sengel, P. (1976a). Tissue interactions in skin morphogenesis. In *Organ culture in biomedical research*, (ed. M. Balls and M. A. Monnickendam), pp. 114–47. Cambridge University Press, Cambridge.

Sengel, P. (1976b). *Morphogenesis of skin*. Cambridge University Press, Cambridge.

Slavkin, H. C. and Bringas Jr, P. (1976). Epithelial–mesenchymal interactions during odontogenesis. IV. Morphological evidence for direct heterotypic cell–cell contacts. *Developmental Biology*, **50**, 428–42.

Sweeny, P. R. and Hardy, M. H. (1976). Ciliated and secretory epidermis produced from embryonic mammalian skin in organ culture by vitamin A. *Anatomical Record*, **185**, 93–100.

Wessells, N. K. (1977). *Tissue interactions and development*. W. A. Benjamin, Menlo Park.

# 15

# Expression of retinoic acid receptors and dermal–epidermal interactions during mouse skin morphogenesis

Jean P. Viallet, Esther Ruberte, Andrée Krust, Arthur Zelent, and
Danielle Dhouailly

## INTRODUCTION

Vertebrate skin comprises two tissues—the epidermis and the underlying dermis. Much data is available relating to the respective roles of these two skin components for each step of scale, feather, and hair morphogenesis (see Sengel 1976; Dhouailly 1977*b*, 1984; Sawyer 1983 for reviews) but little is known about the language of communication used by the tissues. In particular, two different and successive waves of dermal information are required for the initiation and achievement of morphogenesis of the cutaneous appendages (Dhouailly 1973). The early wave, which can be understood and interpreted by an epidermis from another vertebrate class, is responsible for the formation of epidermal placodes; the later wave, which contains indispensable information for regional structural organization of cutaneous appendages, is responsible for the development of a specific programme of differentiation of the placodes, i.e. the choice between scale and feather in birds, and hair and gland formation in mammals.

Exogenous retinoic acid (RA) has marked effects on embryonic skin morphogenesis, leading to the exchange of one developmental pathway for another, e.g. abnormal location of feather is obtained on the normally scaled feet of chick embryos (Dhouailly *et al.* 1980), while glomerular-type glands develop in mouse upper-lip skin (Hardy 1968, 1983, Chapter 14). Furthermore, the fact that these homeotic-type effects are observed only when RA is added at the placodal stage of cutaneous appendage morphogenesis, led to the hypothesis that RA treatment might alter the dermis in such a way that the developing placodes can be diverted to a different programme of morphogenesis (Hardy 1983).

Recently, the discovery of multiple retinoic acid nuclear receptors (RAR-$\alpha$, $\beta$, and $\gamma$; Giguère *et al.* 1987; Petkovich *et al.* 1987; Zelent *et al.* 1989) and their distinct pattern of expression in different mouse embryonic tissues (Dollé *et al.* 1989, 1990; Ruberte *et al.* 1990, 1991), has suggested that each member of the

RAR subfamily may play specific roles during organogenesis. Furthermore, several isoforms of RAR-α, β, and γ, have been characterized (Krust *et al.* 1989; Giguère *et al.* 1990; Kastner *et al.* 1990; Leroy *et al.* 1991; Zelent *et al.* 1991; for a review see Leroy *et al.* Chapter 2). The differential expression of mRAR-α1 and α2 isoforms in various tissues (Chambon, unpublished data) suggests that at least these two isoforms perform distinct functions. In addition to the nuclear receptors, two cytoplasmic RA binding proteins (CRABP I and II) have been identified (Stoner and Gudas 1989; Giguère *et al.* 1990; Boylan and Gudas 1991). Their function is still controversial (see Ruberte *et al.* Chapter 8).

Thus, the use of RA and investigation of the pattern of expression of its receptors offer new experimental opportunities for the understanding of regional specification in avian and mammalian skin, and for elucidation of the molecular basis of the dermal–epidermal interactions. In this article we report the results of the first part of this investigation, describing the expression of the RA receptors during normal hair vibrissa development, and during RA-induced gland formation in mouse upper-lip skin.

A TWO-STEP MORPHOGENETIC ACTIVITY OF THE DERMIS DURING THE DEVELOPMENT OF THE HAIR FOLLICLE AND OF THE SWEAT GLAND

In the mouse, the size, shape, and distribution pattern of pelage hair follicles differ markedly from those of tactile hairs (vibrissae). Furthermore, the skin of the foot pads does not form hairs, but sweat glomerular-type glands.

Heterotopic dermal–epidermal recombinants were performed, involving

**Fig. 15.1** Different steps of appendage differentiation attained by the mouse embryo epidermis in homo- and heterospecific recombinants, cultured for 3(*c*) or 8(*a, b, d, e, f*) days on the chorioallantoic membrane of the chick embryo. (*a*) and (*b*) are homospecific heterotopic recombinants. (*a*) 12-day upper-lip mouse epidermis/14.5-day dorsal mouse dermis. Formation of a stage 6 pelage-type hair follicle of rather uniform and small diameter. (*b*) 12.5-day dorsal mouse epidermis/12.5-day upper-lip mouse dermis. Formation of a large stage 6 follicle. Note the constriction (arrow) above the bulb, which is characteristic of a vibrissa. (*c*) to (*f*) are xenoplastic recombinants involving a chick dermis. (*c*) 12.5-day dorsal mouse epidermis/7-day dorsal chick dermis. Hair placode that developed in contact with chick dermal condensation. (*d*) 14.5-day plantar mouse epidermis/7-day dorsal chick dermis. Development of an arrested stage 3 hair peg in close contact with a dermal papilla (P) constituted by chick cells. (*e*)12.5-day upper-lip mouse epidermis /12-day tarsometatarsal chick dermis. Formation of an epithelial downgrowth with a rounded base. There is no aggregation of dermal papilla cells (arrow). (*f*) 12.5-day upper-lip mouse epidermis/12-day tarsometatarsal chick dermis. Development of a gland containing a duct (DC) and a glomerule (G). Scale bar. 25 μm.

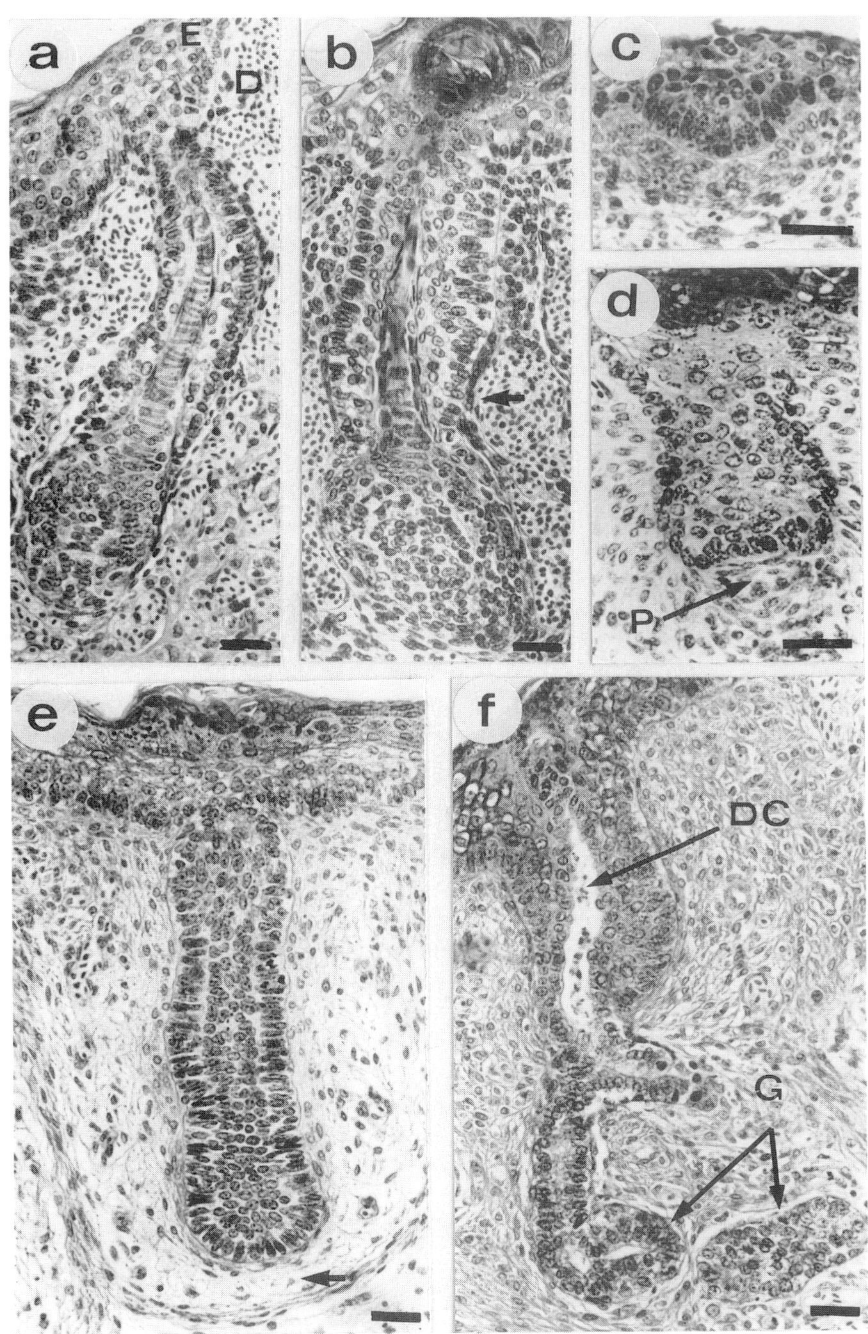
a
E
D
b
c
d
P
e
f
DC
G

embryonic mouse skin tissues from the dorsal (pelage hairs), the upper-lip (vibrissae) and the plantar (sweat glands) regions (Dhouailly, 1977*a*; Delorme and Dhouailly, unpublished data). From these experiments it appears that the specification for the development of whiskers or pelage hairs (Fig. 15.1(a,b), or sweat glands, resided in the dermis. Heterospecific recombinants of mouse epidermis and chick dermis from a feather-forming region (Dhouailly 1973) formed epidermal placodes (Fig. 15.1(c)), which developed into stage 3 (as defined by Hardy, 1969) hair pegs (Fig. 15.1(d)); these hair pegs never developed into hair follicles. In some cases, following recombination of mouse embryonic upper-lip epidermis with chick dermis, hair buds as late as stage 3 were diverted to gland formation: after 8 days of culture, the chick dermal papilla disappeared, while the epithelial downgrowth was abnormally elongated (Fig. 15.1(e)), finally leading to the formation of glandular structures (Fig. 15.1(f)), which are similar to those obtained when mouse upper-lip skin was treated with retinol or RA (Hardy 1968, 1983, Chapter 14; Hardy *et al.* 1990). In this case, the chick dermis acts either as a mouse plantar dermis (Delorme and Dhouailly, unpublished data), or as a mouse upper-lip RA-treated dermis (Hardy *et al.* 1990).

Based on these results, we have investigated the hypothesis that the expression patterns of RA nuclear and cytoplasmic receptors should present sequential changes correlated with the two steps of dermal–epidermal interactions that are required for the initiation and achievement of hair and sweat gland morphogenesis (Fig. 15.2). The early dermal messages, which are responsible for the formation of epidermal placodes, have apparently continued to be based on the same molecular mechanisms during vertebrate evolution. The later messages, however, are responsible for the subsequent programmes of placodal differentiation, which are species-specific, and, within a species, specific to the different body regions. These morphogenetic differences must be correlated with a different pattern of expression of the three RAR genes. Such a variation in the distribution of RA receptors may explain on the one hand the results obtained in xenoplastic dermal–epidermal recombinants, and on the other hand the homeotic-type changes obtained in RA-treated skin explants (Hardy 1968, Chapter 14; Dhouailly *et al.* 1980; Dhouailly 1982).

SPATIAL AND TEMPORAL DISTRIBUTION PATTERNS OF RETINOIC ACID RECEPTORS AND CYTOPLASMIC RETINOIC ACID BINDING PROTEIN I DURING VIBRISSA FOLLICLE DEVELOPMENT

The distribution of RAR-$\alpha$, $\beta$, $\gamma$ and CRABP I transcripts has been studied by *in situ* hybridization with [$^{35}$S]-labelled RNA probes on serial longitudinal sections of upper-lip skin of mouse embryos. As the mysticial pattern appears progressively from the eye region towards the olfactory pit, we have restricted

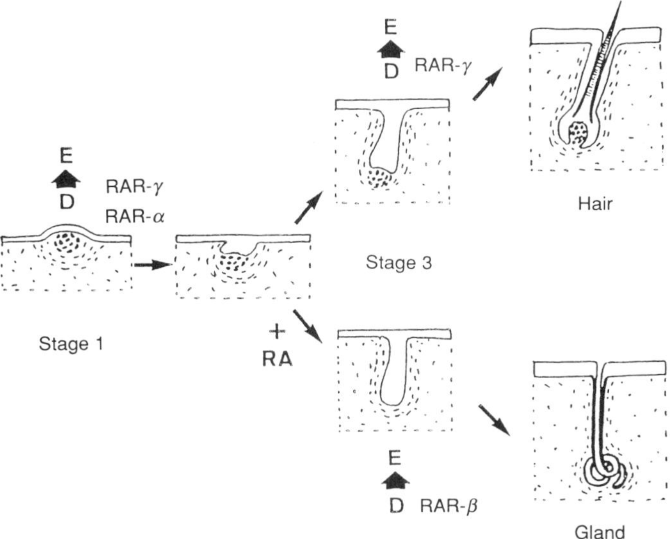


**Fig. 15.2** Diagrammatic representation of the morphogenetic activity of the dermis and of the correlated expression of the RARs during vibrissa follicle development and RA-induced gland morphogenesis.

our observations to the first group of six vibrissae to appear next to the eye. The probes used in this study do not discriminate between the individual isoforms of the three RARs (Kastner *et al.* 1990; Chambon unpublished data), while the CRABP I probe is specific (Ruberte *et al.* Chapter 8).

At stage 1a, on day 12.5 of gestation, the first sign of vibrissa development appears as an epidermal dome associated with a distinctive and round dermal papilla, surrounded by a hemispherical dermal condensation (Fig. 15.3(a)). At this stage, a strong RAR-α signal was observed and was restricted to the cells of the dermal papilla and of the dermal condensation (Fig. 15.3(b)), while an RAR-γ signal of similar intensity was present in both the dermal and the epidermal cells of the hair primordium (Fig. 15.3(d)).

At stage 3, on day 13.5 of gestation, an epithelial column extends into the dermis, the dermal papilla continuing to be associated with the tip of the invaginated peg. The hair peg and the dermal papilla are surrounded by the elongated dermal condensation (Fig. 15.4(a). The level of expression of RAR-α was highest in the epidermal cells of the hair peg (Fig. 15.4(b)), while the RAR-γ transcripts were abundant in both the epidermal and the dermal cells of the hair bud (Fig.15.4(d)).

The CRABP I transcripts were present in the upper-lip skin on day 15.5 of gestation, as already observed by Dollé *et al.* (1990). Furthermore,

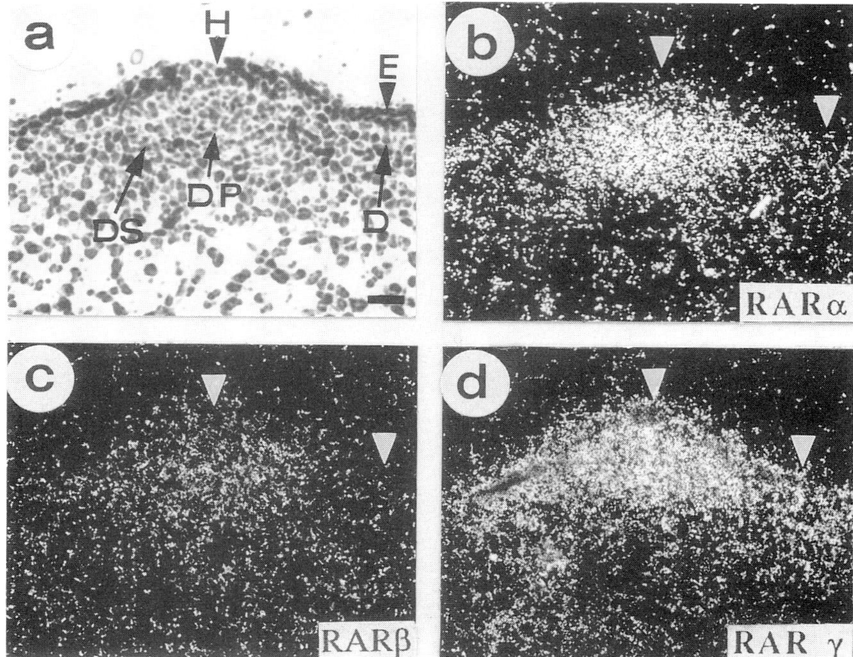



**Fig. 15.3** Distribution of RAR transcripts at stage 1a of mouse vibrissa follicle development (12.5 days of gestation). (*a*) Hair primordium, consisted by an epidermal dome (H), associated with a dermal papilla (DP), surrounded by a dermal condensation (future dermal sheath) (DS) (toluidine blue). (*b*) Expression of the RAR-α gene in the dermal papilla and the dermal condensation. (*c*) No specific labelling over background level is detected with the RAR-β probe. (*d*) Expression of the RAR-γ gene in both the epidermal and the dermal cells of the hair primordium. E, interfollicular epidermis; D, interfollicular dermis. Scale bar: 35 μm.

**Fig. 15.4** Distribution of RAR-α, β, and γ transcripts in 13.5-day mouse embryo upper-lip skin (*a* to *d*) and of RAR-β transcripts in 13.5-day mouse embryo upper-lip skin cultured *in vitro* for 48 hours with added retinoic acid (5 μg/ml)(*e*, *f*). (*a*) Stage 3 hair peg, constituted by an epithelial downgrowth (H), cavitated at its base, where it is in close contact with a dermal papilla (DP), and surrounded by a dermal sheath (DS). (*b*) Expression of the RAR-α gene, both in epidermal and dermal cells which constitute the hair bud (compare with (*a*)). (*c*) No specific labelling over background level is detected with the RAR-β probe. (*d*) Expression of the RAR-γ gene, predominantly in the epithelial downgrowth and its associated dermal papilla. (*e* and *f*) Abnormal asymmetric epithelial downgrowth developed in the presence of RA excess. Note that the RAR-β labelling is found in the dermal cells, with higher levels on the left part of the dermal sheath (black arrows). *a* and *e*: toluidine blue. Scale bar = 35 μm.

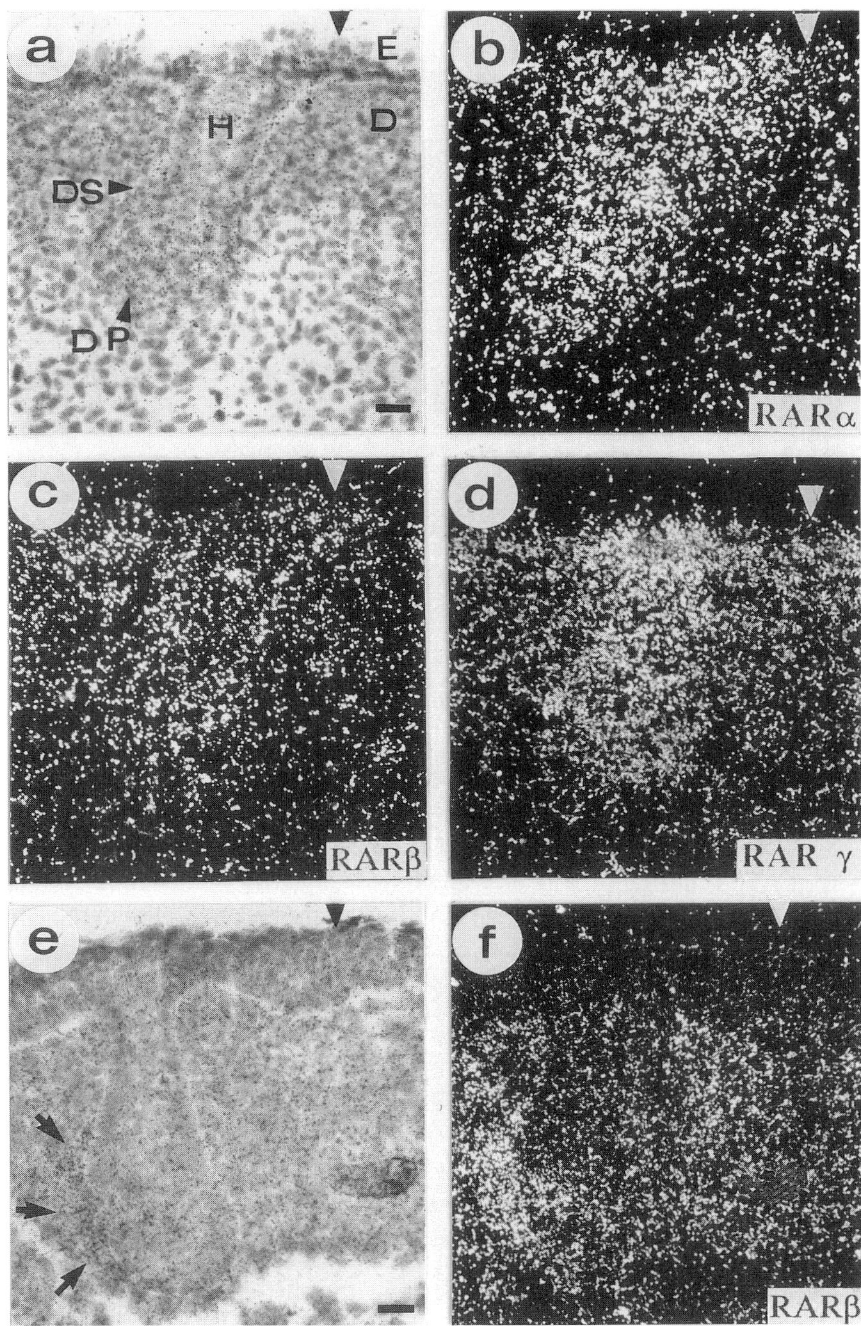
a
E
D
H
DS
DP
b
RARα
c
RARβ
d
RAR γ
e
f
RARβ

longitudinal sections of stage 6 vibrissa follicles showed that these transcripts were strictly localized in the interfollicular dermis and in the upper part of the dermal sheath, corresponding to the infundibulum (Fig. 15.5), while they were absent from the lower part of the dermal sheath and from the dermal papilla.

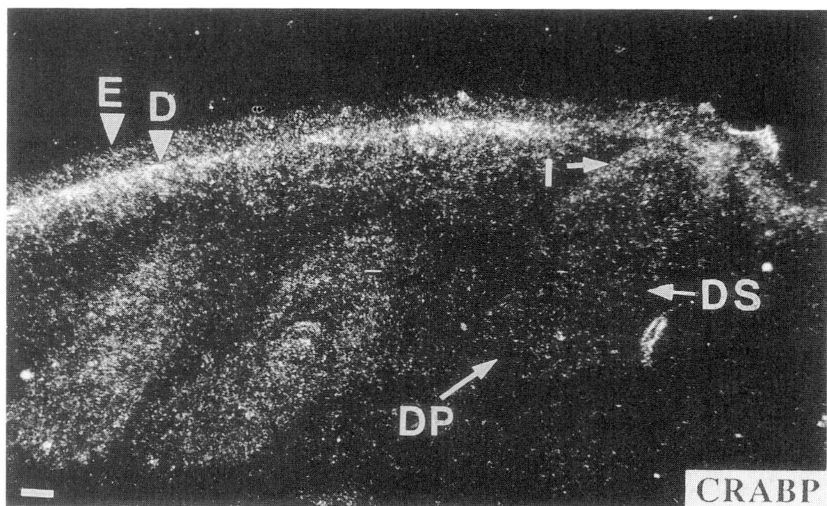


**Fig. 15.5** Distribution of CRABP I transcripts in 15.5-day mouse embryo upper-lip skin. One longitudinal and two oblique sections of vibrissa follicles can be observed. A predominant expression of the receptor occurs in the interfollicular and superficial dermis (D), as well as in the upper part of the dermal sheath of the follicle, i.e. the infundibulum (I). The lower part of the dermal sheath (DS), and the dermal papilla (DP) are not involved in this expression. E, epidermis. Scale bar: 50 $\mu$m.

It should be noticed that the RAR-$\beta$ (Fig.15.3(c) and Fig. 15.4(c)) and the CRABP I genes were not expressed as stages 1a and 3 of hair vibrissa follicle development, i.e. they were not expressed at the two stages during which there is morphogenetic activity in the skin.

DERMAL EXPRESSION OF RAR-$\beta$ DURING RETINOIC ACID-INDUCED GLANDULAR METAPLASIA

It is well known (see Hardy Chapter 14) that RA treatment can interfere with the morphogenesis of embryonic skin, leading to the formation of feathered scales in chick (Dhouailly *et al.* 1980; Dhouailly 1982) and to the development of mouse vibrissa hair buds into glands (Hardy 1968, 1983). We have been interested (Viallet *et al.* 1991) in studying the possible changes in RAR expression that could be correlated with such a switch in skin developmental pathways. Therefore, the 13.5-day mouse upper-lip skin has been cultured *in vitro* with RA at a dose of $16.7 \times 10^{-6}$ M. After 48 hours, most of the explants presented a few abnormally shaped hair buds (Fig. 15.4(e)). Those buds were characterized by an asymmetric bulge end and by the lack of a well defined dermal papilla, as previously shown by Hardy (1983). In contrast with the case of the control explants, where the RAR-$\beta$ gene was not transcribed, the RA-treated explants showed a strong expression of this gene: the RAR-$\beta$ transcripts are abundant overall in the dermis and in particular on one side of the abnormal hair peg, in the dermal sheath (Fig. 15.4(f)). Thus, the RA-induced formation of abnormal hair buds, followed by the differentiation of exocrine-type glands, correlates with a strong RAR-$\beta$ signal in the dermal cells.


CONCLUSION

Our results indicate that the RAR-$\beta$ and the CRABP I genes are not expressed at a detectable level at any stage of vibrissa development. In contrast, the RAR-$\alpha$ and RAR-$\gamma$ genes appear to have specific functions linked to the two steps of the dermal-epidermal interactions which occur during normal mouse upper-lip skin morphogenesis (see Fig. 15.2).

First, the expression of RAR-$\alpha$ and RAR-$\gamma$ genes in the dermal papilla cells at stage 1a of hair development may be involved in the regulation of genes that are responsible for the formation of the epidermal placode. The subsequent appearance of RAR-$\gamma$ transcripts into the ingrowing epidermal cells of the hair peg may be part of their response to the dermal information.

Second, the large amount of RAR-$\gamma$ transcripts, both in the dermal papilla and the dermal sheath cells, indicates that this expression may be correlated with the transmission of the late dermal messages which are responsible for the achievement of hair follicle morphogenesis.

When the RAR-$\gamma$ transcripts are replaced by RAR-$\beta$ transcripts after RA treatment of the upper-lip skin at stage 3, the hair bud epithelium is diverted to gland formation. Thus, genes activated by RAR-$\beta$ may change the nature of the second dermal message responsible for the achievement of hair vibrissa morphogenesis. Such an induction of RAR-$\beta$ gene transcription by RA has been previously shown by many authors in different systems (among them Zelent *et al.* 1989; de Thé *et al.* 1989) and is presumably mediated by either RAR-$\alpha$ or $\gamma$.

Of particular interest is the late appearance and the specific distribution of the CRABP I transcripts in the interfollicular dermis at 15.5 days of gestation. At this stage, the interfollicular skin is known to be stabilized by a large amount of collagen deposit (Mauger *et al.* 1987). Our results are thus in conformity with a previous hypothesis of several authors, according to which the CRABP I gene expression is not necessary for mediating the RA signal and, in contrast, could serve to prevent RA from activating its receptors (Dover and Koeffer, 1982; Darmon *et al.* 1988; Ruberte *et al.* 1991; Boylan and Gudas 1991).

Our working hypothesis has now progressed to two points. First, we presume that the early dermal messages, which can be understood and interpreted by an epidermis from another vertebrate class, may be characterized by a similar RAR-$\alpha$ and RAR-$\gamma$ gene expression, not only during hair development, but also during sweat gland morphogenesis in mammals, as well as during scale and feather formation in birds. Second, the RAR-$\beta$ signal may be correlated with the second step of dermal induction involved in normal mammalian gland development, as well as in avian skin morphogenesis.

Further studies on possible correlations with the expression domains of other genes such as *Hox* or TGF-$\beta$ will be able to provide clues to identify some of the target genes of RAR-$\alpha$, RAR-$\beta$, and RAR-$\gamma$ during morphogenesis of the cutaneous appendages.


ACKNOWLEDGEMENTS

This work is dedicated to Margaret H. Hardy for her pioneering work on hair vibrissa glandular metaplasia.

We gratefully acknowledge the support of Professor Pierre Chambon, for making this study possible; Pr. P. Chambon and Dr Marie Pierre Gaub for helpful discussions; Dr Gillian Morriss-Kay for critical reading of the manuscript; and M. C. Bernard for histology. This research was supported by an ARC grant no. 6233 and a GEFLUC grant to D. Dhouailly.



REFERENCES

Boylan, J. F. and Gudas, L. J. (1991). Overexpression of the cellular retinoic acid binding protein-I (CRABP-I) results in a reduction in differentiation-specific gene expression in F9 teratocarcinoma cells. *Journal of Cell Biology*, **112(5)**, 965–79.

Darmon, M., Rocher, M., Cavey, M. T., Martin, B., Rabilloud, T., Delescluse, C., and Shroot, B. (1988). Biological activity of retinoids correlates with affinity for nuclear receptor but not for cytosolic binding protein. *Skin Pharmacology* **1**, 161–75.

de Thé, H., Marchio, A., Tiollais, P., and Dejean, A. (1989). Differential expression and ligand regulation of the retinoic acid receptor $\alpha$ and $\beta$ genes. *EMBO Journal*, **8**, 429–33.

Dhouailly, D. (1973). Dermo-epidermal interactions between birds and mammals: differentiation of cutaneous appendages. *Journal of Embryology and Experimental Morphology*, **30**, 585–603.

Dhouailly, D. (1977*a*). Regional specification of cutaneous appendages in mammals. *Wilhelm Roux's Archives*, **181**, 3–10.

Dhouailly, D. (1977*b*). Dermo-epidermal interactions during morphogenesis of cutaneous appendages in amniotes. *Frontier Matrix Biology*, **4**, 86–121.

Dhouailly, D. (1982). Effects of retinoic acid on development properties of the foot integument in avian embryo. In *Embryonic development part B: Cellular aspects*, pp. 309–316. Alan R. Liss, New York.

Dhouailly, D., Hardy, M. H., and Sengel, P. (1980). Formation of feathers on chick foot scales: a stage-dependent morphogenetic response to retinoic acid. *Journal of Embryology and Experimental Morphology*, **58**, 63–78.

Dhouailly, D. (1984). Specification of feather and scale patterns. In *Pattern formation* (eds G. M. Malacinski, and S. V. Bryant), pp. 581–601. MacMillan Publications, New York.

Dollé, P., Ruberte, E., Kastner, P., Petkovich, M., Stoner, C., Gudas, M., and Chambon, P. (1989). Differential expression of genes encoding $\alpha$, $\beta$ and $\gamma$ retinoic acid receptors and CRABP in the developing limbs of the mouse. *Nature (Lond.)* **342**, 702–4.

Dollé, P., Ruberte, E., Leroy, P., Morriss-Kay, G., and Chambon, P. (1990). Retinoic acid receptors and cellular binding proteins I. A systematic study of their differential pattern of transcription during mouse organogenesis. *Development*, **110**, 1133–51.

Dover, D. and Koeffer, H. P. (1982). Retinoic acid inhibition of the clonal growth of human myeloid leukemia cells. *Journal of Clinical Investigation*, **69**, 277–83.

Giguère, V., Ong, E. S., Segui, P., and Evans, R. M. (1987). Identification of a receptor for the morphogen retinoic acid. *Nature (Lond.)*, **330**, 624–9.

Giguère, V., Lyn, S., Yip, P., Siu, C. H., and Amin, S. (1990). Molecular cloning of cDNA encoding a second cellular retinoic acid-binding protein. *Proceedings of the National Academy of Sciences USA*, **87**, 6233–7.

Hardy, M. H. (1968). Glandular metaplasia of hair follicles and other responses to vitamin A excess in cultures of rodent skin. *Journal of Embryology and Experimental Morphology*, **19**, 157–80.

Hardy, M. H. (1969). The differentiation of hair follicles and hairs in organ culture. In *Advances in biology of skin*, (ed. W. Montagna), vol. 9, pp. 35–60. Pergamon Press, Oxford.

Hardy, M. H. (1983). Vitamin A and the epithelial-mesenchymal interactions in skin differentiation. In *Epithelial–mesenchymal interactions in development*, (eds R. H. Sawyer, and J. F. Fallon), pp. 163–88. Praeger, New York.

Hardy, M. H., Dhouailly, D., Torma, H., and Valquist, A. (1990). Either chick embryo dermis or retinoid treated mouse dermis can initiate glandular morphogenesis from mammalian epidermal tissue. *The Journal of Experimental Zoology*, **256**, 279–89.

Kastner, P., Krust, A., Mendelsohn, C., Garnier, J. M., Zelent, A., Leroy, P., Staub, A., and Chambon, P. (1990) Murine isoforms of retinoic acid receptor-$\gamma$ with specific patterns of expression. *Proceedings of the National Academy of Sciences USA*, **87**, 2700–4.

Krust, A., Kastner, P., Petkovich, M., Zelent, A., and Chambon, P. (1989). A third human retinoic acid receptor, hRAR-$\gamma$. *Proceedings of the National Academy of Sciences USA*, **86**, 5310–14.

Leroy, P., Krust, A., Zelent, A., Mendelsohn, C., Garnier, J. M., Kastner, P., Dierich, A., and Chambon, P. (1991). Multiple isoforms of the mouse retinoic acid receptor $\alpha$ are generated by alternative splicing and differential induction by retinoic acid. *EMBO Journal*, **10**, 59–69.

Mauger, A., Emonard, H. Hartmann, D. J., Foisdart, J. H., and Sengel, P. (1987). Immunofluorescent localization of collagen types I, III, IV, fibronectin, laminin, and basement membrane proteoglycan in developing mouse skin. *Wilhelm Roux's Archives*, **195**, 295–302.

Petkovich, M., Brand, N. J., Krust, A., and Chambon, P. (1987). A human retinoic acid receptor which belongs to the family of nuclear receptors. *Nature*, **330**, 444–50.

Ruberte, E., Dollé, P., Krust, A., Zelent, A., Morriss-Kay, G., and Chambon, P. (1990). Specific spatial and temporal distribution of retinoic acid receptor gamma transcripts during mouse embryogenesis. *Development*, **108**, 213–22.

Ruberte, E., Dollé, P., Chambon, P., and Morriss-Kay, G. (1991). Retinoic acid receptors and cellular binding proteins. II. Their differential pattern of transcription during early morphogenesis in mouse embryos. *Development*, **111**, 45–60.

Sawyer, R. H. (1983). The role of epithelial-mesenchymal interactions in regulating gene expression during avian scale morphogenesis. In *Epithelial–mesenchymal interactions in development*, (eds R. H. Sawyer, and J. F. Fallon), pp. 115–46. Praeger, New York.

Sengel, P. (1976). Tissue interactions in skin morphogenesis. In *Organ culture in biomedical research*, (eds M. Balls, and M. A. Monnickendam), pp. 111–47. Cambridge University Press, Cambridge.

Stoner, C. M. and Gudas, L. J. (1989). Mouse cellular retinoic acid binding protein: cloning, complementary DNA sequence, and messenger RNA expression during the retinoic acid-induced differentiation of F9 wild type and RA-3-10 mutant teratocarcinoma cells. *Cancer Research*, **49**, 1497–504.

Viallet, J. P., Ruberte, E., du Manoir, S., Krust, A., Zelent, A., and Dhouailly, D. (1991). Retinoic acid-induced glandular metaplasia in mouse skin is linked to the dermal expression of retinoic acid receptor $\beta$ mRNA. *Development Biology*, **144**, 424–8.

Zelent, A., Krust, A., Petkovich, M., Kastner, P., and Chambon, P. (1989). Cloning of murine $\alpha$ and $\beta$ retinoic acid receptors and a novel receptor $\gamma$ predominantly expressed in skin. *Nature*, **339**, 714–17.

Zelent, A., Mendelsohn, C., Kastner, P., Krust, A. Garnier, J. M., Ruffenach, F., Leroy, P., and Chambon, P. (1991). Differentially expressed isoforms of mouse retinoic acid receptor $\beta$ are generated by usage of two promoters and alternative splicing. *EMBO Journal*, **10**, 71–81.

# PART IV

# Retinoids in gene expression

# 16

# Differential regulation by retinoic acid of the homeobox genes of the HOX family in human embryonal carcinoma cells

Edoardo Boncinelli, Antonio Simeone, Dario Acampora, Antonio Faiella,
Maurizio D'Esposito, Anna Stornaiuolo, Maria Pannese, and
Antonio Mallamaci

Morphogens, which convey positional information and induce pathways of differentiation through a gradient of concentration, are believed to play a key role in vertebrate development (Brockes 1989). Retinoic acid (RA) has been implicated as a natural morphogen in chicken development, where it contributes to the specification of the limb anterior–posterior (AP) axis (Tickle *et al.* 1982; Thaller and Eichele 1987). There is also a possible implication in frogs, where alteration of embryonic RA levels dramatically affects specific structures of the developing central nervous system (Durston *et al.* 1989), although these interpretations have recently been reconsidered (Brockes 1991). The identification of specific cellular RA binding proteins and of a number of nuclear RA receptors in mouse and man supports the concept that this substance may act as a morphogen with direct gene control functions. Although the genes involved in interpreting signals provided by morphogens are still unknown, homeobox genes, which specify positional information in *Drosophila* and possibly in vertebrate embryogenesis, are among suggested candidates.

Extensive genetic analysis of early embryogenesis in *Drosophila* has provided insights into the molecular events controlling body plan formation and, in particular, the process of segmentation and the specification of segment identity along the body axis. These processes are controlled by a relatively limited number of genes, which appear to be organized in a complex regulatory network (Akam 1987). Most of them encode DNA-binding proteins with a role in transcriptional control (Levine and Hoey 1988). The DNA-binding domain present in these proteins is often a homeodomain (HD) encoded by a homeobox (Gehring 1987). Genes containing a homeobox code for nuclear proteins with regulatory functions in a wide variety of organisms, from yeast to man (Scott *et al.* 1989). Many of these genes appear to have a role in control of the early developmental programme. Unravelling this regulatory network is critical for our understanding of development.

Sequences encoding HDs most closely related to the archetypal *Antennapedia* (*Antp*) HD have been termed class I homeoboxes. In *Drosophila*, genes containing a class I homeobox are clustered in two complex loci, the *Antennapedia*-complex (ANT-C) and the *Bithorax*-complex (BX-C) (Akam 1987). Mouse and human class I homeobox genes appear to be clustered in a similar way to restricted genomic regions (*Hox* loci) located on at least four chromosomes (Boncinelli *et al.* 1988; Duboule and Dollé 1989; Acampora *et al.* 1989; Graham *et al.* 1989). The four human clusters have been designated HOX-1, 2, 3, and 4 and mapped on chromosomes 7, 17, 12, and 2, respectively. Each cluster contains at least nine genes organized in a homologous linear arrangement whereby genes occupying the same position in each cluster encode protein products showing the highest peptide identity at their amino termini and in extended domains including the HD itself. A similar *Hox* gene organization has been observed in the mouse genome (Duboule and Dollé 1989; Graham *et al.* 1989) and evidence is accumulating for comparable arrangements in several other vertebrate species (Kappen *et al.* 1989).

*Hox* genes are expressed and developmentally regulated in mouse (Holland and Hogan 1988; Dollé *et al.* 1989; Duboule and Dollé 1989; Graham *et al.* 1989; Wilkinson *et al.* 1989), frog (Wright *et al.* 1989) and human (Acampora *et al.* 1991) embryos. Expression of these genes along the embryonic A/P body axis uniformly follows a 5′-posterior/3′-anterior rule, even if their expression domains largely overlap (Gaunt *et al.* 1988; Duboule and Dollé 1989; Akam 1989; Giampaolo *et al.* 1989; Graham *et al.* 1989). This phenomenon was first observed for the homeotic genes in the BX-C of *Drosophila* and was termed *colinearity* (Lewis 1978).

We have so far identified 38 human HOX genes distributed on four homologous chromosomal loci (Fig. 16.1). These genes can be attributed to 13 groups primarily on the basis of the peptide sequence of the encoded homeodomain protein (Fig. 16.2). Genes belonging to the same group occupy corresponding positions in the four loci, supporting the notion of duplication events of a single homeobox gene complex with subsequent dispersion in different chromosomes (Hart *et al.* 1987; Boncinelli *et al.* 1988; Acampora *et al.* 1989). HOX groups 5, 10, 12, and 13 are in turn closely related to four fly homeotic genes in ANT-C and BX-C complex loci (Fig. 16.2) suggesting that mammalian HOX loci are true homologues of the insect homeotic gene complexes (Akam 1989). Furthermore, both fly and mammalian genes show colinearity between relative position in the cluster and anterior boundaries of expression along the AP body axis of the embryo (Duboule and Dollé 1989; Giampaolo *et al.* 1989; Graham *et al.* 1989). HOX and homeotic genes have remained clustered in limited chromosomal regions and in the same physical order for a long evolutionary period. Their orderly diversification must play

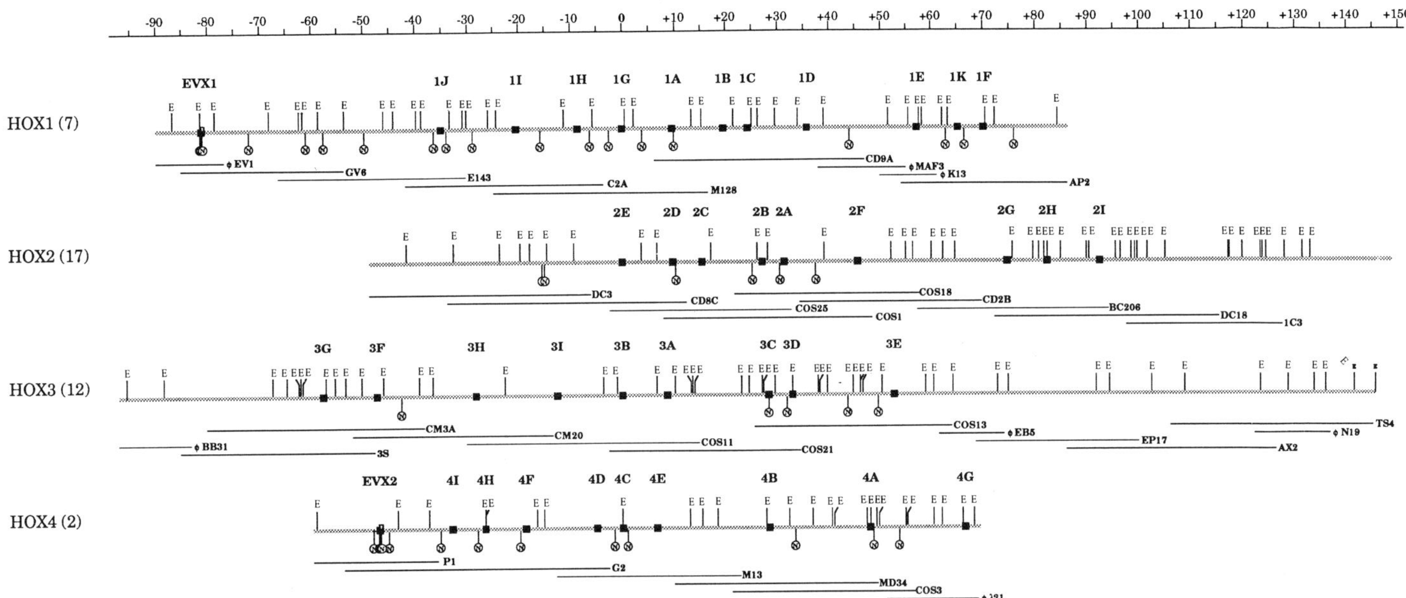

**Fig. 16.1** Genomic organization of mapped regions of the human HOX loci. Identified homeobox sequences are shown as filled boxes. The nomenclature used for HOX genes has been previously reported (Acampora *et al.* 1989; Stornaiuolo *et al.* 1990). Chromosomal location of the locus is indicated in brackets after each HOX designation. We mapped in HOX1 and HOX4—two homeobox related to the *eve* homeobox, EVX1 and EVX2 respectively. Distances in kb are shown. Maps of various loci derive from the analysis of overlapping cosmid or phage ($\phi$) clones. E, Eco RI; N, Not I.

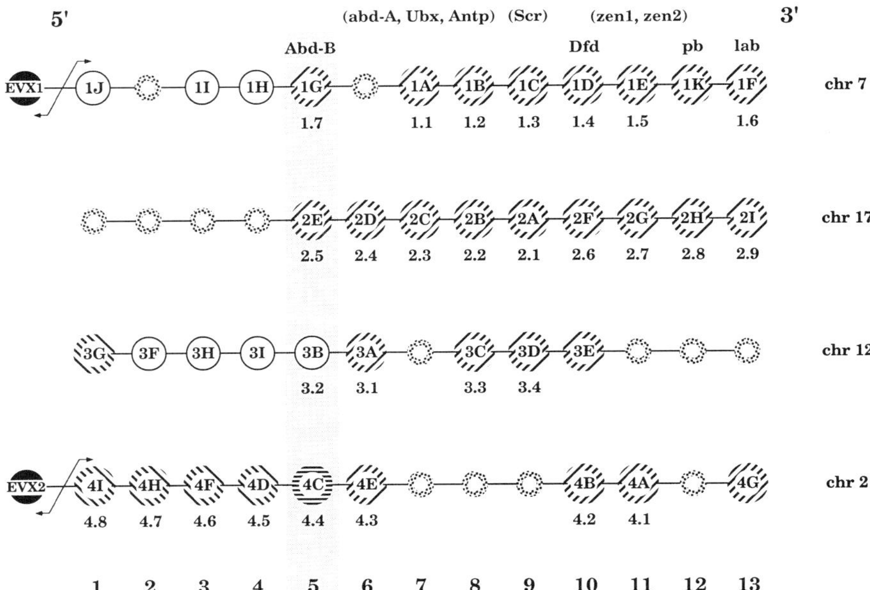

Fig. 16.2 Schematic representation of 40 HDs (circles) in the four chromosomal human HOX clusters. Transcriptional orientation of all HOX genes is from left to right. This alignment identifies 13 HOX HD groups, indicated at the bottom. Below the circles are shown the names of known murine HOX homologues. Stippled empty circles indicate HDs predicted in the scheme but yet unidentified. HDs from *Drosophila* BX-C and ANT-C genes are indicated above the scheme. Each fly HD is placed on top of the group of human HDs to which it is most closely related in sequence. Correspondence of fly HDs in brackets in unclear. *Drosophila* genes are *Abdominal-B* (*Abd-B*), *abdominal-A*, (*abd-A*), *Ultrabithorax* (*Ubx*), *Antp*, *Sex combs reduced* (*Scr*), *Deformed* (*Dfd*), *zerknuellt* (*zen*), *proboscipedia* (*pd*), and *labial* (*lab*). Homology group 5 is highlighted. Striping of HOX circles reflects the HOX gene activation in NT2/D1 cells induced to differentiate by $10^{-5}$ M RA. Ascending (descending) striping indicates genes activated (down-regulated) in time upon RA addition. HOX4C is constitutively expressed at a very low level. Open circles indicate genes whose expression is constantly undetectable in these cells. The scheme also includes the EVX1 and EVX2 HDs. Transcriptional orientation of the corresponding genes is opposite to that of all HOX genes, that is from right to left.

some role in the general regulatory network, which possibly includes their own gene products. The evolutionary implications of such a genomic organization appear to be even more intriguing as we have recently identified two human homologues of the fly *even-skipped* (*eve*) segmentation gene at the 5′ end of the HOX1 and HOX4 loci (D'Esposito *et al.* 1991; our unpublished results) (Figs 16.1 and 16.2). These genes, termed EVX1 and EVX2 respectively, are in turn homologous to *Evx-1* and *Evx-2*, recently identified in the mouse (Bastian and Gruss 1990).

Molecular analysis of expression of these genes in general, and the study of the relationship between expression and physical arrangement in chromosomes in particular, seem crucial in the present phase of the research. Expression analysis of HOX genes in embryonic and extraembryonic tissues has been complemented by studies in differentiating embryonal carcinoma (EC) cells, both in mouse (Colberg-Poley *et al.* 1985*a,b*, 1987; Jackson *et al.* 1985; Breier *et al.* 1986; Deschamps *et al.* 1987*a,b*; Krumlauf *et al.* 1987; Rubin *et al.* 1987; Fibi *et al.* 1988; Graham *et al.* 1988; La Rosa and Gudas 1988*a,b*) and man (Hauser *et al.* 1985; Mavilio *et al.* 1988; Simeone *et al.* 1989, 1990; Stornaiuolo *et al.* 1990; Simeone *et al.* 1991). These cells provide a convenient model to study the molecular basis of developmental gene regulation in response to specific signals.

The cloned human EC cell line NT2/D1 shows a phenotypic pattern of multipotent stem cells of very early embryonic stages and may be induced to differentiate when cultured in a medium containing retinoic acid (RA) (Andrews 1984). Differentiation is characterized by the appearance of several cell types, including neurons, at 7–28 days after induction. Although several morphologically distinct cell types may be found in induced cultures, only neurons were positively identified. The heterogeneity of the cultures increased with time after the addition of RA and recognizable neuronal cells appear relatively late (3–4 weeks), after most morphological changes have already occurred.

We studied the expression of HOX genes in this cell line before and after a time course induction, from 1 to 195 hours, in the continuous presence of $10^{-5}$ M or $10^{-7}$ M RA. Within this period, NT2/D1 cells appear to grow steadily and homogeneously and do not undergo the morphological and phenotypical changes later associated with RA-induced differentiation (Mavilio *et al.* 1988; Simeone *et al.* 1990). Expression of single genes was analysed both by Northern blot analysis and by RNase protection assays.

Northern blot analysis reveals that none of the 38 HOX genes is significantly expressed in untreated stem NT2/D1 cells, whereas 25 of them are activated within a week in differentiating cells upon treatment with $10^{-5}$ M RA (Stornaiuolo *et al.* 1990). Conversely, the expression of 13 HOX genes remains undetectable even after this treatment. These 13 genes are located at the 5′ end of their loci: three in HOX1, five in HOX3 and five in HOX4. The boundary between responding and non-responding genes roughly corresponds to HOX genes of group 5 (see Fig. 16.2).

We further studied in detail the expression of the nine HOX2 genes (Simeone *et al.* 1990). Genes located in the 3′ half of the cluster are induced at peak levels by $10^{-8}$ M RA, whereas a concentration of $10^{-5}$ M RA is required to fully activate 5′ genes. The time course of individual gene expression was monitored by RNase protection assays, at RA concentrations that activate all the genes or just the 3′ genes of the cluster. At $10^{-5}$ M RA, HOX2H and HOX2I were rapidly induced, reaching maximal expression at 24 hours, while the other genes were activated sequentially from HOX2G to HOX2E between 1 and 4 days. At $10^{-7}$ M RA, only the five most 3' genes, HOX2I to HOX2A, are induced, with slower kinetics, but in the same order. Thus, HOX2 genes are sequentially activated in the 3' to 5' direction; this order corresponds with their patterns of expression in the developing central nervous system of humans and mice. It is particularly interesting that HOX2A, the cut-off between the two groups, is the homologue of the murine *Hox-2.1* gene, the expression of which marks the border of hindbrain and spinal cord *in vivo* (Wilkinson *et al.* 1989).

We have now extended this analysis to all 38 HOX genes (Fig. 16.3) in NT2/D1 cells and in a second EC line, namely Tera-2/clone 13 (Simeone *et al.* 1991). Tera-2/clone 13 cells were originally isolated from the human teratoma line, Tera-2 (Thompson *et al.* 1984). Upon induction with RA these cells differentiate in a variety of cell types, relatively similar to RA-induced NT2/D1 cells, but including a lower percentage (1–3 per cent) of neuron-like cells. In all four *Hox* loci, 3' genes are sequentially activated by RA in a 3' to 5' order, colinear with their physical arrangement within the cluster. Activation is relatively rapid in both cell lines and largely precedes the phenotypical and morphological changes associated with RA-induced differentiation. Previous studies on both mouse (Deschamps *et al.* 1987*a*) and human (Mavilio *et al.* 1988) EC cells already showed that homeobox gene activation is strictly correlated with the presence of RA and not with the onset of differentiation, either spontaneously or chemically induced. In the HOX1, HOX3, and HOX4 loci, expression of 5' genes is undetectable in both uninduced and induced EC cells, as analysed by Northern analysis (Stornaiuolo *et al.* 1990). The boundary between activated and silent genes is approximately at the level of the homology group 5. No gene of the first four groups is present in the HOX2 cluster, which is indeed entirely induced by RA in NT2/D1 cells (Simeone *et al.* 1990). The more sensitive RNase protection analysis showed a further distinction (see Fig. 16.2) between: (i) truly silent genes, belonging to groups 1–5 in HOX1 and 2–5 in HOX3; (ii) boundary genes expressed at constant low level, i.e. HOX4C and, to some extent, HOX1A; and (iii) genes expressed at low level in uninduced cells and down-regulated upon RA induction, such as HOX3G and those belonging to groups 1–4 in the HOX4 cluster (Simeone *et al.* 1991).

The general pattern of sequential activation of RA-responding genes and colinear separation of the two sets of responding and non-responding genes was observed in both studied EC cell lines. Significant differences were only observed in the HOX2 cluster, which is activated only in the 3' portion up to HOX2F in Tera-2/clone 13 cells, and in the first four HOX4 genes, which in the same cells are not down-regulated by RA. However, the difference is apparent only when comparing cells induced by high levels ($10^{-5}$ M) of RA. At lower RA concentration ($10^{-7}$ M), partial activation of the HOX2 cluster and continuous expression of the first four HOX4 genes also occur in NT2/D1

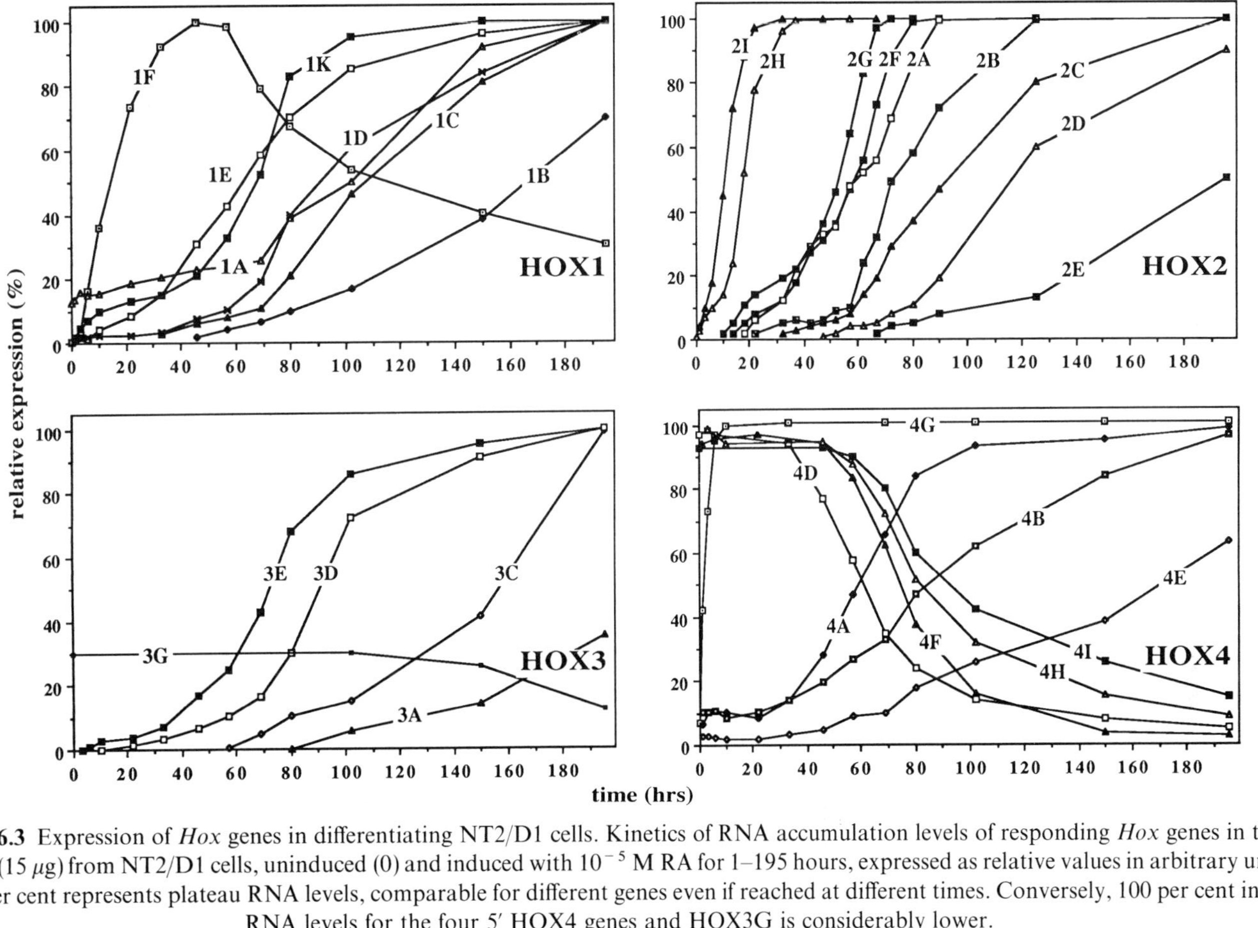



**Fig. 16.3** Expression of *Hox* genes in differentiating NT2/D1 cells. Kinetics of RNA accumulation levels of responding *Hox* genes in total RNA (15 $\mu$g) from NT2/D1 cells, uninduced (0) and induced with $10^{-5}$ M RA for 1–195 hours, expressed as relative values in arbitrary units: 100 per cent represents plateau RNA levels, comparable for different genes even if reached at different times. Conversely, 100 per cent initial RNA levels for the four 5' HOX4 genes and HOX3G is considerably lower.

cells, thus giving rise to an expression pattern almost coincident with that of Tera-2/clone 13. These data indicate a similarity between the expression pattern of Tera-2/clone 13 induced with high RA levels and that of NT2/D1 cells induced with lower RA concentrations. Although the significance of these differences remains obscure, the study of Tera-2/clone 13 cells indicates that the differential regulation of HOX genes by RA and the observed colinearity betwen genomic organization and kinetics of gene activation are not peculiar to a single EC cell line but could rather reflect a phenomenon of general physiological relevance.

In conclusion, we found that HOX genes are differentially activated by RA in human EC cells according to their physical location within the four clusters in a time- and concentration-dependent fashion. The boundaries between RA-responding and non-responding genes are approximately at the level of the same homology group, although not all groups are represented in all clusters (see Fig. 16.2). It is interesting to note that the same boundary also separates the first four groups, which do not correspond to specific *Drosophila* homeotic genes, from the other nine, whose correspondence and colinearity with the *Drosophila* clusters are already well established. The evolutionary origin of genes belonging to the first four groups remains poorly understood. They might have been present in the ancestral cluster predating the divergence between insects and vertebrates and subsequently lost (or heavily rearranged) in the evolutionary lineage leading to flies. Alternatively, these four groups may have arisen specifically in the lineage leading to vertebrates in connection with the complex remodelling of the body plan. Many existing data suggest a specific regulatory pattern for these HOX genes, different from that of downstream genes. For example, known gene products of the genes of the first five HOX groups appear to lack the conserved pentapeptide, termed YPWM homeopeptide (Mavilio *et al.* 1986), present in every HOX gene product of the other groups. Recently reports on involvement of upstream genes of the mouse *Hox-4* locus in limb pattern formation (Dollé *et al.* 1989) represent a first step toward the understanding of this specific pattern.

In both EC cell lines studied, an asymmetry exists between the four loci as far as silent and down-regulated genes of the first four groups are concerned: HOX3G and the four most 5' HOX4 genes exhibit a unique expression pattern in these cells (Figs 16.2 and 16.3).

Expression of *Hox* genes is regionally restricted along the AP axis in the developing CNS. This has probably been best studied for the *Hox-2* locus, whereby 3' genes are expressed more rostrally in the spinal cord and in the hindbrain and 5' genes more caudally in the spinal cord (Giampaolo *et al.* 1989). In the early development of mouse hindbrain, the anterior boundaries of expression of *Hox-2* genes correlate with their relative position in the cluster, with *Hox-2.8* (= HOX2H), *Hox-2.7* (= HOX2G) and *Hox-2.6* (=HOX2F) starting at the level of rhombomeres 3, 5, and 7, respectively (Wilkinson *et al.* 1989). The present data, obtained in EC cells *in vitro*, suggest that RA levels regulated in space and/or time might at least in part underlie a differential activation of *Hox* genes *in vivo* consistent with the observed region-restricted expression pattern. The existence of endogenous sources of retinoids in mammalian central nervous system that could provide concentration gradients analogous to those observed in limb buds has recently been established *in vivo* (Wagner *et al.* 1990). Differential exposure to the morphogen might determine, in cells of different embryonic regions, which genes are to be locally activated, thus providing positional information for the developing embryo. A possible role of RA in connection with homeobox gene function in diverse aspects of vertebrate development has repeatedly been emphasized (Balling *et al.* 1989; Brockes 1989; De Robertis *et al.* 1989; Dollé *et al.* 1989). In this light the anteriorization of the central nervous system observed in frog embryos treated with RA (Durston *et al.* 1989) could be interpreted as a phenomenon correlated to ectopic expression of *Hox* genes. The observed anteriorization implies a considerable reduction in size of the forebrain and midbrain with a relative expansion of spinal cord and hindbrain. The most anterior boundaries of *Hox* gene expression in the developing central nervous system of vertebrates are actually within the hindbrain although ectopic expression of *Hox-2.9* has now been observed in the anterior hindbrain of RA-treated mouse embryos, and correlated with malformation of this region (Murphy *et al.* Chapter 17).

The trouble in extrapolating data from studies in EC cells is that the equation (3′-anterior)-early/(5′-posterior)-late has not been substantiated for the expression of *Hox* genes in embryos, even if circumstantial evidence exists in the literature (see Kessel and Gruss 1990) for a review. In the developing mouse limb a dynamic, temporally restricted, pattern of expression of *Hox*-4 genes was observed with 3′ genes expressed earlier and more proximally and 5′ genes later and more distally (Dollé *et al.* 1989).

We have further investigated the molecular mechanisms underlying the differential activation of HOX genes observed in EC cells. The observed RNA increase appears to be regulated primarily at the transcriptional level as revealed by nuclear run-on transcription assay and comparison of nuclear and cytoplasmic RNA levels during induction. Whatever the underlying activation mechanisms, *de novo* protein synthesis is apparently not required for the induction of early responding genes, as shown by treating differentiating cells with 20 $\mu$g/ml cycloheximide (CHX) (Simeone *et al.* 1991) or puromycin (50 $\mu$g/ml). Furthermore, treatment with CHX alone caused appearance of RNA transcripts from most responding HOX genes, while leaving unaffected the expression levels of silent and down-regulated genes. A significant amount of transcript was observed within hours, with loss of the 3′ to 5′ pattern of sequential activation. This observation appears particularly significant for late responding genes, such as HOX2E. For this gene, 4 hours of CHX treatment, alone or combined with RA, caused RNA accumulation levels comparable to those observed after 90 hours of induction with RA only. These experiments show that synthesis of short-half-life proteins is required to maintain the timing and polarity of expression of responding HOX genes in EC cells. We conclude that the complex control network operating on HOX gene clusters may include active repression exerted by specific gene products. Activation of 5′ HOX genes *in vivo* might involve down-regulation of this latter set of genes, while CHX treatment might play a similar role in the cell culture system.

The number and size of HOX mature transcripts observed in differentiating NT2/D1 cells generally reflects the pattern observed in human embryonic tissues. Nevertheless, for some HOX genes only a subset of the transcripts observed in embryonic and/or adult tissues can be detected in these cells. For example, the HOX3C 2.2 kb (Simeone *et al.* 1987) and the HOX4B 2.5 kb (Mavilio *et al.* 1986) transcripts are not detectable in these cells (Stornaiuolo *et al.* 1990). The absence of the HOX3C 2.2 kb transcripts is particularly relevant. We reported that at least three contiguous HOX3 homeoboxes, namely HOX3C, 3D, and 3E, are transcribed in some embryonic and adult tissues as parts of a single primary transcript transcribed from a major upstream promoter (Simeone *et al.* 1988). Some of their messengers share a common 5′ exon. In particular, we observed two major HOX3C mRNA classes, 2.2 kb and 1.8 kb in length. The 1.8 kb mRNA appears to be transcribed from a proximal specific promoter, whereas the 2.2 kb mRNA derives from a transcript initiated from the distal major upstream promoter. Similar results have been reported for the frog homologue of this gene (Cho *et al.* 1988). Northern blot analysis revealed that the 2.2 kb HOX3C mRNA is not expressed in differentiating NT2/D1 cells. Moreover, RNase protection assays failed to detect any expression in these cells of the common 5′ exon characteristic of this mRNA (Simeone *et al.* 1991). The HOX3C, 3D, and 3E mRNAs observed in these cells must be therefore transcribed from specific promoters proximal to the corresponding homeoboxes and not from the major upstream promoter.

If RA-treated EC cells really mimic the first appearance of HOX gene transcripts in development we conclude that HOX3 genes are first expressed with transcripts starting from individual proximal promoters. As the HOX3 major upstream promoter has been shown to be active in the placenta and embryonic spinal cord, it is conceivable that alternative or additional biological stimuli, e.g. specific growth factors (Ruiz i Altaba and Melton 1989), are required for the activation of this promoter in specific tissues later on in development. With this in mind, we are studying HOX gene expression in NT2/D1 cells upon treatment with growth factors alone or combined with RA. Preliminary data show that EGF (10 ng/ml), bFGF (10 ng/ml), NGF (100 ng/ml) and XTC-MIF (Smith 1987) by no means change the HOX gene expression pattern, whether administered for 6 hours, 1 and 3 days before, or with RA treatment. We are currently testing other peptide growth factors but we should emphasize that so far RA is the only substance that proved to be able to affect HOX gene expression in these cells.

Another possibility is that RA treatment of NT2/D1 cells does not turn on HOX genes for the first time in these cells, but rather it perturbs a previous pattern of expression. Indeed, we know that HOX2I, 2H, 1F, and some HOX4 genes are already expressed in these cells. In addition, we have shown that 5′ genes like HOX2E are actively repressed in NT2/D1 cells. Finally, these cells differentiate in response to other substances, but only RA activates HOX genes.

Several examples have been reported where RA addition *in vivo* changes the pattern of developing structures (see Brockes 1991 for a recent review). It is conceivable that this effect of RA is mediated through a disruption of the pattern of *Hox* gene expression already established in a given developing structure. One might further suggest that this phenomenon results from the activation of individual proximal promoters in the *Hox* loci. This disruption could in turn bring treated cells to a sort of fundamental ground state as far as relative positional information is concerned.

ACKNOWLEDGEMENTS

We wish to thank Vincenzo Nigro for helpful suggestions. This work was supported by Progetti Finalizzati CNR 'Biotecnologia e Biostrumentazione' and 'Ingegneria Genetica', the Third AIDS Project of Ministero della Sanita' and the Italian Association for Cancer Research AIRC. A. St. and A.M. are recipients of an AIRC fellowship.

REFERENCES

Acampora, D., D'Esposito, M., Faiella, A., Pannese, M., Migliaccio, E., Morelli, F., Stornaiuolo, A., Nigro, V., Simeone, A., and Boncinelli, E. (1989). The human HOX gene family. *Nucleic Acids Research*, **17**, 10385–402.

Acampora, D., Simeone, A., and Boncinelli, E. (1991). Human HOX homeobox genes. *Oxford surveys on eukaryotic genes. vol. 7*, (ed. N. Maclean), pp. 1–28. Oxford University Press, Oxford.

Akam, M. (1987). The molecular basis for metameric pattern in *Drosophila* embryo. *Development*, **101**, 1–22.

Akam, M. (1989). HOX and HOM: Homologous gene clusters in insects and vertebrates. *Cell*, **57**, 347–9.

Andrews, P. W. (1984). Retinoic acid induces neuronal differentiation of a cloned human embryonal carcinoma cell line in vitro. *Developmental Biology*, **103**, 285–93.

Balling, R., Mutter, G., Gruss, P., and Kessel, M. (1989). Craniofacial abnormalities induced by ectopic expression of the homeobox gene *Hox-1.1* in transgenic mice. *Cell*, **58**, 337–47.

Bastian, H. and Gruss. P. (1990). A murine *even-skipped* homologue, *Evx 1*, is expressed during early embryogenesis and neurogenesis in a biphasic manner. *EMBO Journal*, **9**, 1839–52.

Boncinelli, E., Somma, R., Acampora, D., Pannese, M., D'Esposito, M., Faiella, A.,

224   *Edoardo Boncinelli* et al.

and Simeone, A. (1988). Organization of the human homeobox genes. *Human Reproduction*, **3**, 880–6.

Breier, G., Bucan, M., Francke, U., Colberg-Poley, A. M., and Gruss, P. (1986). Sequential expression of murine homeobox genes during F9 EC cell differentiation. *EMBO Journal*, **5**, 2209–15.

Brockes, J. P. (1989). Retinoids, homeobox genes, and limb morphogenesis. *Neuron*, **2**, 1285–94.

Brockes, J. P. (1991). We may not have a morphogen. *Nature*, **350**, 15.

Cho, K. W. Y., Goetz, J., Wright, C. V. E., Fritz, A., Hardwicke, J., and De Robertis, E. M. (1988). Differential utilization of the same reading frame in a *Xenopus* homeobox gene encodes two related proteins sharing the same DNA-binding specificity. *EMBO Journal*, **7**, 2139–49.

Colberg-Poley, A. M., Voss, S. D., Chowdhury, K., and Gruss, P. (1985a). Structural analysis of murine genes containing homoeo box sequences and their expression in embryonal carcinoma cells. *Nature*, **314**, 713–18.

Colberg-Poley, A. M., Voss, S. D., Chowdhury, K., Stewart, C. L., Wagner, E. F., and Gruss, P. (1985b). Clustered homeo boxes are differentially expressed during murine development. *Cell*, **43**, 39–45.

Colberg-Poley, A. M., Pueschel, A. W., Dony, C., Voss, S. D., and Gruss, P. (1987). Post-transcriptional regulation of a murine homeobox gene transcript in F9 embryonal carcinoma cells. *Differentiation*, **35**, 206–11.

De Robertis, E. M., Oliver, G., and Wright, V. E. (1989). Determination of axial polarity in the vertebrate embryo: homeodomain proteins and homeogenetic induction. *Cell*, **57**, 189–91.

D'Esposito, M., Morelli, F., Acampora, D., Migliaccio, E., Simeone, A., and Boncinelli, E. (1991). EVX2, a human homeobox gene homologous to the *even-skipped* segmentation gene, is localized at the 5′ end of HOX4 locus on chromosome 2. *Genomics*, **10**, 43–50.

Deschamps, J., De Laaf, R., Joosen, L., Meijlink, F., and Destrée, O. (1987a). Abundant expression of homeobox genes in mouse embryonal carcinoma cells correlates with chemically induced differentiation. *Proceedings of the National Academy of Science USA*, **84**, 1304–8.

Deschamps, J., De Laaf, R., Verrijzer, P., de Gouw, M., Destrée, O., and Meijlink, F. (1987b). The mouse *Hox-2.3* homeobox-containing gene: Regulation in differentiating pluripotent stem cells and expression pattern in embryos. *Differentiation*, **35**, 21–30.

Dollé, P., Izpisúa-Belmonte, J-C., Falkenstein, H., Renucci, A., and Duboule, D. (1989). Coordinate expression of the murine *Hox*-5 complex homoeobox-containing genes during limb pattern formation. *Nature*, **342**, 767–72.

Duboule, E. and Dollé, P. (1989). The structural and functional organization of the murine Hox gene family resembles that of Drosophila homeotic genes. *EMBO Journal*, **8**, 1497–505.

Durston, A. J., Timmermans, J. P. M., Hage, W. J., Hendriks, H. F. J., de Vries, N. J., Heideveld, M., and Nieuwkoop, P. D. (1989). Retinoic acid causes an anteroposterior transformation in the developing central nervous system. *Nature*, **340**, 140–4.

Fibi, M., Zink, M., Kessel, M., Colberg-Poley, A. M., Labeit, S., Lehrach, H., and Gruss, P. (1988). Coding sequence and expression of the homeobox gene *Hox-1.3*. *Development*, **102**, 349–59.

Gaunt, S. J., Sharpe, P. T., and Duboule, D. (1988). Spatially restricted domains of homeogene transcripts in mouse embryos: relation to a segmented body plan. *Development*, (Suppl.), **104**, 169–81.

Gehring, W. J. (1987). Homeoboxes in the study of development. *Science*, **236**, 1245–52.

Giampaolo, A., Acampora, D., Zappavigna, V., Pannese, M., D'Esposito, M., Care', A., Faiella, A., Stornaiuolo, A., Russo, G., Simeone, A., Boncinelli, E., and Peschle, C. (1989). Differential expression of human HOX-2 genes along the anterior-posterior axis in embryonic central nervous system. *Differentiation*, **40**, 191–97.

Graham, A., Papalopulu, N., Lorimer, J., McVey, J. H., Tuddenham, E. G. D., and Krumlauf, R. (1988). Characterization of a murine homeo box gene, *Hox-2.6*, related to the *Drosophila Deformed* gene. *Genes and Development*, **2**, 1424–38.

Graham, A., Papalopulu, N., and Krumlauf, R. (1989). The murine and Drosophila homeobox complexes have common features of organization and expression. *Cell*, **57**, 367–78.

Hart, C. P., Fainsod, A., and Ruddle, F. H. (1987). Sequence analysis of the murine *Hox-2.2, -2.3,* and *-2.4* homeo boxes: evolutionary and structural comparisons. *Genomics*, **1**, 182–95.

Hauser, C. A., Joyner, A. L., Klein, R. D., Learned, T. K., Martin, G. R., and Tjian, R. (1985). Expression of homologous homeo-box-containing genes in differentiated human teratocarcinoma cells and mouse embryos. *Cell*, **43**, 19–28.

Holland, P. W. H. and Hogan, B. L. M. (1988). Spatially restricted patterns of expression of the homeobox-containing gene *Hox-2.1* during mouse embryogenesis. *Development*, **102**, 159–74.

Jackson, I. J., Schofield, P., and Hogan, B. (1985). A mouse homeo box gene is expressed during embryogenesis and in adult kidney. *Nature*, **317**, 745–8.

Kappen, C., Schughart, K., and Ruddle, F. H. (1989). Two steps in the evolution of Antennapedia-class vertebrate homeobox genes. *Proceedings of the National Academy of Science USA*, **86**, 5459–63.

Kessel, M. and Gruss. P., (1990). Murine developmental control genes. *Science*, **249**, 374–9.

Krumlauf, R., Holland, P. W. H., McVey, J. H., and Hogan, B. L. M. (1987). Developmental and spatial patterns of expression of the mouse homeobox gene, *Hox-2.1*. *Development*, **99**, 603–17.

La Rosa, G. J. and Gudas, L. J. (1988*a*). An early effect of retinoic acid: Cloning of an mRNA (Era-1) exhibiting rapid and protein synthesis-independent induction during teratocarcinoma stem cell differentiation. *Proceedings of the National Academy of Science USA*, **85**, 329–33.

La Rosa, G. J. and Gudas, L. J., (1988*b*). Early retinoic acid induced F9 teratocarcinoma stem cell gene ERA-1: alternate splicing creates transcripts for a homeobox-containing protein and one lacking the homeobox. *Molecular Cellular Biology*, **8**, 3906–17.

Levine, M. and Hoey, T. (1988). Homebox proteins as sequence-specific transcription factors. *Cell*, **55**, 537–40.

Lewis, E. B. (1978). A gene complex controlling segmentation in Drosophila. *Nature*, **276**, 565–70.

Mavilio, F., Simeone, A., Giampaolo, A., Faiella, A., Zappavigna, V., Acampora, D., Poiana, G., Russo, G., Peschle, C., and Boncinelli, E. (1986). Differential and stage-related expression in embryonic tissue of a new human homoeobox gene. *Nature*, **324**, 664–8.

Mavilio, F., Simeone, A., Boncinelli, E., and Andrews, P. W. (1988). Activation of four homeobox gene clusters in human embryonal carcinoma cells induced to differentiate by retinoic acid. *Differentiation*, **37**, 73–9.

Rubin, M. R., King, W., Toth, L. E., Sawczuk, I. S., Levine, M., D'Eustachio, P., and Chi Nguyen-Huu, M. (1987). Murine *Hox-1.7* homeobox gene: Cloning, chromosomal location, and expression. *Molecular Cellular Biology*, **7**, 3836–41.

Ruiz i Altaba, A., and Melton, D. A. (1989). Interaction between peptide growth factors and homoeobox genes in the establishment of antero-posterior polarity in frog embryos. *Nature*, **341**, 33–8.

Scott, M. P., Tamkun, J. W., and Hartzell, G. W. III (1989). The structure and function of the homeodomain. *BBA Review of Cancer*, **989**, 25–48.

Simeone, A., Acampora, D., D'Esposito, M., Faiella, A., Pannese, M., and Boncinelli, E. (1988). At least three human homeoboxes on chromosome 12 belong to the same transcription unit. *Nucleic Acids Research*, **16**, 5379–87.

Simeone, A., Acampora, D., D'Esposito, M., Faiella, A., Pannese, M., Scotto, L., Montanucci, M., D'Alessandro, G., Mavilio, F., and Boncinelli, E. (1989). Post-transcriptional control of human homeobox gene expression in induced NTERA-2 embryonal carcinoma cells. *Molecular Reproduction and Development*, **1**, 107–15.

Simeone, A., Acampora, D., Arcioni, L., Andrews, P. W., Boncinelli, E., and Mavilio, F. (1990). Sequential activation of HOX2 homeobox genes by retinoic acid in human embryonal carcinoma cells, *Nature*, **346**, 763–6.

Simeone, A., Acampora, D., Nigro, V., Faiella, A., D'Esposito, M., Stornaiuolo, A., Mavilio, F., and Boncinelli, E. (1991). Differential regulation by retinoic acid of the homeobox genes of the four HOX loci in human embryonal carcinoma cells. *Mechanisms of Development*, **33**, 215–28.

Simeone, A., Mavilio, F., Acampora, D., Giampaolo, A., Faiella, A., Zappavigna, V., D'Esposito, M., Pannese, M., Russo, G., Boncinelli, E., and Peschle, C. (1987). *Proceedings of the National Academy of Sciences USA*, **84**, 4914–18.

Smith, J. C. (1987). A mesoderm-inducing factor is produced by a *Xenopus* cell line. *Development*, **99**, 3–14.

Stornaiuolo, A., Acampora, D., Pannese, M., D'Esposito, M., Morelli, F., Migliaccio, E., Rambaldi, M., Faiella, A., Nigro, V., Simeone, A., and Boncinelli, E. (1990). Human HOX genes are differentially activated by retinoic acid in embryonal carcinoma cells according to their position within the four loci. *Cell Differentiation and Development*, **31**, 119–27.

Thaller, C. and Eichele, G. (1987). Identification and spatial distribution of retinoids in the developing chick limb bud. *Nature*, **327**, 625–8.

Thompson, S., Stern, P. L., Webb, M., Walsh, F. S., Endstrom, W., Evans, E. P., Shi, W-K., Hopkins, B., and Graham, C. F. (1984). Cloned human teratoma cells differentiate into neuron-like cells and other cell types in retinoic acid. *Journal of Cell Science*, **2**, 37–64.

Tickle, C., Alberts, B., Wolpert, L., and Lee, J. (1982). Local application of retinoic acid to the limb bud mimics the action of the polarizing region. *Nature*, **296**, 564–6.

Wagner, M., Thaller, C., Jessel, T., and Eichele, G. (1990). Polarizing activity and retinoid synthesis in the floor plate of the neural tube. *Nature*, **345**, 819–22.

Wilkinson, D. G., Bhatt, S., Cook, M., Boncinelli, E., and Krumlauf, R. (1989).

Segmental expression of Hox-2 homoeobox-containing genes in the developing mouse hindbrain. *Nature*, **341**, 405–9.
Wright, C. V. E., Cho, K. W. Y., Oliver, G., and De Robertis, E. M. (1989). Vertebrate homeodomain proteins: families of region-specific transcription factors. *Trends in Biochemical Science*, **14**, **(2)**, 52–6.

# 17

# Retinoid-induced alterations of segmental organization and gene expression in the mouse hindbrain

Paula Murphy, Duncan R. Davidson, Robert E. Hill, and
Gillian M. Morriss-Kay

## SEGMENTATION OF THE VERTEBRATE HINDBRAIN AND SEGMENTAL EXPRESSION OF POTENTIAL REGULATORY GENES

The organization of the vertebrate body plan during development is not well understood. In invertebrates, such as *Drosophila*, the embryo is divided into repeated units along the anteroposterior (AP) axis; which subsequently develop into distinctive parts of the adult (for review see Akam 1987). The process of segmentation, in which initially similar units diversify according to position, simplifies the task of constructing a complex organism. *Drosophila* segmentation has been well characterized at the genetic level and more than 50 genes that establish or distinguish segments have been identified (Lewis 1978; Nusslein-Volhard and Wieschaus 1980; Perrimon *et al.* 1982; Jurgens *et al.* 1984; Nusslein-Volhard *et al.* 1984; Wieschaus *et al.* 1984; Schupbach and Wieschaus 1986; Akam 1987). Many of these are DNA regulators that belong to conserved families of genes that are represented in vertebrates (for reviews see Holland and Hogan 1988; Kessell and Gruss 1990). However, not surprisingly, vertebrate embryos are more complex than *Drosophila* and are not entirely segmental in form. One part of the vertebrate embryo that is clearly segmented is the hindbrain, where repetitive morphological structures (rhombomeres) have been shown to reflect an underlying segmental organization to neurogenesis (Adelman 1925; Lumsden and Keynes 1989). In mouse embryos, individual hindbrain segments show specific patterns of gene expression. Two of these segment-specific genes show an alteration in their pattern of expression when retinoic acid (RA) levels are raised. They are: *Krox-20*, which contains zinc finger DNA binding motifs and is expressed in rhombomeres 3 and 5 (Wilkinson *et al.* 1989a), and *Hox-2.9*, which is expressed in the intervening rhombomere, rhombomere 4, and contains a homeobox similar to those found in *Drosophila* homeotic genes which confer segment identity (Murphy *et al.* 1989; Wilkinson *et al.* 1989b). Among *Drosophila* genes, *Hox-2.9* is most closely related to *labial*, which identifies an ancestral segment in the *Drosophila* head (Frohman *et al.* 1990; Murphy and Hill 1991).

A study of the expression patterns of *Hox-2.9* and *Krox-20* during segmentation of the hindbrain (8–9 days) revealed that their segmental domains are established progressively, prior to the morphological appearance of rhombomeres (Murphy and Hill 1991). This indicates that these genes are closely linked to the segmentation process, and that their potential role as developmental regulators involves defining and/or identifying individual rhombomeres. In this article we describe the effects of RA excess on the segmental patterns of expression of these two genes, and on the development of morphological segmentation in the hindbrain.


## THE EFFECTS OF EXCESS RETINOIC ACID ON GENES INVOLVED IN EMBRYONIC DEVELOPMENT

There is good evidence that RA plays an important role in organizing the vertebrate embryo. This has been most strikingly demonstrated by the fact that RA can respecify positional identity in the chick limb bud, where it mimics the effect of transplantation of the zone of polarizing activity (Tickle *et al.* 1982; see also Ispisúa-Belmonte *et al.* Chapter 18). The mechanisms by which RA influences development are likely to involve activation of the nuclear RA receptors, which themselves are able to activate (or repress) specific genes or sets of genes (see Chapters 2 and 4). Among the genes found to respond to RA levels in cell culture systems are homeobox-containing genes including *Hox-2.9* (Colberg-Poley *et al.* 1985; Mavilio *et al.* 1988; LaRosa and Gudas 1988*a*,*b*; Murphy 1991; Boncinelli *et al.* Chapter 16). In particular, clustered human homeobox-containing genes respond differentially to raised RA levels depending on their position within the cluster, which is also related to their domain of expression along the AP axis of the embryo (Simeone *et al.* 1990; Boncinelli *et al.* Chapter 16). Further, at least one mouse homeobox-containing gene, *Hox-1.6* (which is most closely related to *Hox-2.9*), shows the characteristics of a direct response to RA (LaRosa and Gudas 1988*a*,*b*). *Hox-2.9* and *Krox-20* are particularly interesting candidate targets for RA because, as described above, they are related to developmental regulators in *Drosophila* that respond to positional information in the embryo.

RA is normally acquired by the mammalian embryo through metabolism of retinol (vitamin A) supplied via the maternal bloodstream. Free RA levels may differ in different parts of the embryo and at different stages of development, possibly through specific spatiotemporal patterns of expression of the cytoplasmic proteins that specifically bind retinol (CRBP I) and RA (CRABP I and II) (Dollé *et al.* 1990; Ruberte *et al.* 1991 and Chapter 8). Maintenance of low RA levels in some embryonic tissues may be crucial for their normal development.

## EFFECTS OF EXCESS RETINOIC ACID ON CRANIOFACIAL DEVELOPMENT

The developmental effects of RA excess depend on embryonic stage at the time of exposure (Shenefelt 1972). Exposure of rodent embryos to retinoid excess at or just before the early neural plate stage leads to the development of a characteristic pattern of craniofacial malformations (Morriss 1972; Morriss and Thorogood 1978). We have used a combination of morphological and molecular techniques to investigate the relationship between RA-induced effects on morphological development of the hindbrain and on the segmental pattern of gene expression in this region (Morriss-Kay *et al.* 1991). Mouse embryos *in utero* were exposed to RA by maternal administration of 10 or 12 mg/kg RA on day $7\frac{3}{4}$, day 8 or day $8\frac{1}{4}$ of pregnancy. Orally administered RA reaches near maximal levels in the embryo within 2 hours and falls off rapidly after 4 hours (Creech Kraft *et al.* 1987), so the time of maternal administration is very close to the time of embryonic exposure to RA. At the three treatment times, the embryos are just beginning neural plate differentiation (day $7\frac{3}{4}$), or are at the late presomite, early neural plate stage (day 8), or are just beginning somitogenesis (day $8\frac{1}{4}$). Embryos examined on day 9, day $9\frac{1}{2}$ or day 10 showed clear cranial abnormalities after treatment on day $7\frac{3}{4}$ or day 8, but were apparently normal after exposure to RA on day $8\frac{1}{4}$.

Otocyst position provides a useful marker for assessing normality or the degree of malformation. In control and day $8\frac{1}{4}$-treated embryos it was positioned level with the second pharyngeal arch; in day 8-treated embryos it was shifted rostrally to be level with the first pharyngeal arch; in day $7\frac{3}{4}$-treated embryos it was on the side of the face, level with the maxillary region. In relation to hindbrain development, the otocyst normally lies alongside rhombomere 5 and the upper part of rhombomere 6; the rostral shift observed in RA-treated embryos was associated with a shortening of the preotic part of the hindbrain. We therefore examined in more detail the effects of RA on the basic segmental organization of the hindbrain, including both morphology and the pattern of expression of genes thought to be involved in the segmentation process. These genes, *Hox-2.9* and *Krox-20*, in addition to being intrinsically interesting as possible target genes for RA (see above), are particularly appropriate in this context because their normal domains of expression (rhombomeres 3, 4, and 5) lie within the affected part of the hindbrain.

## THE EFFECT OF EXCESS RETINOIC ACID ON THE MORPHOLOGY OF THE DEVELOPING BRAIN

Both scanning electron microscopy (Figs 17.2 and 17.3) and histological examination of embryo sections (Figs 17.1 and 17.3) revealed that recognizable rhombomeres do not form following RA treatment at or before 8 days of development. Control embryos at all stages between 8 and 20 somites clearly showed the characteristic pattern of development of 7 rhombomeric sulci (Figs 17.1 and 17.2). In 12-somite day 8-treated embryos an enlarged rhombomere-like structure was visible just posterior to the midbrain in the normal position of rhombomere 1 (Fig. 17.1(d)). In some cases there was a smaller sulcus just caudal to this one but no others could be distinguished. Slightly later in development (16–20 somites), day $7\frac{3}{4}$-treated embryos showed a number of irregular undulations along the neural epithelium extending from the trunk region to the midbrain/hindbrain junction. These however were assymetrical indicating that they do not reflect an underlying segmental organization.

In addition to the lack of rhombomeres, the treated embryos also showed delayed neurulation and a reduction in size of the forebrain and midbrain. Exposure to excess RA at these early stages did not produce malformations in the embryonic trunk or limbs.

THE EFFECT OF EXCESS RETINOIC ACID ON EXPRESSION OF *HOX-2.9* AND *KROX-20*

*In situ* hybridization of control and RA-treated embryo sections with *Hox-2.9* and *Krox-20*-specific probes revealed that a number of features of the normal expression patterns of these genes in the hindbrain between 9 and $9\frac{1}{2}$ days are altered. First, the domains are shifted anteriorly. This is clearly shown by the fact that *Hox-2.9* expression in the treated hindbrain occupies the previously described, enlarged rhombomere-like structure adjacent to the hindbrain/midbrain junction (which is in the normal position of rhombomere 1) (Figs 17.1(e) and 17.4). In control embryos, *Hox-2.9* is expressed in rhombomere 4, which is the fourth rhombomere from the hindbrain/midbrain junction (Figs 17.1(b) and 17.4). Secondly, only one of the two normal *Krox-20* expression domains is detectable in treated embryos. The single remaining *Krox-20* domain is posterior to *Hox-2.9* expression (Figs 17.1(f) and 17.4), and in this respect it is equivalent to normal expression in rhombomere 5 (Fig. 17.1(c)) but is in a more anterior position. There is no expression of *Krox-20* anterior to the *Hox-2.9* domain. Thirdly, the expression domains and the otic vesicle show a similar anterior shift, although the domains are not precisely in register with the otic vesicle. This is shown by *Krox-20* expression, which extends slightly anterior to the otic vesicle in treated embryos but not in controls (Fig. 17.1(c) and (f)). Lastly, in treated embryos, the *Hox-2.9* and *Krox-20* domains lack the sharp planar boundaries that are a characteristic of normal segmental expression. At the single *Hox-2.9*/*Krox-20* mutual expression boundary in the treated embryo there is an irregular alternation of cells expressing the two genes (Morris-Kay *et al.* 1991; Fig. 17.4). As in control embryos, there does not appear to be intermingling of the *Hox-2.9*- and *Krox-20*-expressing cells, indicating that compartmentalization of the cell types is maintained, even

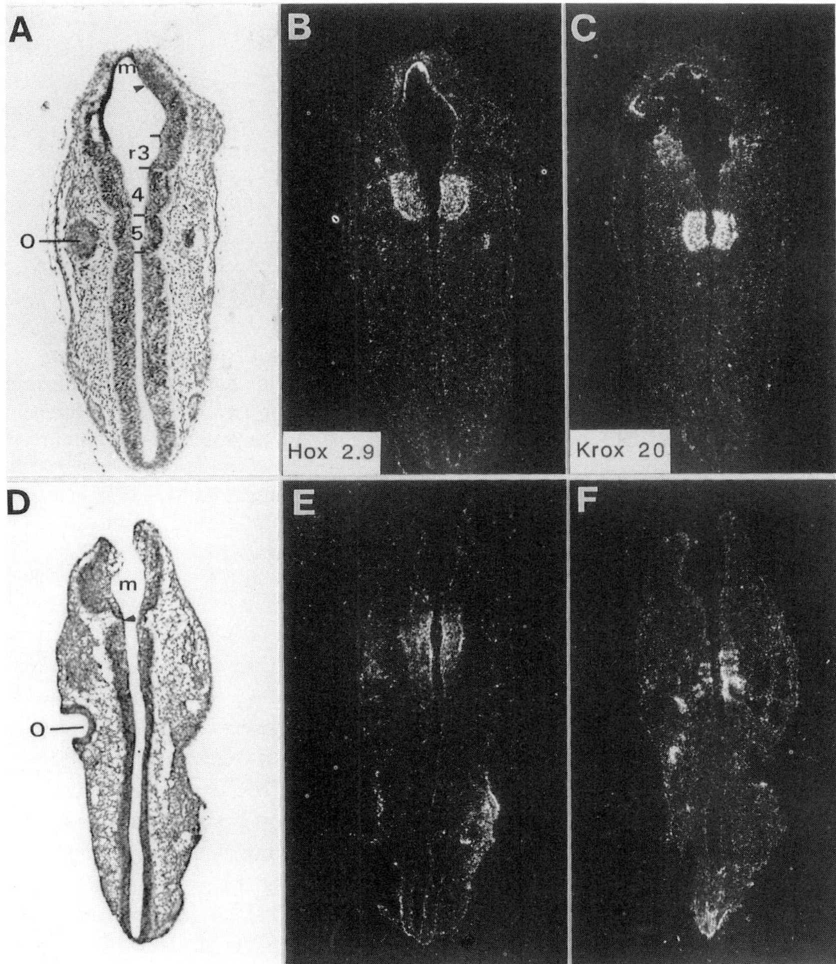



**Fig. 17.1** Coronal sections of control (*a–c*) and RA-treated (*d–f*) day 9 mouse embryos. (*b*) and (*e*) are darkfield exposures of the sections shown by brightfield illumination in (*a*) and (*d*) respectively; (*c*) and (*f*) are adjacent sections. (*b*) and (*e*) show the location of *Hox-2.9* transcripts; (*c*) and (*f*) show *Krox-20*. In the control embryo sections, *Hox-2.9* is expressed in rhombomere 4 and *Krox-20* in rhombomeres 3 and 5, with sharp planar boundaries between them. Rhombomere 5 is level with the otocyst. In the RA-treated embryo sections, the anterior boundary of *Hox-20* expression is the midbrain–hindbrain junction, but both anterior and posterior boundaries are ill-defined; *Krox-20* is expressed in patches alternating with patches of *Hox-2.9* expression, slightly rostral to the level of the otic pit. m, midbrain; o, otocyst/otic pit; r3 to 5: rhombomeres 3 to 5; arrowhead, midbrain–hindbrain junction.

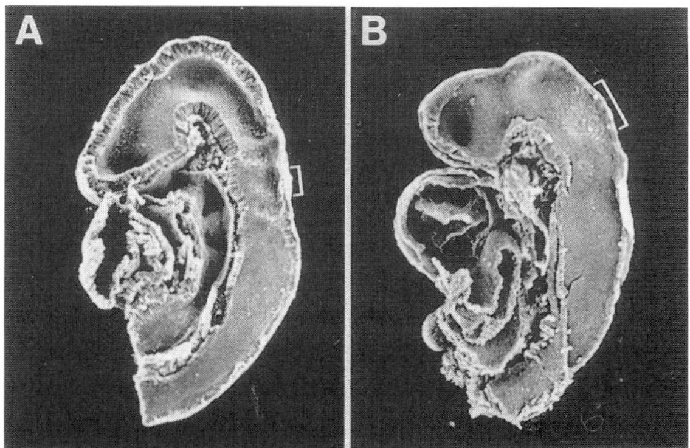

**Fig. 17.2** Scanning electron micrographs of sagitally halved 12-somite stage control (*a*) and RA-treated (*b*) embryos showing the normal and RA-affected hindbrain morphology. (*a*) The normal pattern of rhombomeric sulci and gyri seen on the internal surface of the hindbrain neuroepithelium (*a*) shows as a clear alternation of concave sulci and sharply convex gyri. (*b*) In the RA-treated embryo there is a large sulcus-like concavity adjacent to the midbrain and a less clear second sulcus separated from it by a gentle convex curve. The brackets indicate the domains of *Hox-2.9* expression as observed by *in situ* hybridization, i.e. rhombomere 4 in (*a*) and the large rhombomere-like structure adjacent to the midbrain in (*b*).

though precise regional localization of the cells along the AP axis is not achieved.

In both treated and untreated embryos, specific subsets of neural crest cells that originate within the hindbrain neural epithelium express *Hox-2.9* (Fig. 17.3) and *Krox-20* (Morriss-Kay *et al.* 1991). Like the expression patterns within the epithelium itself, neural crest cell expression is shifted anteriorly in RA-treated embryos. *Hox-2.9* is normally expressed specifically in neural crest cells that emigrate from rhombomere 4 on day 9 (Fig. 17.3(a, b)); these cells populate the acousticofacial ganglion (Hunt *et al.* 1991), so do not migrate far from the neural tube. Cells emigrating from this region at an earlier stage migrate further, to form the mesenchyme of the second pharyngeal arch (Tan and Morriss-Kay 1986); they do not express *Hox-2.9* (Hunt *et al.* 1991). In RA-treated embryos, *Hox-2.9*-expressing neural crest cells migrate from the abnormally extensive and abnormally positioned *Hox-2.9*-expressing area of the hindbrain (Fig. 17.3(d, e)). They form a larger population than those of controls and migrate to a greater than normal distance from the neural tube, into the side of the developing face. They therefore differ from normal *Hox-2.9*-expressing crest cells in terms of their level of origin, their migration pathway and their migration distance. Their mesenchymal organization resembles pharyngeal arch mesenchyme rather than presumptive ganglion cells. One intriguing possibility is that they form the ectopic maxillary

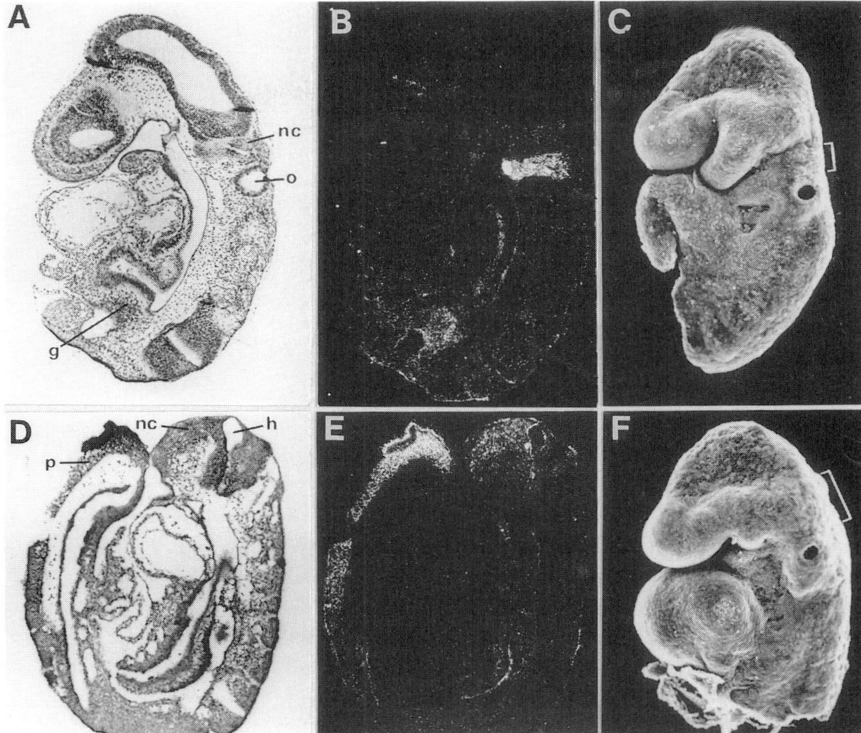

**Fig. 17.3** Expression of *Hox-2.9* in neural crest cells of control and RA-treated embryos. (*a*) and (*b*) are light and darkfield views of a parasagittal section of a day 9 control embryo showing *Hox-2.9* expression in neural crest cells (nc) derived from rhombomere 4 (i.e. immediately anterior to the otocyst, (o)). (*d*) and (*e*) show a section cut parasagittal in the caudal region but obliquely through the cranial region, so that the left side of the head is anterior to the right side. Expression of *Hox-2.9* is seen in the hindbrain neural epithelium (h) on the left side, and in the neural crest cells (nc) extending from it across into the side of the developing face. Other areas of *Hox-2.9* expression, in the gut-associated mesenchyme (g) and in the primary mesenchyme (p) are present in appropriate sections of both control and RA-treated embryos. (*c*) and (*f*) are scanning electron micrographs showing external views of control (*c*) and RA-treated (*f*) embryos, showing the abnormal position of the otic pit and reduced size of the pharyngeal arches. The level of the *Hox-2.9*-expressing neural crest cells, as revealed by *in situ* hybridization, is indicated by the brackets.

cartilage that is seen at later stages (Morriss and Thorogood 1978). *Krox-20* expression is normally detected in the neural crest cells just posterior to rhombomere 5, which also expresses *Krox-20* (Wilkinson *et al.* 1989*a*; Morriss-Kay *et al.* 1991). In RA-treated embryos neural crest cells expressing

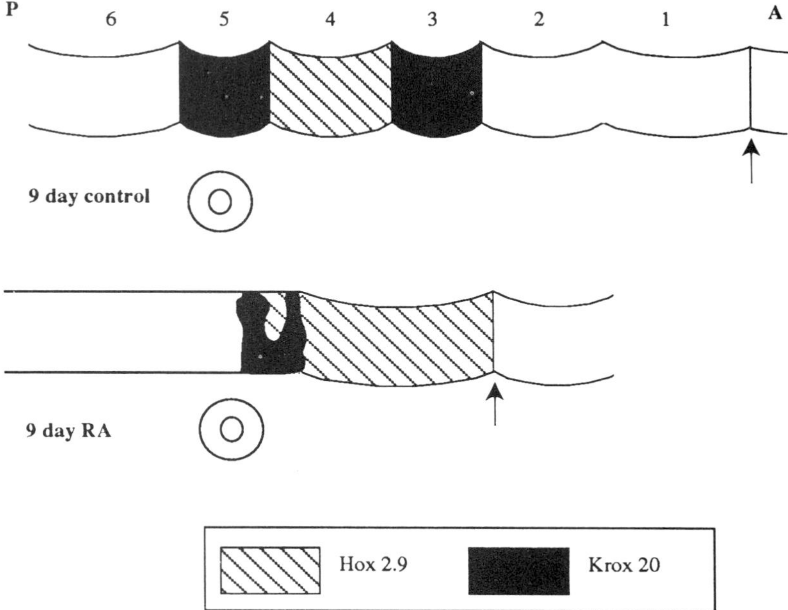



**Fig. 17.4** Diagrammatic representation of the expression of *Hox-2.9* and *Krox-20* in the control and RA-treated day 9 embryonic hindbrain, showing the RA-induced disruption of the segmental expression domains. The otic vesicles have been aligned so that the anterior shift in expression domains is not represented. A, anterior (rostral); P, posterior (caudal); 1–6, rhombomeres 1–6. Arrows: midbrain/hindbrain junction.

*Krox-20* are present adjacent and posterior to the *Krox-20* hindbrain domain. Their level of expression is higher than that of controls.

HINDBRAIN SEGMENTATION IS SUPPRESSED BY EXPOSURE TO EXCESS RETINOIC ACID

The experiments described here show that the process of hindbrain segmentation is disrupted following *in vivo* exposure to excess RA. In the normal embryo the segments are defined gradually in a rostrocaudal direction between 8 and 9 days of development. This is illustrated by the expression of *Hox-2.9* and *Krox-20*, which become localized to specific segments, with boundaries that become progressively sharper as the segments are defined (Murphy and Hill 1991; Fig. 17.4). Following treatment with RA during the initial stages of this process (8 days), the expression of *Hox-2.9* fails to become sharply defined, showing that full definition of the segmental units is not achieved (Fig. 17.4).

The behaviour of *Hox-2.9*- and *Krox-20*-expressing cells at their mutual expression boundary in treated embryos may indicate the method by which segmental domains are maintained in the normal hindbrain. It has previously been suggested (Lumsden 1990) that segmental compartments may be maintained by two alternative methods, either by the existence of cells at segment boundaries that do not allow cell migration or by cells in each segment recognizing cellular segmental differences and mixing only with cells of their own segment. The result of the present study, where *Hox-2.9*- and *Krox-20*-expressing cells do not intermingle in the absence of a segmental boundary, indicate that cellular recognition plays a role in segment definition, but is not sufficient by itself to establish sharp segmental boundaries. Cell recognition mechanisms are therefore more likely to be involved in the maintenance than in the establishment of cell compartments.

Maximal sensitivity of the hindbrain to raised RA levels is at or just before 8 days of development, which is before the definition of hindbrain segments. Immediately after this 'critical period', *Krox-20* expression begins in the area that becomes rhombomere 3 and *Hox-2.9* is expressed throughout the neural tube and posterior hindbrain as far as the posterior boundary of the *Krox-20* domain. This boundary is morphologically defined by the preotic sulcus. Subsequently, *Hox-2.9* expression is restricted to rhombomere 4, *Krox-20* expression begins in rhombomere 5, and the morphological features of segmentation become apparent (Murphy and Hill 1991). The timing of hindbrain sensitivity is therefore consistent with involvement of RA in spatial organization of the hindbrain.

The effects of RA excess in mammalian embryos at this stage are slightly different from those observed in amphibian embryos, in which treatment during and shortly after gastrulation causes truncation of hindbrain structures (Durston *et al.* 1989; Sive *et al.* 1990). However, similar effects on the anterior hindbrain have also been observed in amphibians (Papalopulu *et al.* 1991). The ability of RA to interfere with the AP axis of both mammalian and amphibian embryos, together with the response of other systems to RA, suggests that RA may be an endogenous signalling substance within the cranial neural ectoderm. The results of exposure to excess RA can be interpreted as a distortion of this normal function.

It has been suggested that segmentation of the embryonic head may extend outside the neural ectoderm, and that the head mesenchyme may be segmentally patterned via the neural crest cells, which arise from the segmented hindbrain (Couly and Le Douarin 1990). The existence of segmental differences between cranial neural crest cells is supported by observations of their segment-specific patterns of gene expression (Frohman *et al.* 1990; Hunt *et al.* 1991; Murphy and Hill 1991) and by transplantation studies showing that they have different potentials for differentiation according to their axial level of origin (Noden 1983). The migration of segmentally patterned neural crest cells to inappropriate sites is clearly an important element in the genesis of RA-induced craniofacial malformations. The abnormal pattern appears to be initiated while the crest cells lie within the hindbrain neural epithelium.

ACKNOWLEDGEMENTS

We thank Martin Barker for histology and Colin Beesley and Sandy Bruce for photographic work. The *Krox*-20 probe was received from David Wilkinson. The work was supported by a grant from Hoffmann-La Roche to GMM-K and by the Medical Research Council.

REFERENCES

Adelman, H. B. (1925). The development of the neural folds and cranial ganglia of the rat. *Journal of Comparative Neurology*, **39**, 19–171.

Akam, M. (1987). The molecular basis of metameric pattern in the Drosophila embryo. *Development*, **101**, 1–22.

Colberg-Poley, A. M., Voss, S. D., Chowdhury, K., and Gruss, P. (1985). Structural analysis of murine genes containing homeobox sequences and their expression in embryonal carcinoma cells *Nature*, **314**, 713–18.

Couly, G. and LeDouarin, N. M. (1990). Head morphogenesis in embryonic avian chimeras: evidence for a segmental pattern in the ectoderm corresponding to the neuromeres. *Development*, **108**, 543–58.

Creech Kraft, J., Kochhar, D. M., Scott, W. J., and Nau, H. (1987). Low teratogenicity of 13-cis-retinoic acid (isotretinoin) in the mouse corresponds to low embryo concentrations during organogenesis: comparison to the all-*trans* isomer. *Toxicology and Applied Pharmacology*, **87**, 474–82.

Dollé, P., Ruberte, E., Leroy, P., Morriss-Kay, G., and Chambon, P. (1990). Retinoic acid receptors and cellular retinoid binding proteins. I. A systematic study of their differential pattern of transcription during mouse organogenesis. *Development*, **110**, 1133–51.

Durston, A. J., Timmermans, J. P. M., Hage, W. J., Hendricks, H. F. J., de Vries, N. J., Heidelveld, M., and Nieuwkoop, P. D. (1989). Retinoic acid causes an anteroposterior transformation in the developing central nervous system. *Nature*, **340**, 140–4.

Frohman, M. A., Boyle, M., and Martin, G. R. (1990). Isolation of the mouse *Hox-2.9* gene; analysis of embryonic expression suggests that positional information along the anterior-posterior axis is specified by mesoderm. *Development*, **110**, 589–607.

Holland, P. W. H. and Hogan, B. L. M. (1988). Expression of homeobox genes during mouse development: a review. *Genes and Development*, **2**, 773–82.

Hunt, P., Wilkinson, D., and Krumlauf, R. (1991). Patterning of the vertebrate head: *Hox*-2 genes mark distinct subpopulations of premigratory and migrating neural crest. *Development*, **112**, 43–50.

Jurgens, G., Wieschaus, E., Nusslein-Volhard, C., and Kluding, H. (1984). Mutations affecting the pattern of the larval cuticle in *Drosophila melanogaster*. II Zygotic loci on the third chromosome. *Wilhelm Roux's Archives of Developmental Biology*, **193**, 283–95.

Kessel, M. and Gruss, P. (1990). Murine developmental control genes. *Science*, **249**, 374–9.

LaRosa, G. J. and Gudas, L. J. (1988*a*). An early effect of retinoic acid: Cloning of an mRNA (Era-1) exhibiting rapid and protein synthesis independent induction during teratocarcinoma stem cell differentiation. *Proceedings of the National Academy of Science USA*, **85**, 329–33.

LaRosa, G. J. and Gudas, L. J. (1988*b*). Early retinoic acid induced F9 teratocarcinoma stem cell gene ERA-1: alternate splicing creates transcripts for a homeobox-containing protein and one lacking the homeobox. *Molecular and Cellular Biology*, **8**, 3906–17.

Lewis, E. B. (1978). A gene complex controlling segmentation in *Drosophila*. *Nature*, **276**, 565–70.

Lumsden, A. (1990). The cellualr basis of segmentation in the developing hindbrain. *TINS*, **13**, 329–35.

Lumsden, A. and Keynes, R. (1989). Segmental patterns of neuronal development in the chick hindbrain. *Nature*, **337**, 424–8.

Mavilio, F., Simeone, A., Boncinelli, E., and Andrews, P. W. (1988). Activation of four homeobox gene clusters in human embryonal carcinoma cells induced to differentiate by retinoic acid. *Differentiation*, **37**, 73–9.

Meijlink, F., de Laaf, L., Verrikzer, P., de Graaf, W., and Deschamps, J. (1989). Regulation of expression of the *Hox-2.3* gene. In *Cell to cell signals in mammalian development*, vol. H26, (ed. S. W. de Laat), pp. 23–41, NATO ASI Series, Springer-Verlag, Berlin.

Morriss, G. M. (1972). Morphogenesis of the malformations induced in rat embryos by hypervitaminosis A. *Journal of Anatomy*, **113**, 241–50.

Morriss, G. M. and Thorogood, P. V. (1978). An approach to cranial neural crest migration and differentiation in mammalian embryos. In *Development in mammals*, vol. 3, (ed. M. H. Johnson). pp. 387–400. Elsevier North-Holland, Amsterdam.

Morriss-Kay, G. M., Murphy, P., Hill, R. E., and Davidson, D. R. (1991). Effects of retinoic acid excess on expression of *Hox 2.9* and *Krox 20* and on morphological segmentation in the hindbrain of mouse embryos. *EMBO Journal*, **10**, 2985–95.

Murphy, P. (1991), Mouse labial-like homeobox-containing genes: structure and expression during embryogenesis. Ph.D. thesis, Edinburgh University.

Murphy, P., Davidson, D. R., and Hill, R. E. (1989). Segment specific expression of a homeobox-containing gene in the mouse hindbrain. *Nature*, **341**, 156–9.

Murphy, P. and Hill, R. E. (1991). Expression of the mouse *labial*-like homeobox containing genes, *Hox-2.9* and *Hox-1.6*, during segmentation of the hindbrain. *Development*, **111**, 61–74.

Noden, D. M. (1983). The role of the neural crest in patterning of avian cranial skeletal, connective and muscle tissues. *Developmental Biology*, **96**, 144–65.

Nusslein-Volhard, C. and Wieschaus, E. (1980). Mutations affecting segment number and polarity in *Drosphila*. *Nature*, **287**, 795–801.

Nusslein-Volhard, C., Wieschaus, E., and Kluding, H. (1984). Mutations affecting the pattern of the larval cuticle in *Drosophila melanogaster*. I. Zygotic loci on the second chromosome. *Wilhelm Roux's Archives of Developmental Biology*, **193**, 267–82.

Papalopulu, N., Clarke, J. D. W., Wilkinson, D., Krumlauf, R., and Holder, N. (1991).

Retinoic acid causes abnormal development and segmental patterning of the anterior hindbrain in *Xenopus* embryos. *Development*, (in press).

Perrimon, N., Engstrom, L., and Mahowald, A. P. (1982). The effects of zygotic lethal mutations on female germ line functions in *Drosophila Developmental Biology*, **105**, 404–14.

Ruberte, E., Dollé, P., Chambon, P., and Morriss-Kay, G. (1991). Retinoic acid receptors and cellular retinoid binding proteins: II. Their differential pattern of transcription during early morphogenesis in mouse embryos. *Development*, **111**, 45–60.

Shenefelt, R. E. (1972). Morphogenesis of malformations in hamsters caused by retinoic acid: relation to dose and stage of treatment. *Teratology*, **5**, 103–18.

Schupbach, T. and Wieschaus, E. (1986). Maternal effect mutations affecting the segmental pattern of *Drosophila*. *Wilhelm Roux's Archives of Developmental Biology*, **195**, 302–7.

Simeone, A., Acampora, D., Arcioni, L., Andrews, P. W., Boncinelli, E., and Mavilio, F. (1990). Sequential activation of *HOX-2* homeobox genes by retinoic acid in human embryonal carcinoma cells. *Nature*, **346**, 763–6.

Sive, H. L., Draper, B. W., Harland, R. M., and Weintraub, H. (1990). Identification of a retinoic acid-sensitive period during primary axis formation in *Xenopus laevis*. *Genes and Development*, **4**, 932–42.

Tan, S. S. and Morriss-Kay, G. M. (1986). Analysis of cranial neural crest cell migration and early fates in postimplantation rat chimaeras, *Journal of Embryology and Experimental Morphology*, **98**, 21–58.

Tickle, C., Alberts, B., Wolpert, L., and Lee, J. (1982). Local application of retinoic acid to the limb bud mimics the action of the polarising region. *Nature*, **296**, 564–6.

Webster, W. S., Johnston, M. C., Lammer, E. J., and Sulik, K. K. (1986). Isotretinoin embryopathy and the cranial neural crest: an in vivo and in vitro study. *Journal of Craniofacial Genetics and Developmental Biology*, **6**, 211–22.

Wieschaus, E., Nusslein-Volhard, C., and Jurgens, G. (1984). Mutations affecting the pattern of the larval cuticle in *Drosophila melanogaster*. I. Zygotic loci on the X chromosome and the fourth chromosome. *Wilhelm Roux's Archives of Developmental Biology*, **193**, 267–82.

Wilkinson, D. G., Bhatt, S., Chavrier, P., Bravo, R., and Charnay, P. (1989*a*). Segment specific expression of a zinc finger gene in the developing nervous system of the mouse. *Nature*, **337**, 461–4.

Wilkinson, D. G., Bhatt, S., Cook, M., Boncinelli, E., and Krumlauf, R. (1989*b*). Segmental expression of *Hox 2* homeobox genes in the developing mouse hindbrain. *Nature*, **341**, 405–9.

18

# *Hox*-4 genes, retinoic acid, and the specification of positional information during chick wing morphogenesis


Juan-Carlos Izpisúa-Belmonte, Pascal Dollé, Cheryll Tickle,
Lewis Wolpert, and Denis Duboule


The morphogenesis of the vertebrate limb is a widely used model system to study the cellular and molecular basis of positional signalling processes during development. A conceptual framework for pattern formation was provided by Wolpert (1969) who proposed that various cells within a developing system could have their positions specified relative to one or more reference points or organizing regions. Accordingly, each cell would then interpret its own positional value and respond by differentiating in a particular way according to its developmental history. Cells highly related, phenotypically and functionally, would therefore appear as 'non-equivalent' (Lewis and Wolpert 1976) as far as patterning is concerned. The nature of the genetic control of 'non-equivalence' is largely unknown, but the past few years have provided us with serious candidate genes for encoding positional information. Among them are the homeobox-containing (HOX) genes, which are the vertebrate gene cognates of the *Drosophila* homeotic genes. In insects, these genes specify the identities of the various parasegmental units in a position-dependent manner. In vertebrates, functional and circumstantial evidence suggest that the *Hox* genes encode positional information. *Hox* genes are responsive to retinoic acid (RA) treatment *in vitro* (Simeone *et al.* 1990) and it is well established that retinoids can respecify patterns in the developing vertebrate limb (Tickle *et al.* 1982, 1985). Here we discuss evidence that genes from the vertebrate *Hox-4* complex might play a role in setting-up positional cues during mouse and chick limb development and that the expression of these genes can be regulated by RA *in vivo*.

MORPHOGENESIS OF THE CHICKEN LIMB

The chick wing bud arises as an outgrowth of flank mesenchyme opposite somites 15 to 20. It comprises a thin peripheral layer of ectoderm surrounding a uniform mass of mesodermal cells. After a mesodermal induction of an apical ectodermal ridge (AER), this stratified epithelium will control the growth of the limb bud by maintaining an apical zone of high mitotic activity, the progress zone. Correct anterior–posterior (AP) patterning seems to depend upon the 'polarizing' activity of a particular group of cells, which are located at the posterior margin of the limb. This area is therefore believed to be responsible for the AP asymmetry of the developing limbs (Saunders and Gasseling 1968). When transplanted anteriorly, the polarizing region induces a duplication of structures along the AP axis resulting in a digit pattern 4, 3, 2, 2, 3, 4 in place of the normal 2, 3, 4 (2 being anterior and 4 posterior). This respecification of the digit pattern in the anterior cells takes place in the progress zone, where undifferentiated mesenchymal cells are thought to acquire their positional information. The signalling of the polarizing region can be understood in terms of the release of a polarizing factor that would be differentially distributed across the limb bud with the highest concentration at the posterior margin (Tickle *et al.* 1975). Cells would therefore need to read the local concentration of this factor for instruction regarding their anteroposterior location (their positional values).

Retinoic acid induces digit pattern duplications when applied to the anterior margin of the developing wing bud. Such duplications are dose-, time- and position-dependent, similar to duplications produced by grafting polarizing regions (Tickle *et al.* 1982; Eichele *et al.* 1985; Tickle *et al.* 1985). Moreover, the observation that RA is present in the chick wing in concentrations similar to those that induce duplications (Tickle *et al.* 1985, Thaller and Eichele 1987) suggests that RA could be the polarizing factor (or part of it), or alternatively could induce surrounding cells to acquire this morphogenetic activity (Wanek *et al.* 1991). In both cases, the question remains open as to how to translate a rather simple and localized signal into a complicated network of positional information required to establish patterns in the limb. Our recent work on the genes of the *Hox-4* complex suggests that the *Hox* genes are used to read this simple morphogenetic signal and to translate it into a more complex plurimolecular network of positional information.

COORDINATE EXPRESSION OF THE MURINE *HOX-4* COMPLEX

A systematic analysis of the expression of five gene members of the murine *Hox-4* (formerly referred to as *Hox-5*) complex has shown that each gene has a characteristic spatial and temporal expression pattern in the limb bud (Dollé *et al.* 1989). The sequence of temporal activation is colinear with the position of the genes in the complex so that 3'-located genes are transcribed first and 5' genes last (the temporal colinearity). Genes located 3' in the complex are expressed more proximally at the base of the limb, whereas genes located at more 5′ positions are expressed more distally. The transcript domains of all five genes encompass the posterior part of the limb bud and a given expression domain is always included in the domain of the gene located immediately 3′ to it. As the area where these expression domains overlap coincides with the signalling region (the ZPA) we proposed that *Hox-4* gene could be involved in the patterning of the limb by responding to the polarizing signal of the ZPA (Dollé *et al.* 1989). Cells could therefore be non-equivalent in the combination of *Hox* genes that they would express. Because these genes encode transcription factors, various combinations of homeoproteins could differentially affect target genes.

A possible approach to investigate the validity of this proposal is to modify patterns and examine whether changes in the expression domains of the *Hox-4* genes can be observed before morphological changes occur. If true, such observations of the rearranged *Hox-4* expression domains would allow predictions on the novel wing phenotypes to be obtained. The developing chick wing bud is an appropriate system in this context due to its experimental accessibility and because of the high reproducibility of the changes in pattern formation obtained after either grafting of polarizing regions or local release of RA.

## *HOX-4* GENES, RETINOIC ACID, AND PATTERN FORMATION IN THE CHICK WING BUD

The cloning, sequencing, and physical mapping of the chicken *Hox-4*-complex genes revealed an overall organization practically identical to that found in mammals (Izpisúa-Belmonte *et al.* 1991*a*). The intergenic spaces and transcriptional direction of the various genes in the complex are conserved (Fig. 18.1). Sequence analysis revealed similarities ranging from 77 to 92 per cent at the nucleotide level, and a total conservation at the protein level of all the homeodomain proteins (Fig. 18.1). This indicates that both the general structure and the individual coding units have been highly conserved between rodents and birds.

The subsequent analysis of the expression domains of these genes in the buds during the development of the wing revealed transcript distribution similar to those described in mice (Dollé *et al.* 1989; Izpisúa-Belmonte 1991*b*). These genes are expressed in different but overlapping cellular domains. The more 5′ a gene is located in the *Hox-4* complex, the more its transcript domain will be restricted posteriorly and distally (Fig. 18.2). As for the mouse, the timing of the expression of the chicken genes is dependent upon their position within the complex: genes located at 5′ positions in the complex are expressed, in the limb bud, several hours after the onset of expression of those located in more 3′ positions (Izpisúa-Belmonte 1991*b*). We therefore conclude that the

244    *J-C. Izpisúa-Belmonte* et al.

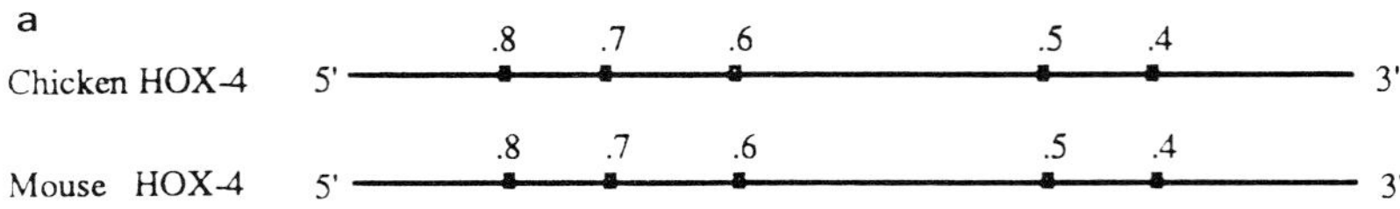

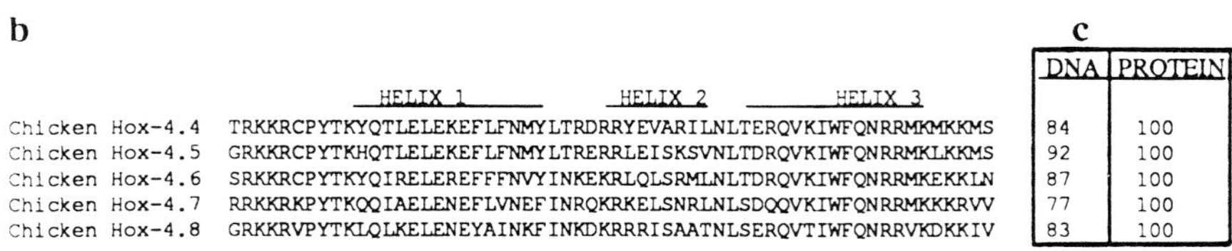



**Fig. 18.1** (*a*) Scheme of the organization of the upstream region of the chicken *Hox-4* complex and comparison with the mouse cognate complex. The genes (black squares) have the same orientations of transcription, 5′ to 3′ from left to right and the average distance between two genes is about 5 to 10 kb. The mouse and chicken *Hox-4* complexes are very similar to each other. (*b*) Protein sequences of the five chicken homeodomains from the genes belonging to the *Hox-4* complex are shown under (*a*). The positions of the three alpha helices (1, 2, 3) are shown on the top. (*c*) Percentages of identity between mouse and chicken DNA and protein sequences. Only the homeobox (homeodomain) sequences are compared. The homeodomain protein sequences are identical.

temporal and spatial features of *Hox-4* gene expression are very similar, if not identical, in chick and mouse limb buds (Fig. 18.2).

Beads soaked in 0.1 mg/ml RA implanted at the anterior margin of early chick wing buds lead to mirror image duplication of the digit pattern (4, 3, 2, 2, 3, 4) so that cells in the anterior part of the bud now form posterior structures (Tickle *et al.* 1982). This pattern respecification process takes place in a biphasic manner (Eichele *et al.* 1985), as the removal of the bead during the first 12 hours following implantation (the priming phase) invariably produces a normal 2, 3, 4 digit pattern. If the bead is removed after this, additional digits are formed but always in an anterior–posterior sequence (extra digit 2 first, digit 3, and digit 4 last). By 24 hours of RA treatment, the reprogramming is irreversible and full digit duplications will develop.

If the *Hox-4* genes encode positional information, one might expect that any changes in patterning are preceded by a reorganization of the *Hox-4* expression domains. In this case, the 'posteriorization' of the anterior part of the limb should occur in parallel with the ectopic expression of the 'posterior' *Hox* genes in anterior domains. This is indeed the case, as exemplified in Fig. 18.3, which shows an *in situ* hybridization analysis of the *Hox-4.6* transcription domain, 24 hours after implantation of a RA-soaked bead in the anterior margin. It shows the appearance of a novel ectopic *Hox-4.6* expression domain in the anterior part of the limb bud, a region that would

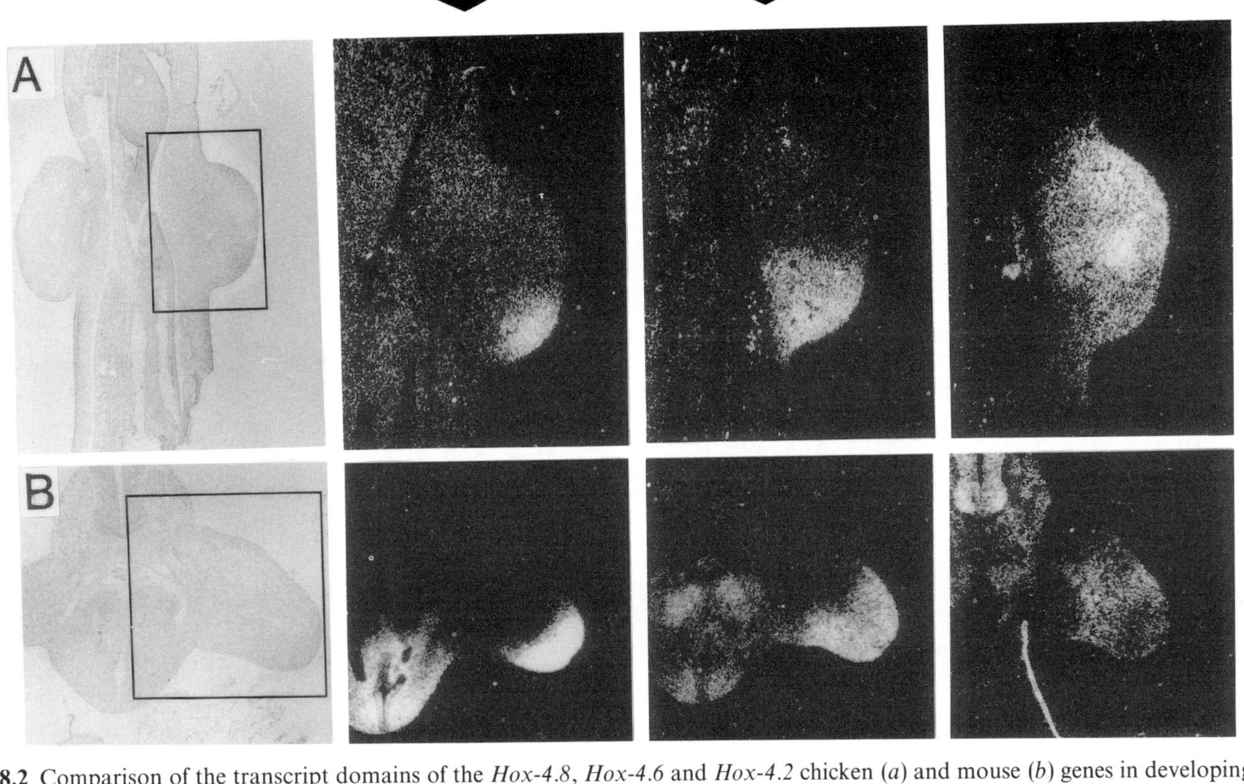



**Fig. 18.2** Comparison of the transcript domains of the *Hox-4.8*, *Hox-4.6* and *Hox-4.2* chicken (*a*) and mouse (*b*) genes in developing limb buds (stage 21 for the chick (*a*) and day 12 postcoitum for the mouse (*b*)). In both mouse and chicken, the expression domain of the *Hox-4.8* gene overlaps with that of *-4.6* while *-4.6* overlaps with that of *-4.4*. The smallest domain (*-4.8*) has a posterodistal localization.

normally never express this gene. Similar observations are made for the other members of the *Hox-4* complex.

*In situ* analysis of the transcript domains at different times after RA treatment shows that the strict temporal sequence of expression of the *Hox-4* genes, which is observed in non-manipulated buds, is reproduced in cells located anteriorly while they are being reprogrammed to form posterior structures (Izpisúa-Belmonte *et al.* 1991*b*). Genes lying at the 3′ end of the complex are switched on before genes located at the 5′ end. This is in agreement with the data of Simeone *et al.* (1990) (see also Boncinelli *et al.* Chapter 16) showing that the genes of the HOX-2 cluster located in the 3′ region respond to RA *in vitro* more rapidly than genes located towards the 5′ end of the complex. The molecular mechanism involved in such a 'colinear' response to RA is unknown. Among other possibilities, the various *Hox-4* genes could respond to similar doses of RA but would require first to be accessible ('open') to the (direct or indirect) stimulation by RA. A similar temporal colinearity is observed for these genes during the development of the trunk (Izpisúa-Belmonte, 1991*a*).

When a bead impregnated with RA is implanted in the limb region of stage 10–14 chicken embryos, it must remain until stage 17 to induce a duplicated limb skeleton (Wilde *et al.* 1987). Stage 17 is the stage when the bud and the AER develop and, moreover, is the stage at which the *Hox-4* genes are sequentially activated (Dollé *et al.* 1989; Izpisúa-Belmonte *et al.* 1991*b*). In buds treated with RA, the pattern along the proximodistal (PD) axis is established in a comparable way as in the normal limb bud. This correlates well with the observation that changes in pattern do not involve the most proximal structures in the limb, but only the forearm and digits. It has been suggested that patterning along the PD axis results from the age of the dividing cells in the progress zone (Summerbell *et al.* 1973). The time spent in this zone could thus allow different cells to express different *Hox* genes. The sequential activation of the genes along the cluster would therefore be an essential factor in this process.


CONCLUSION

Homeobox-containing (*Hox*) genes are expressed along the craniocaudal axis during development according to the same principles as we describe here in the limbs (Duboule and Dollé 1989; Graham *et al.* 1989). This suggests that patterning along the different body axes is achieved by similar mechanisms, the aims of which are to translate an initial signal (e.g. a monotonic gradient of retinoid) into a more complicated and informative network. This would provide the cells with the necessary positional values (the homeogene products) and would allow the cells to interpret and express these values in terms of modulating the transcription of sets of target genes. This pathway

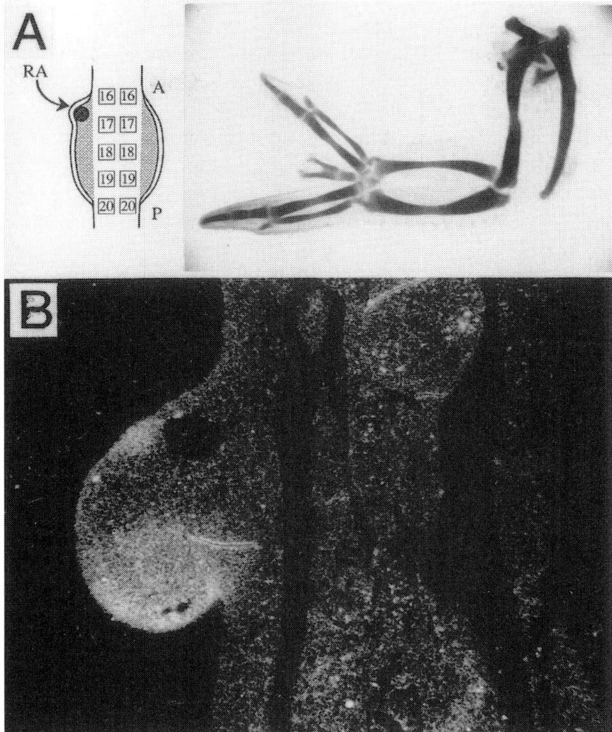


**Fig. 18.3** The *Hox-4.6* transcript domains in chicken wing buds treated with retinoic acid. (*a*) Scheme of the experiment in which a bead soaked in retinoic acid is implanted at the anterior margin of the limb (left). The wing pattern resulting from such an experiment is shown on the right. The bead soaked in a 0.1 mg/ml retinoic acid solution was introduced at the margin of a stage 18 wing bud and removed after 24 hours. (*b*) Shows the *de novo* transcription of the *Hox-4.6* gene in the anterior part of the wing after the treatment. The ectopic domain is restricted to a subectodermal region.

would, in principle, allow the transmission from basic information (e.g. a graded morphogen) to the establishment of complex cell–cell interaction mechanisms leading to the proper spatial arrangement of the structures.


ACKNOWLEDGEMENTS

We wish to thank Dr C. Murphy for her comments and suggestions on the manuscript and H. Davies and the EMBL photolab for preparing the manuscript.

REFERENCES

Dollé, P., Izpisúa-Belmonte, J.-C., Falkenstein, H., Renucci, A., and Duboule, D. (1989). Coordinate expression of the murine Hox-5 complex homeobox-containing genes during limb pattern formation. *Nature*, **342**, 767–72.

Duboule, D. and Dollé, P. (1989). The structural and functional organization of the murine HOX gene family resembles that of Drosophila homeotic genes. *EMBO Journal*, **8**, 1497–505.

Eichele, G., Tickle, C., and Alberts, B. (1985). Studies on the mechanisms of retinoid-induced pattern duplications in the early chick limb bud: temporal and spatial aspects. *Journal of Cell Biology*, **101**, 1913–20.

Graham, A., Papalopulu, N., and Krumlauf, R. (1989). The murine and drosophila homeobox gene complexes have common features of organization and expression. *Cell*, **57**, 367–78.

Izpisúa-Belmonte, J.-C., Falkenstein, H., Dollé, P., Renucci, A., and Duboule, D. (1991*a*). Murine genes related to the Drosophila *AbdB* homeotic gene are sequentially expressed during development of the posterior part of the body. *EMBO Journal*. **10**, 2279–89.

Izpisúa-Belmonte, J.-C., Tickle, C., Dollé, P., Wolpert, L., and Duboule, D. (1991*b*). Expression of the homeobox HOX-4 genes and the specification of positional information in chick wing development. *Nature*, **350**, 585–9.

Lewis, J. H. and Wolpert, L. (1976). The principle of nonequivalence in development. *Journal of Theoretical Biology*, **62**, 479–90.

Saunders, J. W. and Gasseling, M. T. (1968). Ectodermal–mesenchymal interactions in the origin of limb symmetry. In *Epithelial–mesenchymal interactions*. Williams and Wilkins, Baltimore.

Simeone, A., Acampora, D., Arcioni, L., Andrews, P. W., Boncinelli, E., and Mavilio, F. (1990). Sequential activation of HOX 2 homeobox genes by retinoic acid in human embryonal carcinoma cells. *Nature*, **346**, 763–6.

Summerbell, D., Lewis, J. H., and Wolpert, L. (1973). Positional information in chick limb morphogenesis. *Nature*, **244**, 492–6.

Thaller, C. and Eichele, G. (1987). Identification and spatial distribution of retinoids in the developing chick limb bud. *Nature*, **397**, 625–8.

Tickle, C., Summerbell, D., and Wolpert, L. (1975). Positional signalling and specification of digits in chick limb morphogenesis. *Nature*, **254**, 199–202.

Tickle, C., Alberts, B., Wolpert, L., and Lee, J. (1982). Local application of retinoic acid to the limb bud mimics the action of the polarizing region. *Nature*, **296**, 564–5.

Tickle, C., Lee, J., and Eichele, G. (1985). A quantitative analysis of the effect of all-trans-retinoic acid on the pattern of chick wing development. *Developmental Biology*, **109**, 82–5.

Wanek, N., Gardiner, D. M., Muneoka, K., and Bryant, S. V. (1991). Conversion by retinoic acid of anterior cells into ZPA cells in the chick wing bud. *Nature*, **350**, 81–3.

Wilde, S. M., Wedden, S. E., and Tickle, C. (1987). Retinoids reprogramme pre-bud mesenchyme to give changes in limb pattern. *Development*, **100**, 723–33.

Wolpert, L. (1969). Positional information and the spatial pattern of cellular differentiation. *Journal of Theoretical Biology*, **25**, 1–47.

# 19

# Retinoid-regulated expression of transglutaminases: links to the biochemistry of programmed cell death

Peter J. A. Davies, Joseph P. Stein, E. Antonio Chiocca,
James P. Basilion, Vittorio Gentile, Vilmos Thomazy, and Laszlo Fesus

## INTRODUCTION

Transglutaminases are enzymes that catalyse the covalent cross-linking of proteins by promoting the formation of $\varepsilon$-($\gamma$-glutaminyl)lysine isopeptide bonds between polypeptide chains (Fig. 19.1; for reviews see Folk 1983; Lorand and Conrad 1984). These enzymes contain an active-site cysteine that reacts with the $\gamma$-carboxamide of a specific protein-bound glutamine residue, displacing the amide nitrogen and forming a transient thiolester enzyme–substrate intermediate. This intermediate can react with a number of naturally occurring and synthetic amines, such as polyamines (for a review see Davies *et at.* 1988) but the preferred substrate is a protein bound lysine residue (as shown in Fig. 19.1), in which case the two proteins become cross-linked by an isopeptide bond. A striking feature of the transglutaminase reaction is that the cross-link introduced into the substrate proteins is both mechanically and chemically stable. Cells do not contain enzymes that hydrolyse isopeptide bonds. The cross-linked proteins must be completely degraded to liberate the $\varepsilon$-($\gamma$-glutaminyl)lysine dipeptide, which can be broken down to lysine and 5-oxo-proline by the action of $\gamma$-glutamylamine cyclotransferase (Fink *et al.* 1980).

Transglutaminases are a family of enzymes that catalyse protein cross-linking in a number of different physiological compartments. Several of these enzymes, factor XIII (Grundmann *et al.* 1986; Ichinose *et al.* 1986; Takahashi *et al.* 1986), tissue transglutaminase (Chiocca *et al.* 1988; Ikura *et al.* 1988; Gentile *et al.* 1991), keratinocyte transglutaminase (Phillips *et al.* 1990; Kim *et al.* 1991), and epidermal transglutaminase (Kim *et al.* 1990) have now been cloned and sequenced and it is clear that they represent distinct but homologous proteins encoded by unique genes. The transglutaminases that

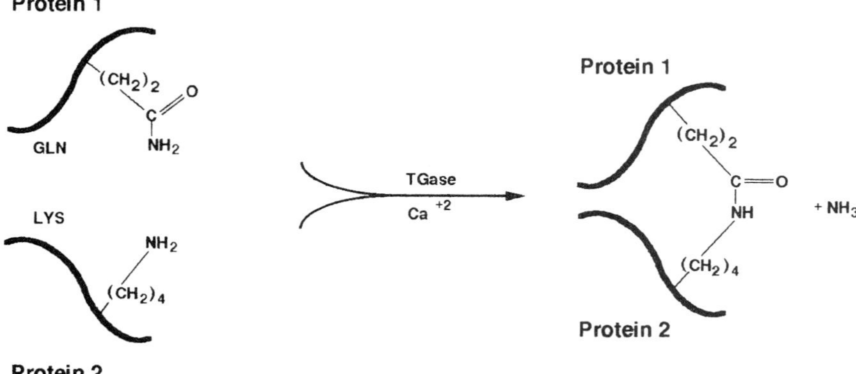



**Fig. 19.1** The transglutaminase reaction.

are found predominantly in extracellular compartments, such as plasma transglutaminase (factor XIII) and seminal plasma transglutaminase, are involved in cross-linking proteins during clotting reactions (Schwartz *et al.* 1973; Porta *et al.* 1991). The function of the intracellular transglutaminases is less well understood. Keratinocyte transglutaminase is a membrane-bound intracellular transglutaminase that is specifically induced during the terminal differentiation of keratinocytes (Parenteau *et al.* 1986). The enzyme becomes activated by increased intracellular calcium and cross-links a series of specific substrate proteins, involucrin (Rice and Green 1979; Simon and Green 1988), loricrin (Mehrel *et al.* 1990), and keratolinin (Zettergren *et al.* 1984) to form an envelope that invests the keratins and contributes to the mechanical stability of the cell (Rice and Green 1978).

We have been interested in another intracellular transglutaminase, tissue transglutaminase, a cytosolic enzyme that is expressed in many cells and tissues. Tissue transglutaminase is particularly abundant in some terminally differentiated cells such as erythrocytes and lens epithelia (Lorand and Conrad 1984) as well as endothelial cells (Greenberg *et al.* 1987; Nara *et al.* 1989). In senescent cells, such as erythrocytes and corneal cells, the enzyme is involved in the cross-linking of intracellular proteins. In endothelial cells the primary function of the enzyme is different and involves the cross-linking of matrix proteins (e.g. fibrinogen, fibronectin, and vitronectin) that contribute to cellular adhesivity (Fesus *et al.* 1986; Martinez *et al.* 1989; Sane *et al.* 1991). A striking feature of tissue transglutaminase is that expression of the enzyme is frequently linked to cellular differentiation. Several years ago Birckbichler and his colleagues showed that tissue transglutaminase activity was low in transformed and rapidly proliferating cells and it increased dramatically if the proliferation of the cells was suppressed or their differentiation was induced (Birckbichler and Patterson 1978). We have been interested in investigating the molecular mechanisms involved in these changes in tissue transglutaminase expression in both normal and transformed cells.

## ROLE OF RETINOIDS IN TISSUE TRANSGLUTAMINASE EXPRESSION

We have been particularly interested in the regulation of tissue transglutaminase in myeloid cells because physiological activation of these cells is frequently associated with large increases in tissue transglutaminase expression (Schroff *et al.* 1981; Kannagi *et al.* 1982; Leu *et al.* 1982). Figure 19.2

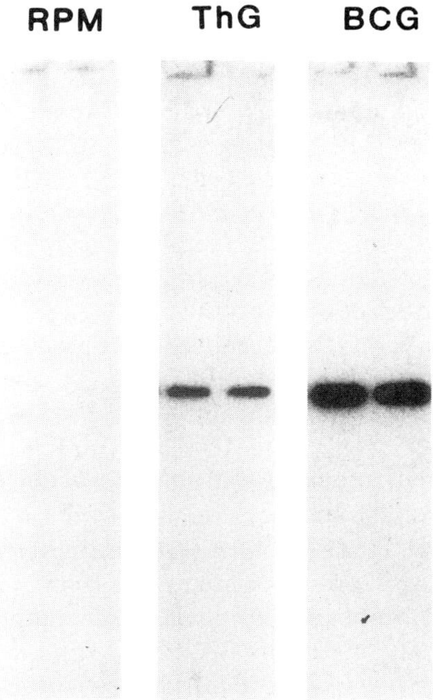


**Fig. 19.2** Tissue transglutaminase in macrophages. Extracts of resident mouse peritoneal macrophages (RPM), thioglycollate-elicited (ThG) or BCG-activated (BCG) macrophages were fractionated by SDS–PAGE on a 5 per cent gel, transferred to nitrocellulose and Western blotted with a purified antibody to tissue transglutaminase.

compares the level of the enzyme (as detected by Western blot analysis) in resident peritoneal macrophages (RPM) and in macrophages recovered from an intraperitoneal inflammatory reaction that had been elicited by injection of either thioglycollate broth (ThG) or heat-killed mycobacteria (BCG). We found that this large *in vivo* induction of the enzyme could be reproduced *in vitro* by exposing the unstimulated macrophages to homologous serum (Murtaugh *et al.* 1983) indicating that serum contained factors that regulated transglutaminase expression.

To identify the transglutaminase-inducing activity of serum, we fractionated serum and found that delipidization of the serum abolished the induction of the enzyme and reconstitution with physiological levels of either retinol (1–5 $\mu$M) or retinoic acid (1–10 nM) restored it (Moore *et al.* 1984). Subsequent studies carried out with cloned mouse tissue transglutaminase cDNA probes showed that retinoic acid (RA) increased tissue transglutaminase expression within minutes of its addition to cells and this effect was attributable to increased expression of the tissue transglutaminase gene (Chiocca *et al.* 1988). The induction of the enzyme was a direct effect of RA because inhibitors of protein synthesis (such as cycloheximide) did not block the induction of tissue transglutaminase mRNA (Chiocca *et al.* 1988, 1989). Although the effects of RA on tissue transglutaminase expression were most striking in mouse peritoneal macrophages (Murtaugh *et al.* 1983; Mehta *et al.* 1987; Roth *et al.* 1988), similar inductions could be detected in other cells *in vitro* (Murtaugh *et al.* 1984; Davies *et al.* 1985; Melino *et al.* 1988; Ceru *et al.* 1989; Cai *et al.* 1991) and *in vivo* (Lichti and Yuspa 1988; Piacentini *et al.* 1988).

The ability of RA to regulate tissue transglutaminase expression appears to be under complex endocrine control. Analogues of cyclic-AMP can increase RA-induced expression of tissue transglutaminase in both normal and leukaemic myeloid cells (Davies *et al.* 1985; Murtaugh *et al.* 1986). This effect is not restricted to myeloid leukaemia cells: Fig. 19.3 demonstrates the dramatic effect of dibutyryl-cyclic AMP (DBC) on RA-induced expression of the enzyme in a cultured fibroblastic cell line (CHO cells). Agents that activate protein kinase C (such as phorbol esters) can block the induction of the enzyme. Fig. 19.4 shows the effect of RA (1 $\mu$M), dibutyryl-cyclic AMP (DBC, 1 mM) or teradecanoyl phorbol acetate (TPA, 10 nM) on transglutaminase induction in mouse peritoneal macrophages. The data demonstrates clearly that TPA can completely block both RA-induced and RA plus DBC-induced expression of the enzyme. These observations suggest that the retinoid responsiveness of cells (as measured by transglutaminase expression) is subject to regulation by other hormones. Thus the biological effects of RA in a specific cell or tissue will depend in part on the nature of other endocrine signals simultaneously regulating the function of that cell.


RETINOIC ACID RECEPTORS AND TRANSGLUTAMINASE EXPRESSION


Our understanding of the molecular basis of retinoid action has undergone considerable revision with the recent identification of a family of nuclear RA

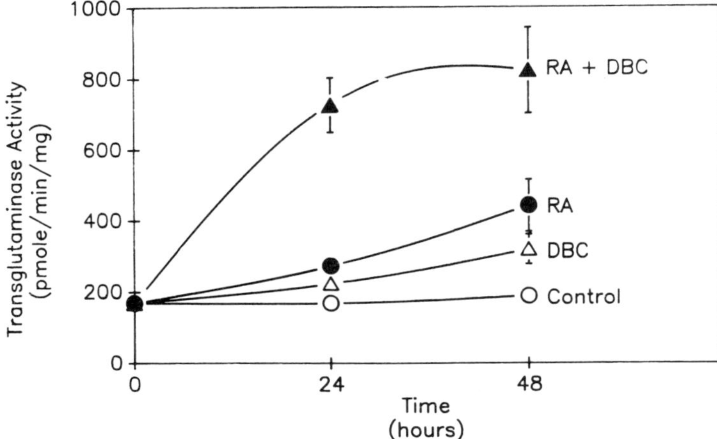


**Fig. 19.3** Effect of retinoic acid (RA) and Dibutyryl cyclic-AMP (DBC) on transglutaminase activity in CHO cells. CHO cells were exposed to medium alone (control) or medium supplemented with RA (1 $\mu$M), DBC (1 mM) or both (RA + DBC). At the indicated times cells were lysed and the transglutaminase activity was assayed.

receptors (RARs) that have been shown to mediate the effects of RA on gene expression (Evans 1988). Three distinct RA receptor genes (RAR-$\alpha$, $\beta$, and $\gamma$) have been identified (Giguère *et al.* 1987; Petkovich *et al.* 1987; Brand *et al.* 1988; Krust *et al.* 1989) and we have been interested in determining whether these receptors mediated the induction of tissue transglutaminase by retinoic acid. Our approach has been to establish cell lines stably transfected with eukaryotic expression vectors that support the expression of either RAR-$\alpha$, $\beta$, or $\gamma$ and then to examine the effect of each receptor on the activation of endogenous tissue transglutaminase gene expression. Balb-C 3T3 cells were chosen for this study because they have very low basal levels of RAR and tissue transglutaminase expression. We were unable to establish cell lines that expressed high levels of RAR-$\alpha$, but we were able to maintain lines with high levels of RAR-$\beta$ and $\gamma$ transcripts. These cell lines showed increased nuclear RA binding activity, RA-induced expression of synthetic retinoid-reporter constructs and retinoid-induced growth inhibition. The transfected cells also acquired a dramatic increase in RA-induced expression of tissue transglutaminase. Figure 19.5 shows the large increase in RA-induced transglutaminase expression in RAR-$\beta$ and RAR-$\gamma$-transfected cells compared to their wild type counterparts. Denning and Verma (1991) have also reported that rat tracheal cells transfected with RAR-$\alpha$ and RAR-$\beta$ also have increased RA-induced expression of tissue transglutaminase. It appears that all three of the RARs can mediate expression of the enzyme.

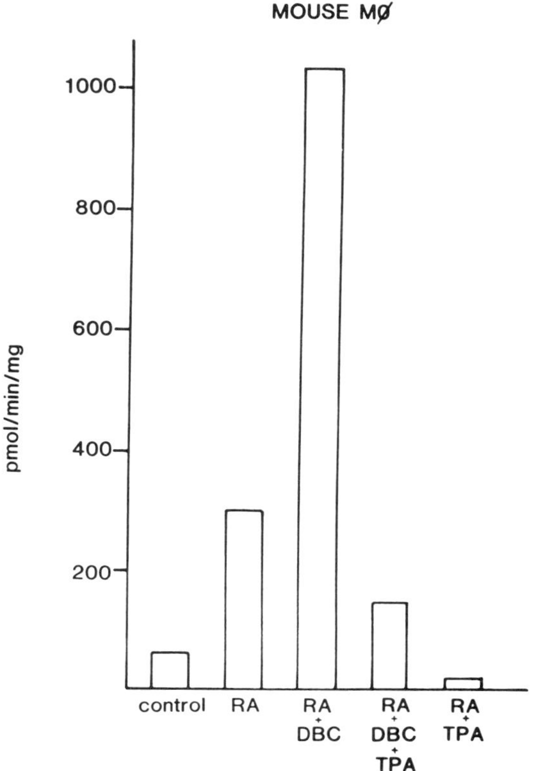



**Fig. 19.4** Effect of phorbol ester (TPA) on transglutaminase induction in mouse macrophages. Mouse resident peritoneal macrophages were cultured for 24 hours in medium alone (RPMI with 10 per cent delipidized mouse serum–control) or media supplemented with retinoic acid (RA—1 $\mu$M), dibutyryl cyclic-AMP (DBC—1 mM), both (RA + DBC) or RA, DBC, and tetradecanoyl phorbol acetate (TPA—10 nM) or RA and TPA. Transglutaminase activity was assayed in cell lysates.

## TRANSGLUTAMINASE EXPRESSION IN THE LIMB BUD

The studies with transfected cells suggested that the three RARs could stimulate tissue transglutaminase expression in cultured cells but they did not address the issue of whether they had the same activity *in vivo*. Recently Chambon and his co-workers have reported on the tissue-specific expression of RARs in the mouse embryo limb bud (Dollé *et al.* 1989; Ruberte *et al.* 1990, Chapter 8). RAR-$\alpha$ was found to be almost ubiquitously distributed, RAR-$\beta$ was concentrated in the interdigital webs and RAR-$\gamma$ was detected in the precartilaginous tissue and cartilages. To determine whether the expression of tissue transglutaminase matched the expression of specific RARs we

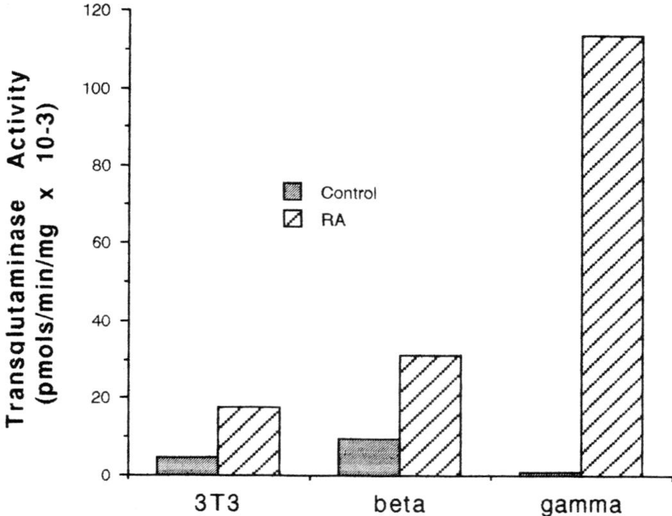


**Fig. 19.5** Retinoic acid-induced expression of tissue transglutaminase activity in retinoid receptor—transfected 3T3 cells. Control 3T3 cells and cells stably transfected with an RAR-$\beta$ and RAR-$\gamma$ expression vector were exposed to retinoic acid (RA— 1 $\mu$M) for 24 hours and the transglutaminase activity was assayed.

used immunohistochemistry to localize the expression of tissue transglutaminase in the limb bud of 7.5 day chick embryos. We found that the expression of the enzyme was focal, it was concentrated in the hypertrophic chondrocytes of the developing bone as well as in apoptotic cells associated with the interdigital webs and the primordial musculature. Figure 19.6 demonstrates the presence of transglutaminase immunoreactivity in apoptotic cells associated with the immature muscle beds. The expression of the enzyme in chondrogenic cells and in regions of apoptosis (programmed cell death) falls within the regions of expression of RAR-$\gamma$ and RAR-$\beta$, respectively (Ruberte *et al.* 1990).

## RETINOIDS, TRANSGLUTAMINASE, AND APOPTOSIS

The induction of tissue transglutaminase in apoptotic cells in the limb bud may explain the close link between retinoids and transglutaminase gene expression. Several studies have demonstrated that the biological effects of retinoids may be due to their ability to induce or expand zones of programmed cell death (Abbott and Pratt 1987; Sulik *et al.* 1988; Alles and Sulik 1989; Yasuda *et al.* 1989; Alles and Sulik 1990; Martin *et al.* 1990; Sacks *et al.* 1990; Piacentini *et al.* 1991*b*). Certainly in the limb bud the induction of apoptosis in

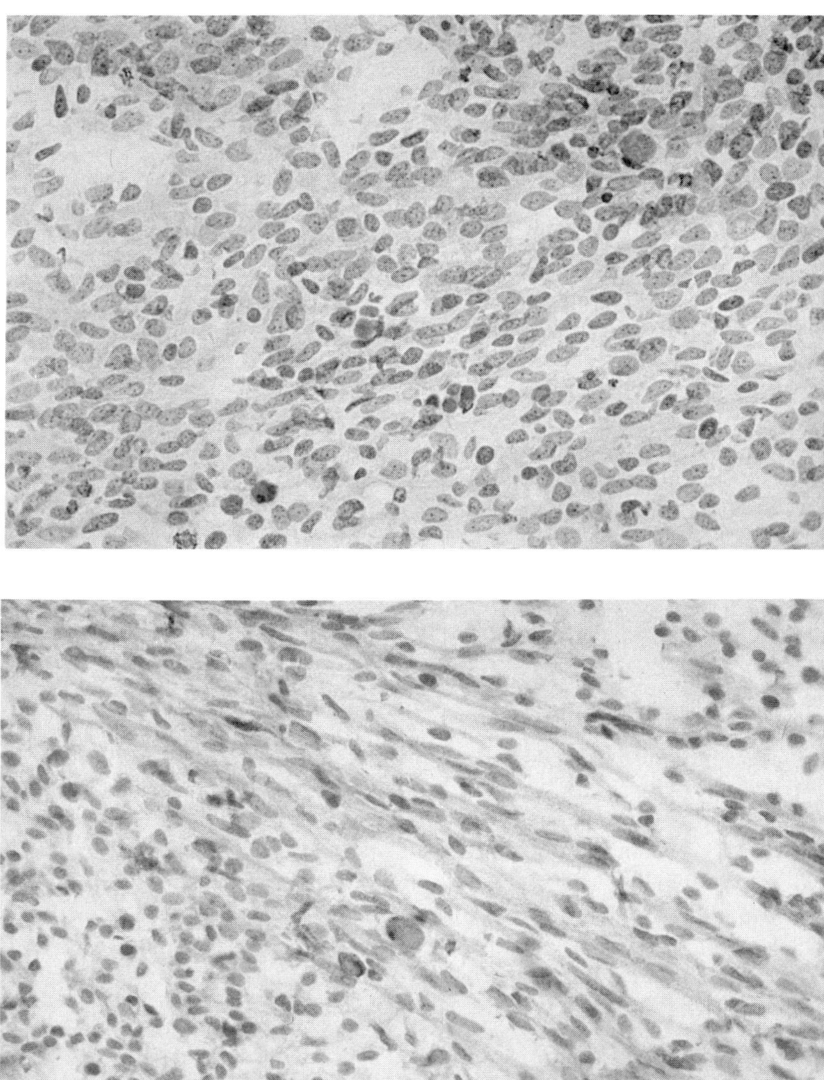

**Fig. 19.6** Transglutaminase expression in apoptotic cells in chick limb bud. Oblique (*a*) and longitudinal (*b*) cryostat sections from the metatarsophalangeal region of 7.5-day chick embryos were paraformaldehyde fixed, briefly trypsinized and reacted with a purified anti-tissue transglutaminase antibody. Immune reactive cells were detected by ABC-peroxidase. The field shows several apoptotic myoblasts in the differentiating muscle strongly reactive with the anti-transglutaminase antibody. The slightly immunopositive ribbons are normal myoblasts. ($\times$ 325).

the interdigital webs is closely linked to retinoid action (Sulik and Dehart 1988; Alles and Sulik 1989). Transglutaminases, on the other hand, have frequently been shown to be both induced and activated in cells undergoing apoptosis (Fesus and Thomazy 1988; Fesus *et al.* 1989; Piacentini *et al.* 1991*a,b*). In some of these instances, such as neuroblastoma cells and adenocarcinoma cells, these effects have been directly linked to the action of RA (Piacentini *et al.* 1991*b*). We believe that the induction and activation of tissue transglutaminase is an integral part of the apoptotic programme of many cell types and RA-induced expression of the enzyme is one example of how an apoptotic factor may increase expression of the enzyme. Figure 19.7

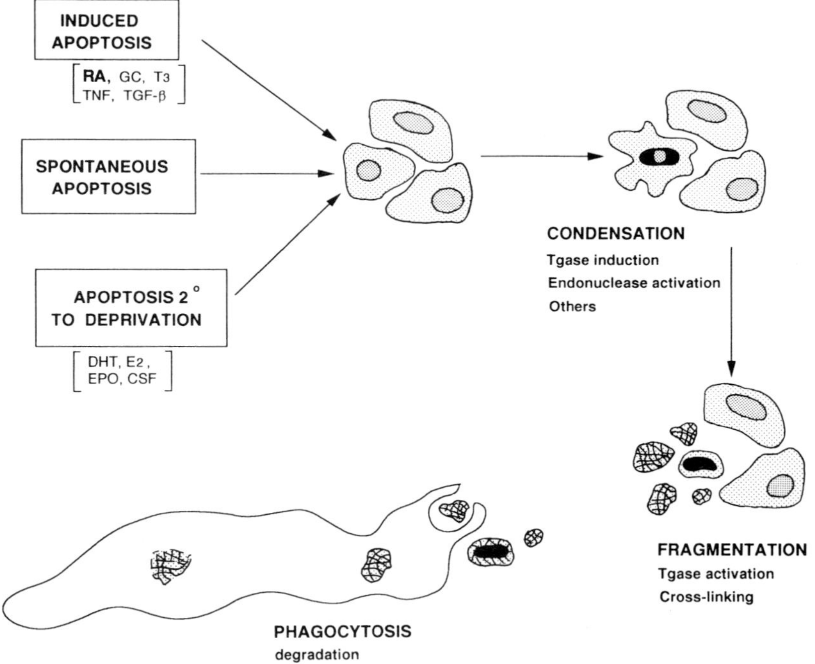


**Fig. 19.7** A simplified model of the role of tissue transglutaminase in apoptotic cell death.

presents a speculative model on the links between retinoid action and transglutaminase expression in apoptotic cells. The induction of apoptosis seems to be a multifactorial process (for recent reviews see (Arends and Wyllie 1991; Fesus *et al.* 1991). In some cells it may be a spontaneous event, genetically programmed into the developmental pathway of that cell lineage, in others it appears to be secondary to deprivation of a cell of critical trophic stimuli, e.g. androgen deprivation of the prostate. A third form of cell death is the apoptosis that is actively induced in susceptible cells following exposure to hormones, such as retinoids or TNF, that can function as specific apoptotic factors.

One of the earliest events in the apoptotic pathway is the condensation of the apoptotic cell. The cytoplasm becomes concentrated, there is condensation of chromatin and activation of specific intranuclear endonucleases. Several studies have demonstrated that the expression of tissue transglutaminase is also induced at this stage of apoptosis and the enzyme accumulates specifically in the cytoplasm of the apoptotic cell (Fesus and Thomazy 1988; Fesus *et al.* 1989; Piacentini *et al.* 1991*a*, *b*). Condensation of the apoptotic cell is rapidly followed by fragmentation. The intracellular transglutaminase is activated at this stage (possibly by increased intracellular $Ca^{2+}$) and there is extensive protein cross-linking. Fesus and his co-workers have shown that these cellular fragments (apoptotic bodies) contain highly cross-linked 'envelope-like' structures analogous to the envelopes that form in cornifying keratinocytes (Fesus *et al.* 1989). The apoptotic bodies are recognized by adjacent phagocytic cells, then ingested and completely degraded in phagolysosomes. Although we do not know the exact role of transglutaminase-mediated cross-linking in this process, it is likely that the cross-linking of intracellular proteins stabilizes the apoptotic body and prevents the leakage of intracellular proteins into the extracellular space. In addition, cross-linking of surface proteins may be associated with increased recognition of the apoptotic bodies by phagocytic cells.

CONCLUSION

Our studies suggest that RA can directly regulate the expression of the tissue transglutaminase gene. The retinoid responsiveness of these cells in turn is linked to multiple factors; it appears to be enhanced by agents that elevate intracellular cyclic AMP and suppressed by factors that activate protein kinase C. The induction of tissue transglutaminase, which appears to be mediated by RA receptors, is linked to the ability of retinoids to induce apoptosis. A detailed understanding of the factors that regulate transglutaminase expression should shed new light on the mechanisms involved in retinoid-induced cell death in both normal and transformed cells, and in embryos.

ACKNOWLEDGEMENTS

Much of the research reported in this review was supported by a grant DK27078 from the National Institutes of Health. The authors would like to express their appreciation to Ms N. Alban-Reeves and M. Sobieski for their excellent technical assistance and Ms J. Jennings for her assistance in preparation of the manuscript.


## REFERENCES

Abbott, B. D. and Pratt, R. M. (1987). Retinoids and epidermal growth factor alter embryonic mouse palatal epithelial and mesenchymal cell differentiation in organ culture. *Journal of Craniofacial Genetics and Developmental Biology*, **7**, 219–40.

Alles, A. J. and Sulik, K. K. (1989). Retinoic-acid-induced limb-reduction defects: perturbation of zones of programmed cell death as a pathogenetic mechanism. *Teratology*, **40**, 163–71.

Alles, A. J. and Sulik, K. K. (1990). Retinoic acid-induced spina bifida: evidence for a pathogenetic mechanism. *Development*, **108**, 73–81.

Arends, M. J. and Wyllie, A. H. (1991). Apoptosis. Mechanisms and role in pathology. *International Review of Experimental Pathology*, **32**, 223–54.

Birckbichler, P. J. and Patterson, M. K. (1978). Cellular transglutaminase, growth and transformation. *Annals of the New York Academy of Science*, **312**, 354–65.

Brand, N., Petkovich, M., Krust, A., Chambon, P., de Thé, H., Marchio, A., Tiollais, P., and Dejean, A. (1988). Identification of a second human retinoic acid receptor. *Nature*, **332**, 850–3.

Cai, D., Ben, T., and De Luca, L. M. (1991). Retinoids Induce Tissue Transglutaminase in NIH-3T3 Cells. *Biochemical and Biophysical Research Communications*, **175**, 1119–24.

Ceru, M. P., Piacentini, M., Piredda, L., Farace, M. G., Autuori, F., and Fesus, L. (1989). Modulation of transglutaminase by retinoic acid in liver cells. *Italian Journal of Biochemistry*, **38**, 278A–280A.

Chiocca, E. A., Davies, P. J., and Stein, J. P. (1988). The molecular basis of retinoic acid action. Transcriptional regulation of tissue transglutaminase gene expression in macrophages. *Journal of Biological Chemistry*, **263**, 11584–9.

Chiocca, E. A., Davies, P. J., and Stein, J. P. (1989). Regulation of tissue transglutaminase gene expression as a molecular model for retinoid effects on proliferation and differentiation. *Journal of Cellular Biochemistry*, **39**, 293–304.

Davies, P. J. A., Murtaugh, M. P., Moore, W. T., Johnson, G. S., and Lucas, D. (1985). Retinoic acid-induced expression of tissue transglutaminase in human promyelocytic leukemia (HL-60) cells. *Journal of Biological Chemistry*, **260**, 5166–74.

Davies, P. J. A., Chiocca, E. A., Basilion, J. P., Poddar, S., and Stein, J. P. (1988). Transglutaminase and their regulation: implications for polyamine metabolism. In *Progress in polyamine research*. (eds, V. Zappia and E. A. Pegg), pp. 399–401. Plenum Publishing Corporation, New York.

Denning, M. F. and Verma, A. K. (1991). Involvement of Retinoic Acid Nuclear Receptors in Retinoic Acid-induced Tissue Transglutaminase Gene Expression in Rat Tracheal 2C5 Cells. *Biochemical and Biophysical Research Communications*, **175**, 344–50.

Dollé, P., Ruberte, E., Kastner, P., Petkovich, M., Stoner, C. M., Gudas, L. J., and Chambon, P. (1989). Differential expression of genes encoding alpha, beta and gamma retinoic acid receptors and CRABP in the developing limbs of the mouse. *Nature*, **342**, 702–5.

Evans, R. M. (1988). The steroid and thyroid hormone receptor superfamily. *Science*, **240**, 889–95.

Fesus, L., Metsis, M. L., Muszbek, L., and Koteliansky, V. E. (1986). Transglutaminase-sensitive glutamine residues of human plasma fibronectin revealed by studying its proteolytic fragments. *European Journal of Biochemistry*, **154**, 371–4.

Fesus, L., and Thomazy, V. (1988). Searching for the function of tissue transglutaminase: its possible involvement in the biochemical pathway of programmed cell death. *Advances in Experimental Biology and Medicine* **231**, 119–34.

Fesus, L., Thomazy, V., Autuori, F., Ceru, M. P., Tarcsa, E., and Piacentini, M. (1989). Apoptotic hepatocytes become insoluble in detergents and chaotropic agents as a result of transglutaminase action. *Febs Letters*, **245**, 150–4.

Fesus, L., Davies, P. J. A., and Piacentini, M. (1991). Apoptosis: Molecular Mechanisms in Programmed Cell Death. *European Journal of Cell Biology*, In Press.

Fink, M. L., Chung, S. I., and Folk, J. E. (1980). Gamma-glutylamine cyclotransferase: Specificity toward epsilon- (L-gamma-glutamyl)-L-lysine and related compounds. *Proceedings of the National Academy of Science USA*, **77**, 4564–8.

Folk, J. E. (1983). Transglutaminases. *Advances in Enzymology*, **54**, 1–56.

Gentile, V., Saydak, M., Chiocca, E. A., Akande, O., Birckbichler, P. J., Lee, K. N., Stein, J. P., and Davies, P. J. (1991). Isolation and characterization of cDNA clones to mouse macrophage and human endothelial cell tissue transglutaminases. *Journal of Biological Chemistry*, **266**, 478–83.

Giguère, V., Ong, E. S., Segui, P., and Evans, R. M. (1987). Identification of a receptor for the morphogen retinoic acid. *Nature*, **330**, 624–9.

Greenberg, C. S., Achyuthan, K. E., Borowitz, M. J., and Shuman, M. A. (1987). The transglutaminase in vascular cells and tissues could provide an alternate pathway for fibrin stabilization. *Blood*, **70**, 702–9.

Grundmann, U., Amann, E., Zettlmeissl, G., and Kupper, H. A. (1986). Characterization of cDNA coding for human factor XIIIa. *Proceedings of the National Academy of Sciences USA*, **83**, 8024–8.

Ichinose, A., Hendrickson, L. E., Fujikawa, K., and Davie, E. W. (1986). Amino acid sequence of the subunit of human factor XIII. *Biochemistry*, **25**, 6900–6.

Ikura, K., Nasu, T., Yokota, H., Tsuchiya, Y., Sasaki, R., and Chiba, H. (1988). Amino acid sequence of guinea pig liver transglutaminase from its cDNA sequence. *Biochemistry*, **27**, 2898–905.

Kannagi, R., Teshigawara, K., Noro, N., and Masuda, T. (1982). Transglutaminase Activity during the Differentiation of Macrophages. *Biochemical and Biophysical Research Communications*, **105**, 164–71.

Kim, H. C., Lewis, M. S., Gorman, J. J., Park, S. C., Girard, J. E., Folk, J. E., and Chung, S. I. (1990). Protransglutaminase E from guinea pig skin. Isolation and partial characterization. *Journal of Biological Chemistry*, **265**, 21971–8.

Kim, H. C., Idler, W. W., Kim, I. G., Han, J. H., Chung, S. I., and Steinert, P. M. (1991). The complete amino acid sequence of the human transglutaminase K enzyme deduced from the nucleic acid sequences of cDNA clones. *Journal of Biological Chemistry*, **266**, 536–9.

Krust, A., Kastner, P., Petkovich, M., Zelent, A., and Chambon, P. (1989). A third human retinoic acid receptor, hRAR-gamma. *Proceedings of the National Academy of Science USA*, **86**, 5310–14.

Leu, R. W., Herriott, M. J., Moore, P. E., Orr, G. R., amd Birkbichler, P. J. (1982).

Enhanced Transglutaminase Activity Associated with Macrophage Activation. *Experimental Cell Research*, **141**, 191–9.

Lichti, U. and Yuspa, S. H. (1988). Modulation of tissue and epidermal transglutaminases in mouse epidermal cells after treatment with 12-O-tetradecanoylphorbol-13-acetate and/or retinoic acid in vivo and in culture. *Cancer Research*, **48**, 74–81.

Lorand, L. and Conrad, S. M. (1984). Transglutaminases. *Molecular and Cellular Biochemistry*, **58**, 9–35.

Martin, S. J., Bradley, J. G., and Cotter, T. G. (1990). HL-60 cells induced to differentiate towards neutrophils subsequently die via apoptosis. *Clinical and Experimental Immunology*, **79**, 448–53.

Martinez, J., Rich, E., and Barsigian, C. (1989). Transglutaminase-mediated cross-linking of fibrinogen by human umbilical vein endothelial cells. *Journal of Biological Chemistry*, **264**, 20502–8.

Mehrel, T., Hohl, D., Rothnagel, J. A., Longley, M. A., Bundman, D., Cheng, C., Lichti, U., Bisher, M. E., Steven, A. C., and Steinert, P. M. (1990). Identification of a major keratinocyte cell envelope protein, loricrin. *Cell*, **61**, 1103–12.

Mehta, K., Claringbold, P., and Lopez Berestein, G. (1987). Suppression of macrophage cytostatic activation by serum retinoids: a possible role for transglutaminase. *Journal of Immunology*, **138**, 3902–6.

Melino, G., Farrace, M. G., Ceru, M. P., and Piacentini, M. (1988). Correlation between transglutaminase activity and polyamine levels in human neuroblastoma cells. Effect of retinoic acid and alpha-difluoromethylornithine. *Experimental Cell Research*, **179**, 429–45.

Moore, W. T., Murtaugh, M. P. , and Davies, P. J. A. (1984). Retinoic Acid-induced Expression of Tissue Transglutaminase in Mouse Peritoneal Macrophages. *Journal of Biological Chemistry*, **259**, 12794–802.

Murtaugh, M. P. Mehta, K., Johnson, J., Myers, M., Juliano, R. L., and Davies, P. J. A. (1983). Induction of tissue transglutaminase in mouse peritoneal macrophages. *Journal of Biological Chemistry*, **258**, 11074–81.

Murtaugh, M. P., Arend, W. P., and Davies, P. J. A. (1984). Induction of tissue transglutaminase in human peripheral blood monocytes. *Journal of Experimental Medicine*, **159**, 114–25.

Murtaugh, M. P., Moore, W. T. Jr., and Davies, P. J. (1986). Cyclic AMP potentiates the retinoic acid-induced expression of tissue transglutaminase in peritoneal macrophages. *Journal of Biological Chemistry*, **261**, 614–21.

Nara, K., Nakanishi, K., Hagiwara, H., Wakita, K., Kojima, S., and Hirose, S. (1989). Retinol-induced morphological changes of cultured bovine endothelial cells are accompanied by a marked increase in transglutaminase. *Journal of Biological Chemistry*, **264**, 19308–12.

Parenteau, N. L., Pilato, A., and Rice, R. H. (1986). Induction of keratinocyte type-I transglutaminase in epithelial cells of the rat. *Differentiation*, **33**, 130–41.

Petkovich, M., Brand, N. J., Krust, A., and Chambon, P. (1987). A human retinoic acid receptor which belongs to the family of nuclear receptors. *Nature*, **330**, 444–50.

Phillips, M. A., Stewart, B. E., Qin, Q., Chakravarty, R., Floyd, E. E., Jetten, A. M., and Rice, R. H. (1990). Primary structure of keratinocyte transglutaminase. *Proceedings of the National Academy of Sciences of the USA*, **87**, 9333–7.

Piacentini, M., Fesus, L., Sartori, C., and Ceru, M. P. (1988). Retinoic acid-induced

262    *Peter J. A. Davies* et al.

modulation of rat liver transglutaminase and polyamines in vivo. *Biochemical Journal*, **253**, 33–8.

Piacentini, M., Autuori, F., Dini, L., Farrace, M. G., Ghibelli, L., Piredda, L., and Fesus, L. (1991*a*). 'Tissue' transglutaminase is specifically expressed in neonatal rat liver cells undergoing apoptosis upon epidermal growth factor-stimulation. *Cell and Tissue Research*, **263**, 227–35.

Piacentini, M., Fesus, L., Farrace, M. G., Ghibelli, L., Pirreda, L., and Melino, G. (1991*b*). The Expression of Tissue Transglutaminase in Two Human Cancer Cell Lines is Related to Programmed Cell Death (Apoptosis). *European Journal of Cell Biology*, **54**, 246–54.

Porta, R., Esposito, C., Metafora, S., Malorni, A., Pucci, P., Siciliano, R., and Marino, G. (1991). Mass spectrometric identification of the amino donor and acceptor sites in a transglutaminase protein substrate secreted from rat seminal vesicles. *Biochemistry*, **30**, 3114–20.

Rice, R. H. and Green, H. (1978). Relation of protein synthesis and transglutaminase activity to formation of the cross-linked envelope during terminal differentiation of cultured human epidermal keratinocytes. *Journal of Cell Biology*, **76**, 705–11.

Rice, R. H. and Green, H. (1979). Presence in human epidermal cells of a soluble protein precursor of the cross-linked envelope: activation of the cross-linking by calcium ions. *Cell*, **18**, 681–94.

Roth, W. J., Chung, S. I., Raju, L., and Janoff, A. (1988). Macrophage transglutaminases: characterization of molecular species and measurement of enzymatic modification by cigarette smoke components. *Advances in Experimental Biology and Medicine*, **231**, 161–73.

Ruberte, E., Dollé, P., Krust, A., Zelent, A., Morriss-Kay, G., and Chambon, P. (1990). Specific spatial and temporal distribution of retinoic acid receptor gamma transcripts during mouse embryogenesis. *Development*, **108**, 213–22.

Sacks, P. G., Oke, V., Calkins, D. P., Vasey, T., and Terry, N. H. (1990). Effects of beta-all-trans retinoic acid on growth, proliferation, and cell death in a multicellular tumor spheroid model for squamous carcinomas. *Journal of Cell Physiology*, **144**, 237–43.

Sane, D. C., Moser, T. L., and Greenberg, C. S. (1991). Vitronectin in the substratum of endothelial cells is cross-linked and phosphorylated. *Biochemical and Biophysical Research Communications*, **174**, 465–9.

Schroff, G., Neumann, C., and Sorg, C. (1981). Transglutaminase as a Marker for Subsets of Murine Macrophages. *European Journal of Immunology*, **11**, 637–42.

Schwartz, M. L., Pizzo, S. V., Hill, R. L., and McKee, P. A. (1973). Human factor XIII from plasma and platelets. Molecular weight, subunit structures, proteolytic activation and cross-linking of fibrinogen and fibrin. *Journal of Biological Chemistry*, **248**, 1395–1407.

Simon, M. and Green, H. (1988). The glutamine residues reactive in transglutaminase-catalyzed cross-linking of involucrin. *Journal of Biological Chemistry*, **263**, 18093–8.

Sulik, K. K., Cook, C. S., and Webster, W. S. (1988). Teratogens and craniofacial malformations: relationships to cell death. **103** (Suppl.), *Development* 213–31.

Sulik, K. K. and Dehart, D. B. (1988). Retinoic-acid-induced limb malformations resulting from apical ectodermal ridge cell death. *Teratology*, **37**, 527–37.

Takahashi, N., Takahashi, Y., and Putnam, F. W. (1986). Primary structure of blood coagulation factor XIIIa (fibrinoligase, transglutaminase) from human placenta. *Proceedings of the National Academy of Science USA*, **83**, 8019–23.

Yasuda, Y., Konishi, H., Matsuo, T., Kihara, T., and Tanimura, T. (1989). Aberrant differentiation of neuroepithelial cells in developing mouse brains subsequent to retinoic acid exposure in utero. *American Journal of Anatomy*, **186**, 271–84.

Zettergren, J. G., Peterson, L. L., and Wuepper, K. D. (1984). Keratolinin: The soluble substrate of epidermal transglutaminase from human and bovine tissue. *Proceedings of the National Academy of Science USA*, **81**, 238–42.

# PART V

# Teratogenesis of 13-*cis*-retinoic acid

20

# Pharmacokinetics, placental transfer, and teratogenicity of 13-*cis*-retinoic acid, its isomer and metabolites

Joan Creech Kraft

## INTRODUCTION

Retinoids, which normally have important physiological functions, are also used in clinical medicine. 13-*cis*-retinoic acid (Accutane, isotretinoin, 13-*cis*-RA) and all-*trans*-retinoic acid (tretinoin, Retin A, all -*trans*-RA) are effective drugs for the treatment of dermatological disorders. Unfortunately, the high teratogenicity of the retinoids is a major drawback to their broad use. A characteristic pattern of congenital cardiac and central nervous system problems was found in human infants who had been exposed to 13-*cis*-RA during the first trimester (Rosa 1983; Braun *et al.* 1984; Lammer *et al.* 1985). This response was elicited by relatively low, clinically therapeutic doses of 0.5–1.5 mg/kg/day. Although 13-*cis*-RA is a potent human teratogen, it is only marginally teratogenic in the mouse at doses 100 times higher than those used in human therapy (Kochhar *et al.* 1984). The all-*trans*-RA isomer of 13-*cis*- RA is a potent teratogen in all laboratory species tested (Willhite *et al.* 1989).

## *IN VIVO* PHARMACOKINETICS AND PLACENTAL TRANSFER OF 13-CIS- AND ALL-TRANS-RA IN THE MOUSE

It has been shown that the low teratogenicity of 13-*cis*-RA in the mouse corresponds to low embryo concentrations during organogenesis (Creech Kraft *et al.* 1987). A single oral dose of 100 mg all-*trans*-RA or 13-*cis*-RA per kg body weight was administered to mice (NMRI: Han, Zentralinstitut für Versuchstierkunde, Hannover, FRG) at two sensitive stages of organogenesis, i.e. day 9 or day 11 of gestation. All-*trans*-RA (100 mg/kg) caused 90 per cent total defects and 13-*cis*-RA (100 mg/kg), only 3 per cent.

The concentrations of the parent drugs, their respective isomers, as well as their 4-oxo-metabolites (Fig. 20.1) were measured, using high performance liquid chromatography (HPLC) in maternal plasma and in the embryo (Creech Kraft *et al.* 1988). All-*trans*-RA and 4 oxo-all-*trans*-RA were transferred to the embryo to a much greater extent (embryo/maternal plasma concentration ratios, 0.4) (Fig. 20.2(a)) than the 13-*cis*-RA and its 4-oxo metabolite (embryo/maternal plasma concentration ratios, 0.03) (Fig. 20.2(b)). Embryo concentrations of all-*trans*-RA on day 9 of gestation exceeded those on day 11 and were in the micromolar range, whereas the embryo levels of 13-*cis*-RA were in the nanomolar range at both gestational stages.

Similar studies were done at 10-fold lower doses of 13-*cis*-RA, all-*trans*-RA and the 4-oxo derivatives to elucidate the metabolism, pharmacokinetics, placental transfer, and teratogenicity of each individual compound (Creech Kraft *et al.* 1989*a*). It was shown that all-*trans*-RA and 4-oxo-all-*trans*-RA were extremely teratogenic, whereas their corresponding *cis* isomers caused only 2 per cent cleft palate. Exposure of the embryo to the *trans* isomers (Figs 20.3 and 20.5) was higher than to the *cis* isomers (Figs 20.4 and 20.6), as was shown by the far higher embryonic peak concentrations of *trans* isomers and by the corresponding 30-fold higher areas under the concentration time curve (AUC). At 8 hours, embryo/maternal plasma ratios were higher than 1 after the administration of the all-*trans* compounds. Concentrations were measured in the placenta and yolk sac and found to be higher for the *trans* forms than for the *cis* forms. At this time, a model of facilitated transport was proposed, supported by the existence of binding proteins that have a high affinity for the *trans* forms of the retinoids but not for the *cis* forms, and it was suggested that the conversion to the *trans* isomer and *trans* metabolite could play a major role in the teratogenicity of 13-*cis*-RA in humans.

In NMRI mice, three 100 mg doses of 13-*cis*-RA/kg, 4 hours apart, caused 84 per cent cleft palate and 50 per cent limb defects (Creech Kraft *et al.* 1991*a*). Multiple high doses of 13-*cis*-RA enhanced its teratogenicity. HPLC measurements showed that the major plasma metabolite of 13-*cis*-RA accumulating in the mouse was 13-*cis*-retinoyl-*β*-glucuronide (13-*cis*-RAG) and not 4-oxo-13-*cis*-RA as in the monkey and human (Brazzel *et al.* 1983; Creech Kraft *et al.* 1991*b*). 13-*cis*-RAG, 13-*cis*- and all-*trans*-RA each were found in the mouse embryo and placenta (Fig. 20.7) after the first and third doses of 13-*cis*-RA at the time points measured. The embryo/maternal plasma ratios for 13-*cis*-RAG were low, indicating that its transfer to the embryo was poor. It was concluded that any of the three retinoids or combinations thereof could have been involved in the induction of malformations. However, all-*trans*-RA, a known potent teratogen detected in the embryo at concentrations between 590 and 80 ng/g for 10 critical hours of gestation, seemed the likely proximate dysmorphogen.


*IN VIVO* PLASMA PHARMACOKINETICS OF 13-CIS- AND ALL-TRANS-RA IN CYNOMOLGUS MONKEY

As the mouse may not be an appropriate model for the human, the metabolism and distribution of 13-*cis*- and all-*trans*-RA in the non-human primate was

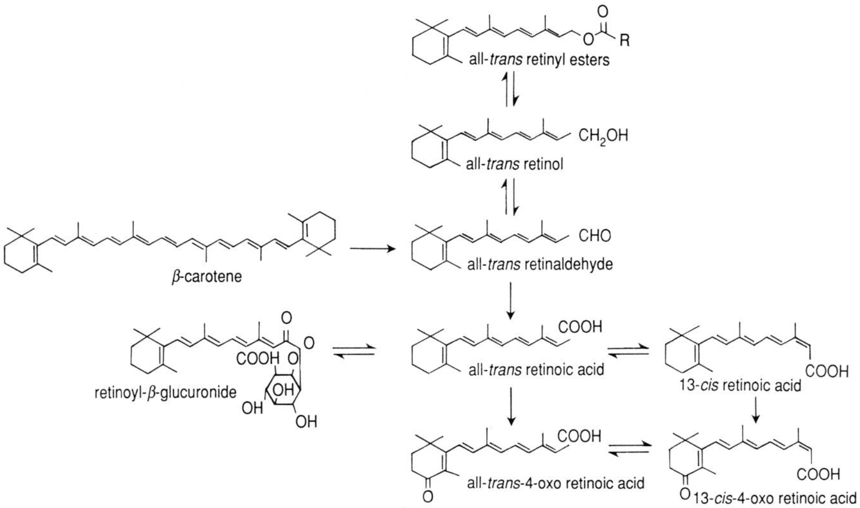

**Fig. 20.1** Metabolic pathways of $\beta$-carotene, all-*trans*-retinol and all-*trans*-RA.

investigated (Creech Kraft *et al.* 1991*b*). Plasma concentrations of parent compounds and metabolites were determined in the cynomolgus monkey on day 1 and day 10 after oral dosing of 2 and 10 mg 13-*cis*- or all-*trans*-RA/kg/day. Comparisons of the two doses revealed that the AUCs were proportional to the dose administered. After the administration of all-*trans*-RA the main metabolite was all-*trans*-RAG; after the administration of 13-*cis*-RA, the main metabolite was 4-oxo-13-*cis*-RA. Evaluation of these studies (Figs 20.8 and 20.9) suggested that there was increased metabolism (involving oxidation, isomerization, and glucuronidation of the retinoic acids) on day 10 compared to day 1. After 10 days of treatment, the parent compounds or metabolites did not accumulate and a trend for rapid elimination of the parent compounds was observed.


HUMAN SERUM CONCENTRATIONS DURING THERAPY

Serum retinoid concentrations of one individual obtained during treatment with 13-*cis*-RA were investigated using our HPLC method to obtain a direct comparison with our comprehensive studies in the mouse and non-human primate (Creech Kraft *et al.* 1991*b*). These data demonstrated that, by day 27 of therapy (0.7 mg/kg/day), the parent compound and metabolites had reached a steady state (Fig. 20.10). One metabolite (4-oxo-13-*cis*-RA) was at concentrations higher than the parent compound. The novelty of this study was that both glucuronides were detected in human serum, as well as the two

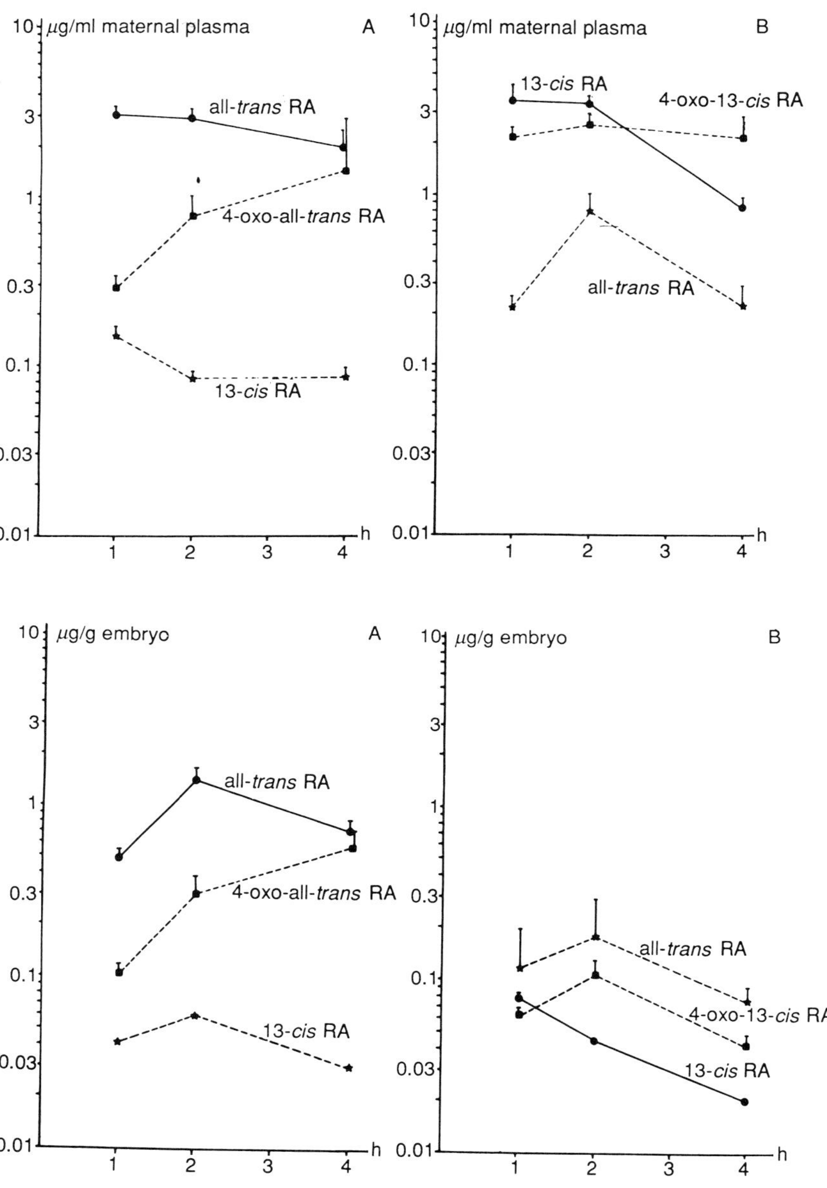



**Fig. 20.2** (*a*) Concentrations of all-*trans*-RA and its transformation products in maternal plasma (top) and embryo (bottom) after NMRI mice were orally dosed with 100 mg/kg all-*trans*-RA on day 11 of gestation (plug day = day 0). (*b*) Concentrations of 13-*cis*-RA and its transformation products in maternal plasma (top) and embryo (bottom) after NMRI mice were orally dosed with 100 mg/kg 13-*cis*-RA on day 11 of gestation (plug day = day 0).

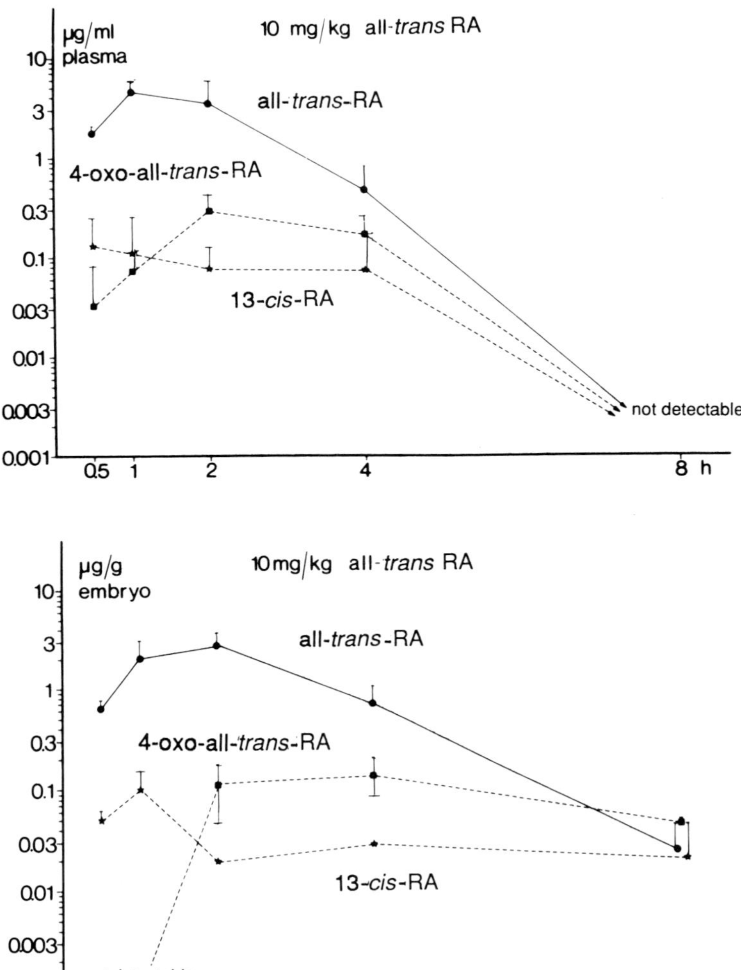


**Fig. 20.3** Concentrations of all-*trans*-RA and its transformation products in maternal plasma (top) and embryo (bottom) after NMRI mice were orally dosed with 10 mg/kg all-*trans*-RA on day 11 of gestation (plug day = day 0).

potent teratogens, all-*trans*-RA and 4-oxo-all-*trans*-RA. Approximately a 10-fold greater dose of 13-*cis*-RA was required in the monkey to produce AUC values comparable to the human values. The human serum concentrations 6 days after discontinuation of treatment with 13-*cis*-RA were: 4-oxo-13-*cis*-RA, 50 ng/ml; 4-oxo-all-*trans*-RA, 5 ng/ml; 13-*cis*-RA, 6 ng/ml and all-*trans*-RA, 1.3 ng/ml. Assuming a first order of elimination the estimated half-lives

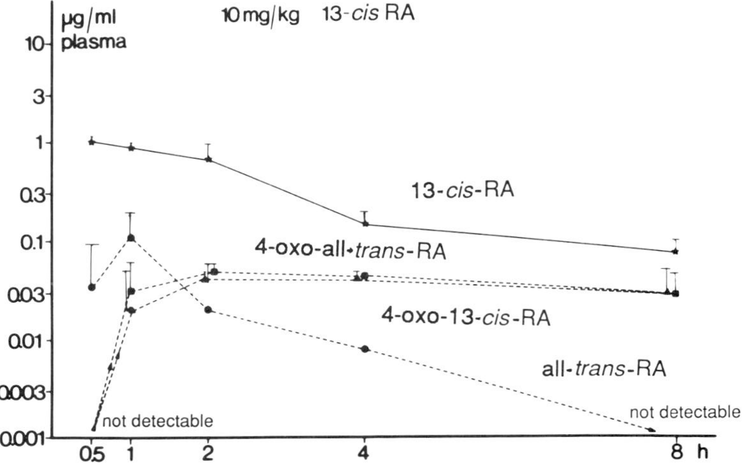


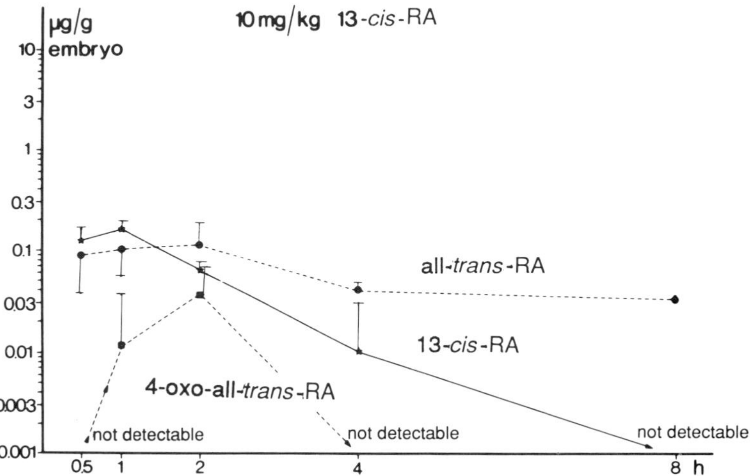


**Fig. 20.4** Concentrations of 13-*cis*-RA and its transformation products in maternal plasma (top) and embryo (bottom) after NMRI mice were orally dosed with 10 mg/kg 13-*cis*-RA on day 11 of gestation (plug day = day 0).

are 29 hours for 4-oxo-all-*trans*-RA, 52 hours for 4-oxo-13-*cis*-RA, 30 hours for 13-*cis*-RA and 34 hours for all-*trans*-RA. In another human case study (Creech Kraft *et al.* 1989*b*), we had the opportunity to study the disposition of 13-*cis*-RA in embryonic tissues obtained from a pregnancy termination of a woman who unintentionally took 40 mg/day 13-*cis*-RA from days 8–28 of gestation. Interestingly, in the embryo, all-*trans*-RA concentrations were

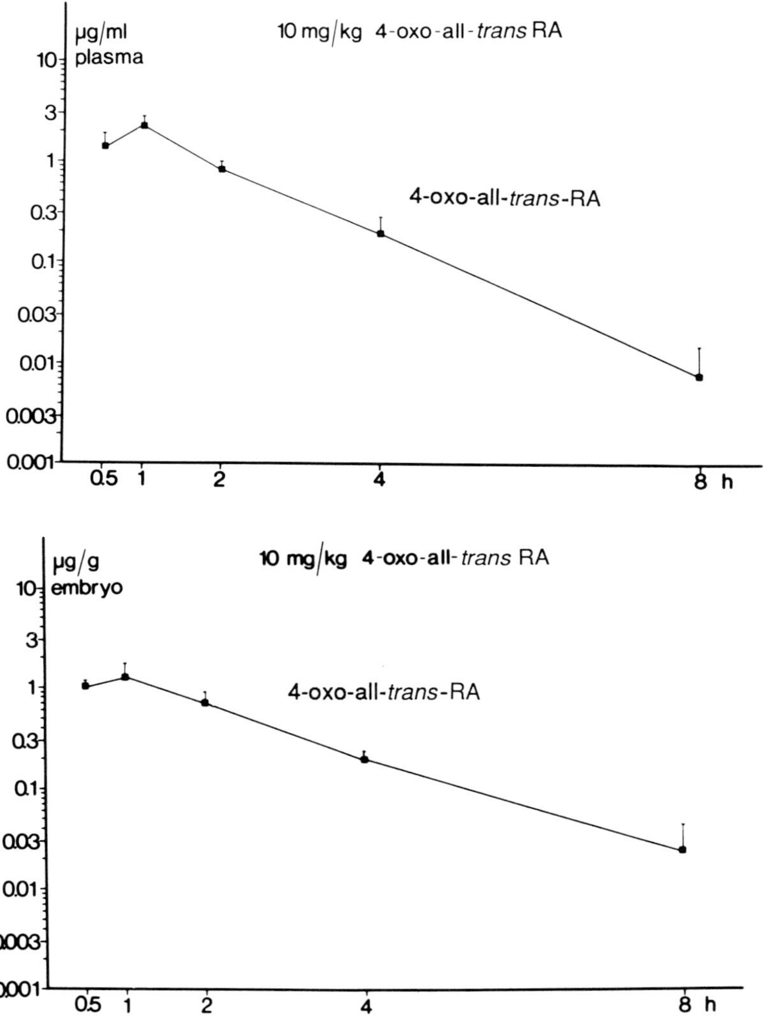


**Fig. 20.5** Concentrations of 4-oxo-all-*trans*-RA in maternal plasma (top) and embryo (bottom) after NMRI mice were orally dosed with 10 mg/kg 4-oxo-all-*trans*-RA on day 11 of gestation (plug day = day 0).

more than twice those for 13-*cis*-RA and were in the micromolar range. Until now there have been no measurements of the concentrations reached in the non-human primate after treatment with 13-*cis*-RA, but, it was recently reported that 13-*cis*-RA was teratogenic in the cynomolgus monkey when 2.5 mg/kg/day were administered during early gestation (Hummler *et al.* 1990).

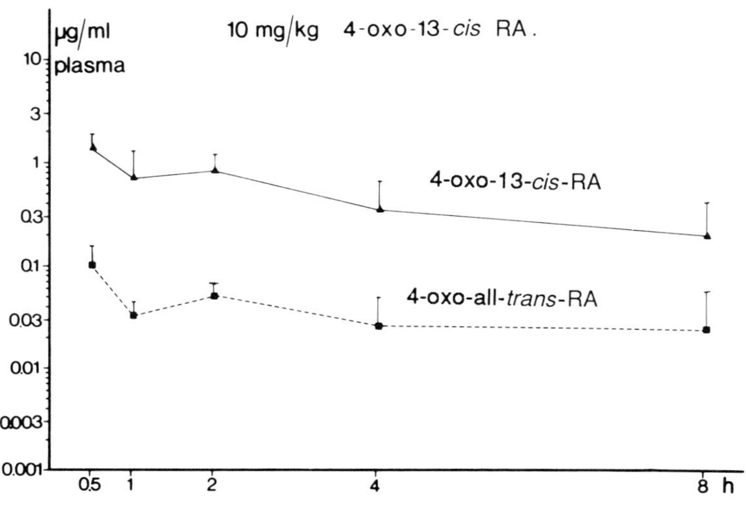



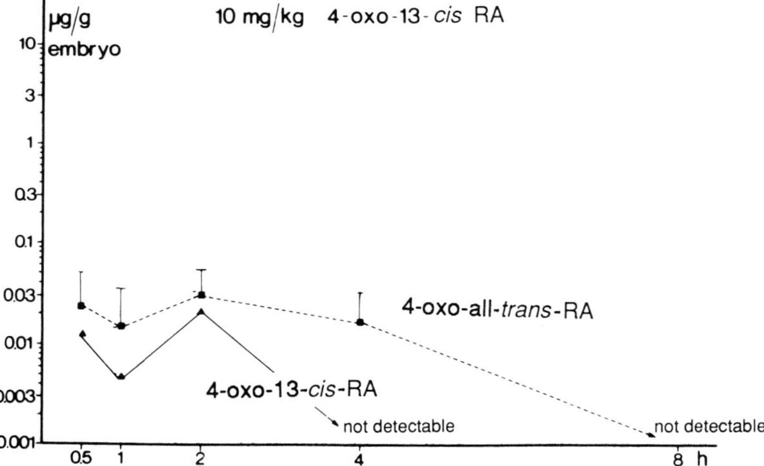



**Fig. 20.6** Concentrations of 4-oxo-13-*cis*-RA and its isomer in maternal plasma (top) and embryo (bottom) after NMRI mice were orally dosed with 10 mg/kg 4-oxo-13-*cis*-RA on day 11 of gestation (plug day = day 0).

## *IN VITRO* EXPERIMENTS USING RAT WHOLE EMBRYO CULTURE

It is obvious that interpretation concerning the teratogenic potency of individual compounds is complicated in testing *in vivo* where maternal metabolism and placental transfer play such important roles. Experiments *in vitro* have the advantage that the compounds can be directly added to the

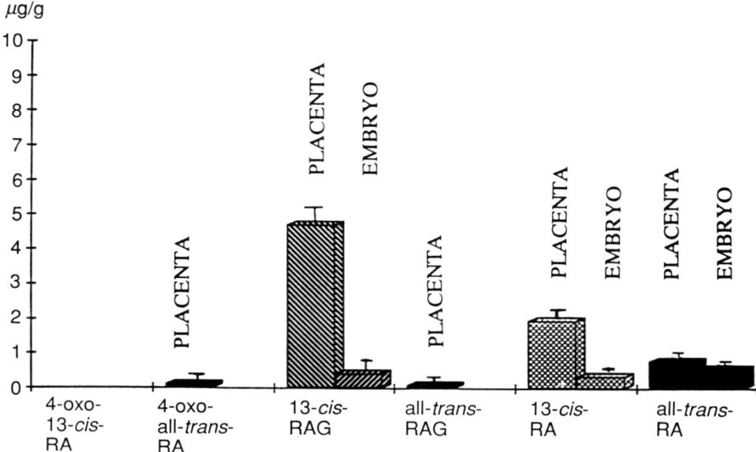


**Fig. 20.7** Concentrations reached in placenta and embryo of NMRI mice, 2 hours after the administration of the first of three oral doses of 13-*cis*-RA (100 mg/kg) on day 11 of gestation.

culture medium and can reveal the direct effects of the agent on the conceptus without interactions with maternal metabolism. *In vitro* experiments with rat or mouse whole-embryo culture showed that both isomers are teratogenic (Goulding and Pratt 1986; Steele *et al.* 1987). Interestingly, the observed pattern of malformations *in vitro* was similar to the malformations seen *in vivo*, and both could be related to the effects of 13-*cis*-RA in humans (Morriss 1972; Morriss and Steele 1977; Webster *et al.* 1986). An answer to the discrepancies in the teratogenic potency between the findings *in vivo* and *in vitro* and even species differences has not yet been found. All previous *in vitro* studies neglected the fact that there is a spontaneous isomerization from *cis* to *trans* and vice versa. This isomerization may be important for the interpretation of the dose–response relationship established with these two substances.

*IN VITRO* EXPERIMENTS ADDING THE COMPOUNDS TO THE CULTURE MEDIUM

In 1989, Klug *et al.* performed an *in vitro* study setting up an extensive dose–relationship for the isomers 13-*cis*-RA and all-*trans*-RA, and verified the role of isomerization of these compounds using measurements with HPLC. *In vitro* experiments using whole rat embryo showed that all-*trans*-RA administered at low concentrations (30 mg/ml culture medium) is 10 times more active than 13-*cis*-RA and 3 times more active when administered at high concentrations (1000 mg/ml culture medium). Morphological investigation of

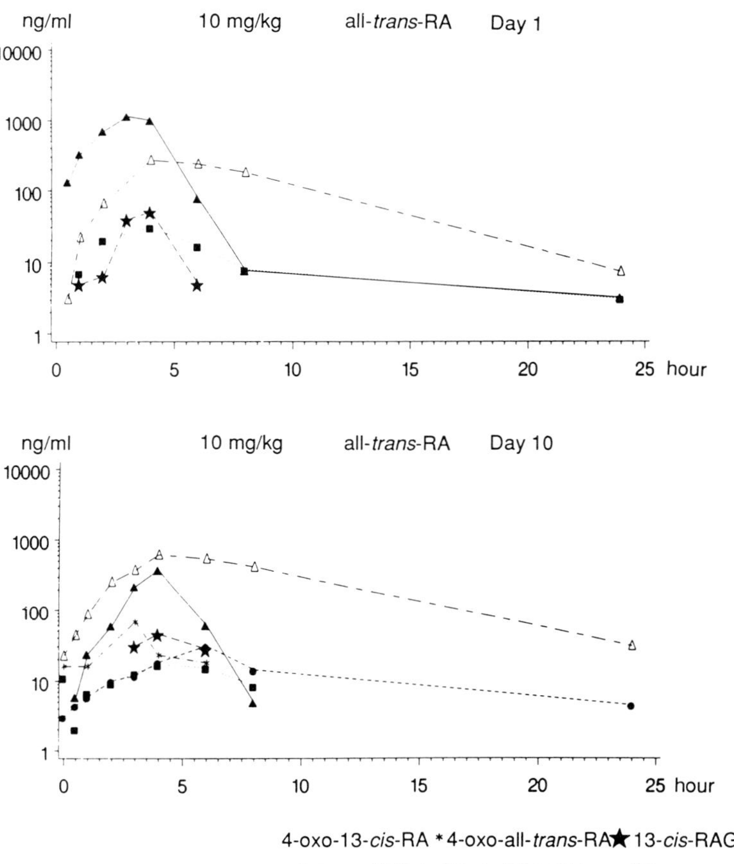



**Fig. 20.8** The time course of concentrations (median) of all-*trans*-RA and its metabolites on day 1 (top) and day 10 (bottom) in four nonpregnant monkeys that had been administered 10 mg all *trans*-RA/kg/day for 10 days.

the embryos showed that both substances directly influenced embryonic development in an identical manner. Isomerization products of the administered compounds (all-*trans*-RA from 13-*cis*-RA and vice versa) were detected by HPLC both in the culture medium and in the embryo. Correlation of the embryonic retinoid concentration with the observed effects led us to suggest that the isomerization to all-*trans*-RA is crucial in regard to the 13-*cis*-RA induced abnormal development. A 100 per cent effect was induced *in vitro* with very low amounts of all-*trans*-RA (7.2 ng/g) in the embryo. These findings were in good agreement with the research in the field of molecular biology concerning RA's role as morphogen in which the gradient across the chick limb bud was 20 to 50 nanomolar (Thaller and Eichele 1987).

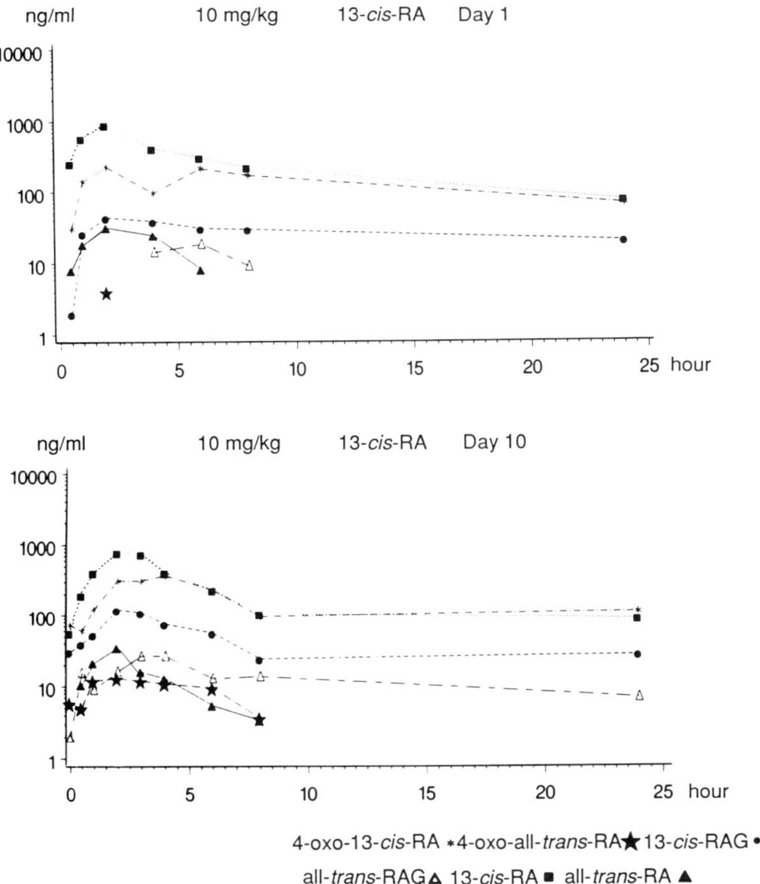



**Fig. 20.9** The time course of concentrations (median) of 13-*cis*-RA and its metabolites on day 1 (top) and day 10 (bottom) in four nonpregnant monkeys that had been administered 10 mg 13-*cis*-RA kg/day for 10 days.

### INTRA-AMNIOTIC MICROINJECTION OF RETINOIDS IN RAT WHOLE EMBRYO CULTURE

Intra-amniotic microinjection is now being used to elucidate the intrinsic dysmorphogenic potential of several retinoids. Retinol, all-*trans*-RAG, 13-*cis* and all-*trans*-RA were microinjected intra-amniotically in rat embryos on day 10 of gestation and cultured until day 11.5. A comparison of the concentration–effect relationships of the retinoids showed qualitatively similar dysmorphogenic effects for all four retinoids. These effects were elicited by a low concentration of all-*trans*-RA (250 ng/ml amniotic fluid), a 6–7-fold higher

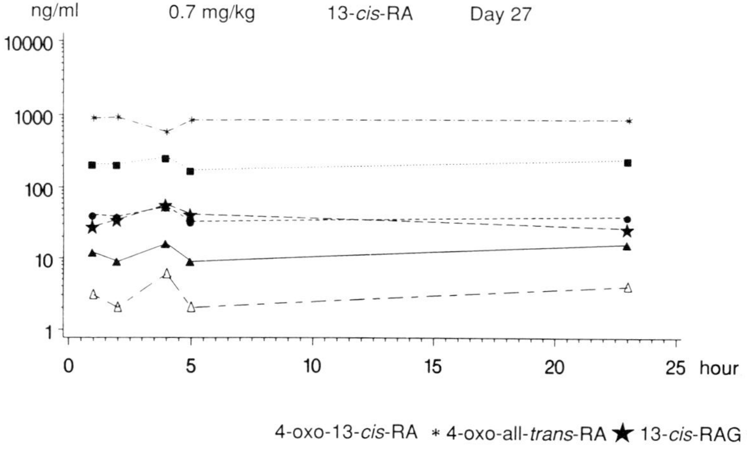



**Fig. 20.10**  Serum concentrations of 13-*cis*-RA and its metabolites in serum of a woman who was therapeutically being treated with 0.7 mg 13-*cis*-RA (isotretinoin, Accutane)/kg/day.

concentration of 13-*cis*-RA and approximately 16-fold higher concentration of retinol or all-*trans*-RAG (Lee *et al.* 1991). HPLC analyses of whole conceptuses and embryos at day 11.5 of gestation revealed quantitatively similar concentrations (in the nanomolar range) of the all-*trans*-isomer after microinjection of either 250 ng/ml all-*trans*-RA, 1500 ng/ml 13-*cis*-RA, retinol (4000 ng/ml) or all-*trans*-RAG (4000 ng/ml) (Creech Kraft *et al.* 1991*c*; Creech Kraft and Juchau 1991).

CONCLUSION

The research reported here has been primarily concerned with the question of whether the teratogenicity of 13-*cis*-RA was associated with the parent compound itself or whether it was the result of metabolic conversion to another metabolite. The original high dose studies, as well as the low dose studies in mice, showed that the *trans* isomers readily reached the embryo and were extremely teratogenic. Multiple high doses of 13-*cis*-RA in mice produced a high percentage of teratogenicity, whereby all-*trans*-RA, 13-*cis*-RA and 13-*cis*-RAG were found in the embryo. In one human case, both 13-*cis*- and all-*trans*-RA were detected in the embryo after a woman had unintentionally taken 13-*cis*-RA during pregnancy. In the human as well as cynomolgus monkey, both teratogens were present in the serum or plasma after treatment with 13-*cis*-RA. The *in vitro* studies in rat suggested that conversion to the *trans* isomer could play a major role in the 13-*cis*-RA-induced abnormal development. Comparisons of the concentration–effect relationships of the retinoids investigated with intra-amniotic microinjections showed that the dysmorphogenic effects were qualitatively similar for all-*trans*-RA, 13-*cis*-RA, all-*trans*-RAG and retinol but were elicited by a low concentration of all-*trans*-RA (250 ng/ml), a 6–7-fold higher concentration of 13-*cis*-RA and an approximately 16-fold higher concentration of retinol. This suggests that bioconversion to the more potent teratogen, all-*trans*-RA, was playing a role. The most decisive evidence that all-*trans*-RA is the ultimate teratogen is that HPLC analysis of the embryos revealed quantitatively similar concentrations, in the nanomolar range, of the all-*trans* isomer after microinjection of either all-*trans*-RA (250 ng/ml), 13-*cis*-RA (1500 ng/ml), retinol (4000 ng/ml) or all-*trans*-RAG (4000 ng/ml).

## ACKNOWLEDGEMENTS

I thank Dr Mont R. Juchau for helpful discussion of this chapter and also for the opportunity of continuing research about retinoids in his laboratory at the University of Washington, Seattle.

## REFERENCES

Braun, J. T., Franciosi, R. A., Mastri, A. R., Drake, R. M., and O'Niel, B. L. (1984). Isotretinoin dysmorphic syndrome. *Lancet*, **1**, 506–7.

Brazzel, R. K., Vane, F. M., Ehmann, C. W., and Colburn, W. A. (1983). Pharmacokinetics of isotretinoin during repetitive dosing to patients. *European Journal of Clinical Pharmacology*, **24**, 695–702.

Creech Kraft, J., Kochhar, D. M., Scott, W. J. Jr., and Nau, H. (1987). Low Teratogenicity of 13-*cis*-retinoic acid (isotretinoin) in the mouse corresponds to low embryo concentrations during organogenesis: comparison to the all-*trans* isomer. *Toxicology and Applied Pharmacology*, **87**, 474–82.

Creech Kraft, J., Eckhoff, C., Kuhnz, W., Löfberg, B., and Nau, H. (1988). Automated determination of 13-*cis* and all-*trans*-retinoic acid, their 4-oxo metabolites and retinol in plasma, serum, amniotic fluid and embryo by reversed-phase high-performance liquid chromatography with a precolumn switching technique. *Journal of Liquid Chromatography*, **11**, 2051–69.

Creech Kraft, J., Löfberg, B., Chahoud, I., Bochert, G., and Nau, H. (1989a) Teratogenicity and placental transfer of all-*trans*, 13-*cis*, 4-oxo-all-*trans*, and 4-oxo-13-*cis*- retinoic acid after a low oral dose during organogenesis in mice. *Toxicology and Applied Pharmacology*, **100**, 162–76.

Creech Kraft, J., Nau, H., Lammer, E., and Olney, A. (1989b). Human embryo retinoid concentrations after maternal intake of Isotretinoin. *New England Journal of Medicine*, **321**, 262.

Creech Kraft, J., Eckhoff, Chr., Kochhar, D. M., Bochert, G., and Nau, H. (1991a). Isotretinoin (13-*cis*-retinoic acid) metabolism, *cis-trans* isomerization, glucuronidation and transfer to the mouse embryo: Consequences for teratogenicity. *Teratogenesis Carcinogenesis and Mutagenesis*, **11**, 21–30.

Creech Kraft, J., Slikker, Jr, W., Bailey, J. R., Roberts, L. G., Fischer, B., Wittfoht, W., and Nau, H. (1991*b*). Plasma pharmacokinetics and metabolism of 13-*cis* and all-*trans* retinoic acid in the cynomolgus monkey and the identification of 13-*cis* and all-*trans*-retinoyl-*β*-glucuronides: A comparison to one human case study with Isotretinoin. *Drug Metabolism and Disposition*, **19**, 317–24.

Creech Kraft, J. and Juchau, M. R. (1991). Correlations between conceptal concentrations of all-trans-retinoic acid and dysmorphogenesis after micro-injections of all-trans-retinoic acid, 13-cis-retinoic acid, all-trans-retinoyl-*β*-glucuronide or retinol in cultured whole or rat embryos. *Drug Metabolism and Disposition*, **20** (in press).

Creech Kraft, J., Lee, Q. P., and Juchau, M. R. (1991*c*). Microinjection of cultured rat embryos: studies with retinoids. *Teratology*, **43**, 442–3A.

Goulding, E. H. and Pratt, R. M. (1986). Isotretinoin teratogenicity in mouse whole embryo culture. *Journal of Craniofacial Genetics and Developmental Biology*, **6**, 99–112.

Hummler, H., Korte, R., and Hendrickx, A. (1990). Induction of malformations in the cynomolgus monkey with 13-*cis* retinoic acid *Teratology*, **42**, 263–72.

Kochhar, D. M., Penner, J. D., and Tellone, C. (1984). Comparative teratogenic activities of two retinoids: Effects on palate and limb development. *Teratogenesis Carcinogenesis and Mutagenesis*, **4**, 377–87.

Lammer, E. J., Chen, D. T., Hoar, R. M., Agnish, N. D., Benke, P. J., Braun, J. T., Curry, C. J., Fernhoff, P. M., Grix, A. W. Jr., Lott, I. T., Richard, J. M., and Sun, S. C. (1985). Retinoic acid embryopathy. *New England Journal of Medicine*, **313**, 837–41.

Lee, Q. P., Juchau, M. R., and Creech Kraft, J. (1991). Microinjection of cultured rat embryos: studies with retinol, 13-*cis*- and all-*trans*-retinoic acid. *Teratology*, **44**, 313–23.

Morriss, G. M. (1972). Morphogenesis of the malformations induced in rat embryos by maternal hypervitaminosis A. *Journal of Anatomy*, **113**, 241–50.

Morriss, G. M. and Steele, C. E. (1977). Comparison of the effects of retinol and retinoic acid on postimplantation rat embryos *in vitro*. *Teratology*, **15**, 109–19.

Rosa, F. W. (1983). Teratogenicity of isotretinoin. *Lancet*, **2**, 513.

Steele, C. E., Marlow, R., Turton, M. J., and Hicks, R. M. (1987). *In vitro* teratogenicity of retinoids. *British Journal of Experimental Pathology*, **68**, 215–23.

Thaller, C. and Eichele, G. (1987). Identification and spatial distribution of retinoids in the developing chick limb bud. *Nature*, **327**, 625–8.

Webster, W. S., Johnston, C., Lammer, E. J., and Sulik, K. K. (1986). Isotretinoin embryopathy and the cranial neural crest: An in vivo and in vitro study. *Journal of Craniofacial Genetics and Developmental Biology*, **6**, 211–22.

Willhite, C. C., Wier, P. J., and Berry, D. L. (1989). Dose response and structural-activity considerations in retinoid induced dysmorphogenesis. *Critical Reviews in Toxicology*, **20**, 113–35.

21

# Malformations of hindbrain structures among humans exposed to isotretinoin (13-*cis*-retinoic acid) during early embryogenesis

Edward J. Lammer and Dawna L. Armstrong

## INTRODUCTION

Increasing evidence indicates that retinoic acid (RA) plays a major role in controlling the morphogenesis of many vertebrate structures. It is metabolically derived from retinol (vitamin A) and can mimic some, but not all, of retinol's properties. For example, RA cannot support retinol's role in reproduction or in the visual system (mediated by conversion of retinol to retinaldehyde), but it can promote cellular differentiation and bone growth as well as retinol does. Both retinol and RA are teratogenic in a number of experimental vertebrate species and humans (Kochhar 1967; Shenefelt 1972; Lammer *et al.* 1985; Irving *et al.* 1986; Webster *et al.* 1986; Hummler *et al.* 1990).

Retinoic acid-induced malformations in humans have resulted from the inadvertent use of isotretinoin (13-cis-RA) during pregnancy (Lammer *et al.* 1985). Isotretinoin is an oral medication taken for the treatment of severe nodulocystic acne. In the United States, isotretinoin was initially marketed in 1982 and is approved only for the treatment of severe nodulocystic acne that is unresponsive to other therapies. It is estimated that 60 000–70 000 reproductive-age women per year obtain new prescriptions for isotretinoin. This number almost certainly exceeds the number of women with severe cystic acne, as this is predominantly a male problem, but the magnitude of over-prescribing has been contentiously debated (Dermatologic Drugs Advisory Committee 1990). Because of the large number of reproductive-age women using isotretinoin and the known failure rates of all contraceptive measures, it is inevitable that use during pregnancy will occur. In the United States, these inadvertent pregnancy exposures have led to a number of adverse outcomes, including spontaneous abortion, premature delivery, and malformed fetuses and infants (Dermatologic Drugs Advisory Committee 1990).

The magnitude of risk for these adverse outcomes of pregnancy is unusually high. Lammer and co-workers (1985) initially found that 18 per cent of 36 prospectively ascertained isotretinoin-exposed pregnancies resulted in delivery of a malformed fetus/infant. Subsequent analysis of a larger cohort of exposed pregnancies identified increased absolute risks for spontaneous abortion (40 per cent), premature delivery (16 per cent), and major malformation (25 per cent) (Lammer *et al.* 1988). Isotretinoin-exposed infants have a characteristic pattern of malformation involving the brain, craniofacies, and the cardiovascular system. Malformations of other organ systems did occur, but much less commonly, so that, with the exception of spontaneous abortions, the most common sequelae of intrauterine isotretinoin exposure were brain malformations, abnormal neurological signs, and neuropsychological dysfunction. In this chapter we report a characteristic pattern of cerebellar and brain stem malformations that we have observed in isotretinoin-exposed fetuses and children.


STUDY POPULATION


Two populations of isotretinoin-exposed fetuses/infants were included in the Longitudinal Study of Infants Exposed to Isotretinoin *in utero* (see Acknowledgements). First, a prospectively ascertained cohort of pregnancies during which isotretinoin was used have been longitudinally followed and a number of outcomes studied. To be eligible for inclusion in this cohort, an exposed pregnancy had to be identified to our study group (or one of several collaborating organizations) before the outcome of the pregnancy was known, and before any antenatal diagnostic procedures took place. The principle benefit of studying this cohort was the ability to determine absolute risks for various adverse outcomes of pregnancy because pregnancies were ascertained without any bias as to outcome. Thus, the findings may be directly applied to the counselling setting. The second population was a retrospectively ascertained case series of isotretinoin-exposed fetuses/infants who have also been longitudinally followed. Most of these fetuses/infants have major malformations and were identified because of them. The data from this case series has led to the formulation of questions that pertain to pathogenetic mechanisms through which RA causes human birth defects.

In this report, cases were drawn from both of the above study populations. For each study participant, relevant maternal and infant medical records, brain computerized tomography (CT) and magnetic resonance imaging (MRI) studies, and autopsy reports were collected and abstracted. The information pertaining to the brain malformations was derived in several ways. The neuropathology of 17 isotretinoin-exposed fetuses/infants was reviewed by one paediatric neuropathologist (D.A.), who was provided with gross pathology descriptions, photographs, and microscopic slides of the nervous system. In some other cases, only an autopsy report was available. In other cases diagnoses were clinical, made on the basis of brain imaging studies. Copies of the original brain imaging tests were obtained and reviewed by one of the authors (E.L.).


## HINDBRAIN MALFORMATIONS

We have observed abnormal neurological signs, functional deficits, and abnormal central nervous system structures of isotretinoin-exposed children. The abnormal neurological signs include hypotonia, hypertonia, and cranial nerve pareses. Functional deficits of the visual and auditory systems have been identified. More subtle development problems detected by neuropsychological performance testing have also been found. At age 5 years, 52 per cent of exposed children, with or without major malformations, had subnormal IQ scores (Adams and Lammer 1991). Among the major malformations of the central nervous system, we identified characteristic anomalies of hindbrain structures. Some of the abnormal neurological signs referred to above may be explained by these malformations.

We observed 26 fetuses and infants with a pattern of hindbrain malformations consisting of partial or complete absence of the cerebellar vermis (Fig. 21.1), cystic dilatation of the roof of the fourth ventricle resulting in elevation of the tentorium cerebelli and prominence of the posterior skull (Fig. 21.2(a, b)), and dilatation of the ventricular system (Fig. 21.2(c)). (Ventricular enlargement was diagnosed as hydrocephalus in some reports and as ventriculomegaly in others; the term hydrocephalus in clinical use usually implies a condition of progressive ventricular dilatation, whereas ventriculomegaly implies that the ventricles are enlarged due to decreased space taken by neural tissue.) This combination of hindbrain malformations is called the Dandy–Walker anomaly (Hart *et al.* 1972). Tables 21.1 and 21.2 illustrate examples of the variety of combinations of these major brain malformations that we observed in the neuropathology review (Table 21.1) and in the review of cases diagnosed by clinical imaging studies (Table 21.2). When hydrocephalus was present, it was frequently secondary to aqueductal stenosis (see Table 21.1). The neuropathology review also defined malformations of the inferior medullary olivary nuclei and/or pontine nuclei (Figs 21.3 and 21.4). Those of the inferior medullary olive were considered to be dysplasias of the nuclei. The normal mature convolutions of the olives were deficient or absent, so that the nuclei appeared as simple rounded or globular spheres (Fig. 21.3(b)). Other microscopic abnormalities included dysplasias of the cerebellar nuclei (Fig. 21.5) and heterotopias of the cerebellar hemispheres (Fig. 21.6).

Among malformed isotretinoin-exposed infants who have survived, a similar pattern of major brain malformations was found. Table 21.2 lists a representative sample of the combination of these defects that were diagnosed by brain CT or MRI techniques. The malformations include cerebral

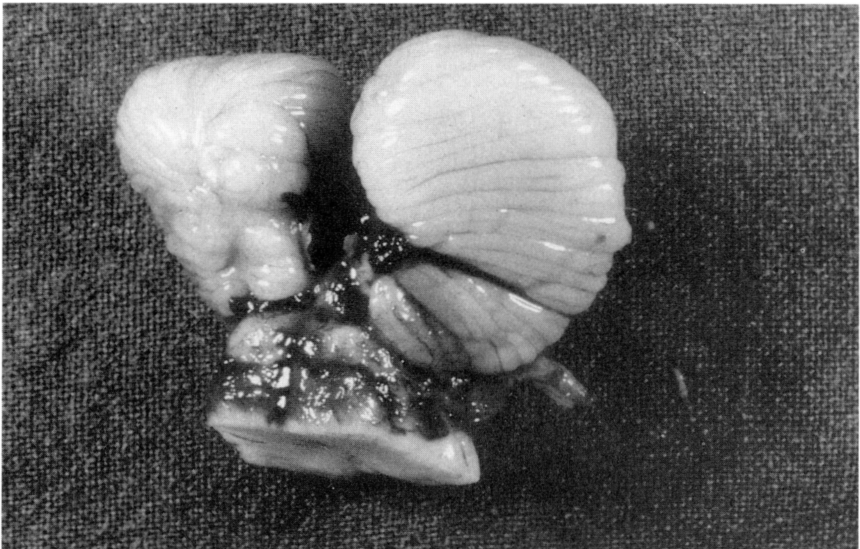

**Fig. 21.1** Agenesis of the cerebellar vermis. Dorsal view of cerebellar hemispheres showing absence of the midline vermis of the cerebellum. The thin, dilated roof the fourth ventricle was torn away during removal of the brain. Courtesy of Dr S. Ayme.

ventriculomegaly or hydrocephalus, deficiencies of the cerebellar vermis and/or small cerebellar hemispheres, and cystic dilation of the roof of the fourth ventricle. The imaging studies also frequently show an enlarged vallecula and cisterna magna (Fig. 21.2(c)).


DISCUSSION

We have described a characteristic pattern of hindbrain malformations in human RA embryopathy that includes cystic dilatation of the roof of the fourth ventricle, cerebellar vermis malformations and/or small cerebellar hemispheres, microscopic cerebellar heterotopias, and malformations of the inferior medullary olivary and pontine nuclei. This pattern of brain malformation, also known as the Dandy–Walker anomaly, is known to occur in several single gene disorders, and is generally regarded as aetiologically heterogeneous (Bordarier and Aicardi 1990). This is the first report of the Dandy–Walker anomaly in humans associated with an environmental agent. A similar cerebellar malformation was described in a pigtail monkey exposed to 10 mg/kg/day of all-*trans*-RA on gestational days 20–44 (Fantel *et al.* 1976). However, most teratology studies of RA that have described central nervous system malformations as an outcome, have focused on neural tube defects.

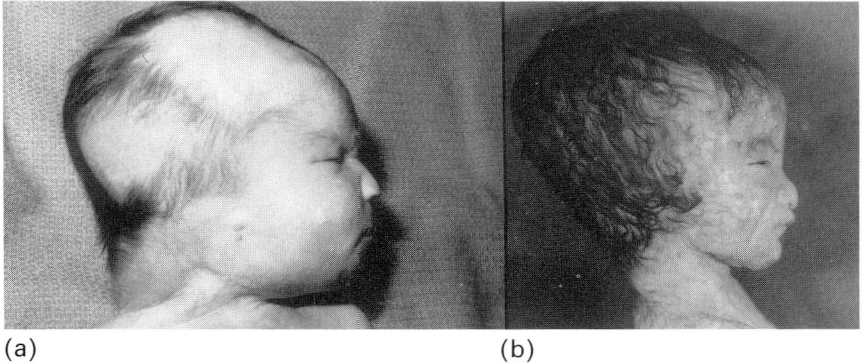

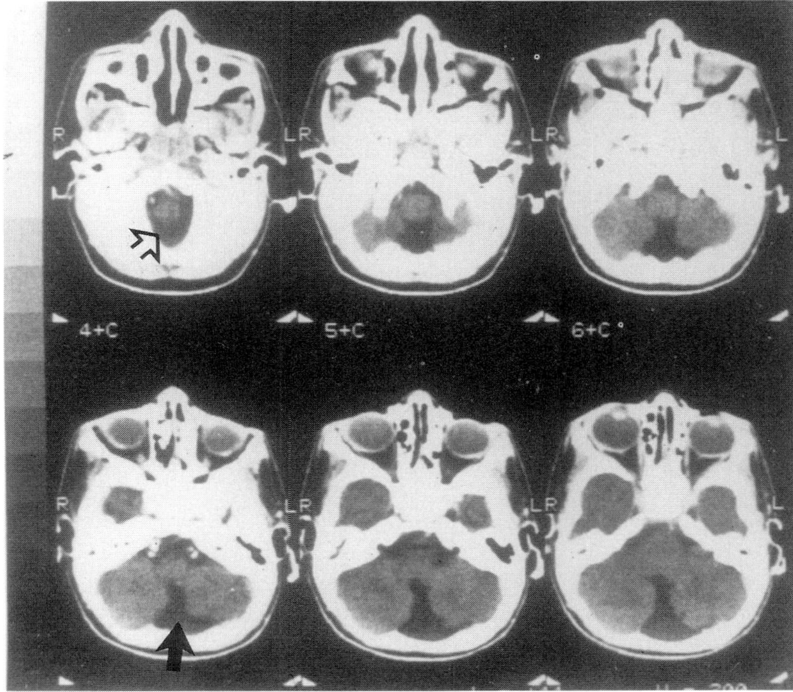

**Fig. 21.2** Abnormal head shape and brain imaging studies. (*a*) Lateral photograph of child with retinoic acid embryopathy. Note the elevated position of the prominent posterior cranial contour caused by the enlarged posterior fossa during fetal development. Note severe microtia and absence of the external ear canal. (*b*) Lateral photograph of a fetus with retinoic acid embryopathy. Note the high prominent posterior cranial contour caused by the enlarged posterior fossa during fetal development, and the complete absence of external ear and canal development. Courtesy of Dr S. Ayme. (*c*) Series of computerized tomography scans from the cisterna magna to the cerebellum of the brain of a child with retinoic acid embryopathy, showing absence of midline cerebellar tissue, enlarged vallecula (arrow), and megacisterna magna.

**Table 21.1** Gross and microscopic abnormalities of rhombencephalic structures among isotretinoin-exposed fetuses/infants

| Case | Exposure interval* (days) | Third and lateral ventricles | Fourth ventricle | Cerebellum | Inferior olivary nuclei | Pons |
|---|---|---|---|---|---|---|
| A | 0–50 | Hydrocephalus, aqueductal stenosis | Cystic dilation | Vermis agenesis | – | Disorganized asymmetric |
| B | 1–32 | Hydrocephalus, type?† | – | Vermis agenesis | Normal | Normal |
| C | 0–52 | Hydrocephalus, aqueductal stenosis | Cystic dilation | Vermis agenesis | Malformed | Normal |
| D | 38–66 | Hydrocephalus, aqueductal stenosis | Cystic dilation | Vermis agenesis | Malformed | |
| E | 1–2 | Hydrocephalus, aqueductal stenosis | Cystic dilation | Vermis agenesis | Normal | Abnormal |
| F | 0–120 | Hydrocephalus, type? | Cystic dilation | Vermis agenesis | Malformed | Abnormal |

* Day 0 = estimated day of conception.

† Cases with hydrocephalus of undetermined type were not included in our neuropathology review.

**Table 21.2** Clinically diagnosed abnormalities of the posterior fossa among isotretinoin-exposed infants

| Case | Exposure interval* (days) | Third and lateral ventricles | Fourth ventricle | Cerebellum | Other |
|---|---|---|---|---|---|
| G | 8–52 | Hydrocephalus | Cystic dilation | Normal | — |
| H | 0–29 | Ventriculomegaly | Cystic dilation | Vermis agenesis | — |
| I | 17–41 | Normal | Normal | Inferior vermis agenesis | — |
| J | 19–26 | Normal | Cystic dilation | Normal | Large cisterna magna |
| K | 25–31 | Ventriculomegaly | Cystic dilation | Vermis agenesis | Asymmetric pons |
| L | 64–68 | Ventriculomegaly | Normal | Vermis agenesis | Large cisterna magna |

* Day 0 = estimated day of conception.

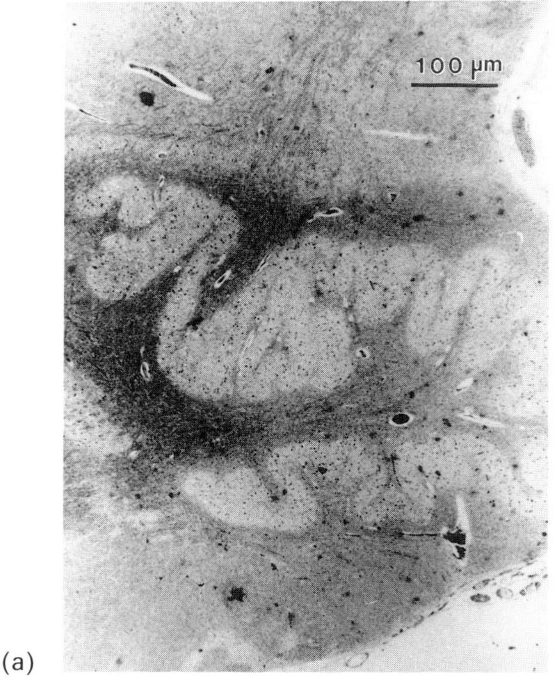


(a)

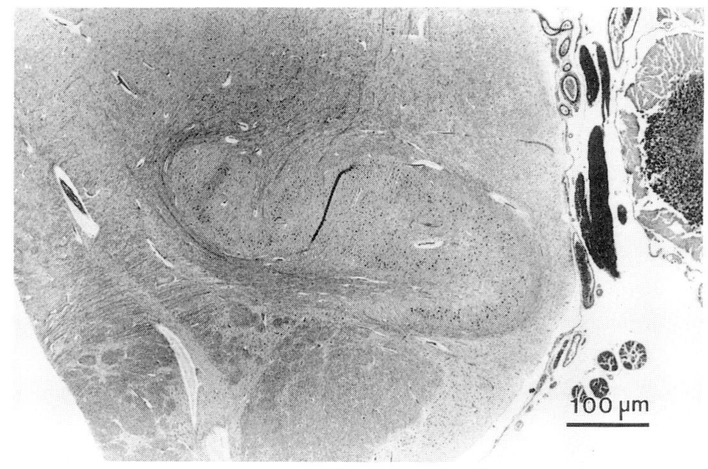


(b)

**Fig. 21.3** Dysplasias of inferior medullary olives. (*a*) Normal appearance of right inferior medullary olive at 3 months postnatal age. The left edge of the photograph is medial. Note the normal convoluted olive with its right-sided reverse-C-shape, and the medial accessory and dorsal accessory olivary nuclei with their normal oval shapes. × 40. (*b*) The inferior medullary olive shows incomplete undulations of the lamellae. The anterior lamella is short and narrow. × 40.

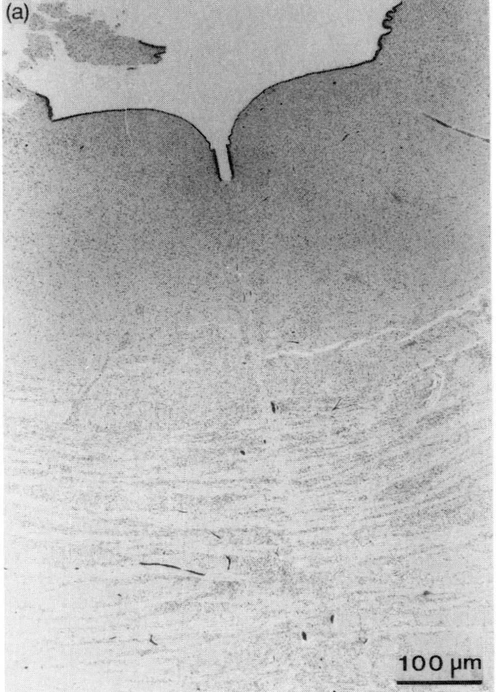


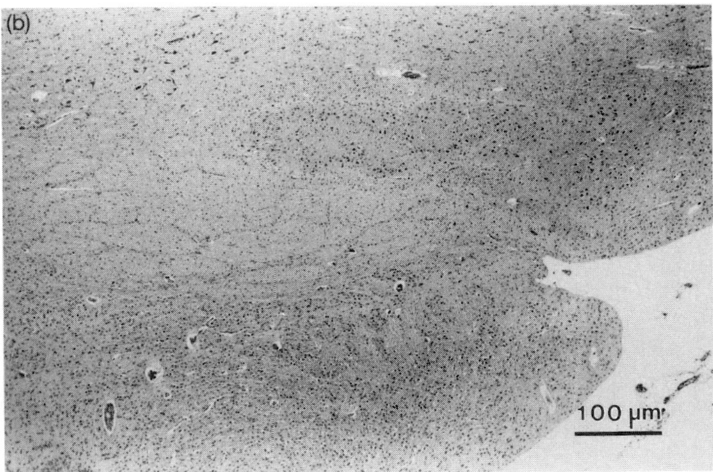


**Fig. 21.4** Abnormalities of pontine relationships. (*a*) Microscopic features of the pons of a 23-week gestation fetus showing normal basis pontis with crossing fibres and the tegmentum above them. (*b*) Basis pontis with nuclear aggregates that are incompletely separated by pontine tracts. × 31.

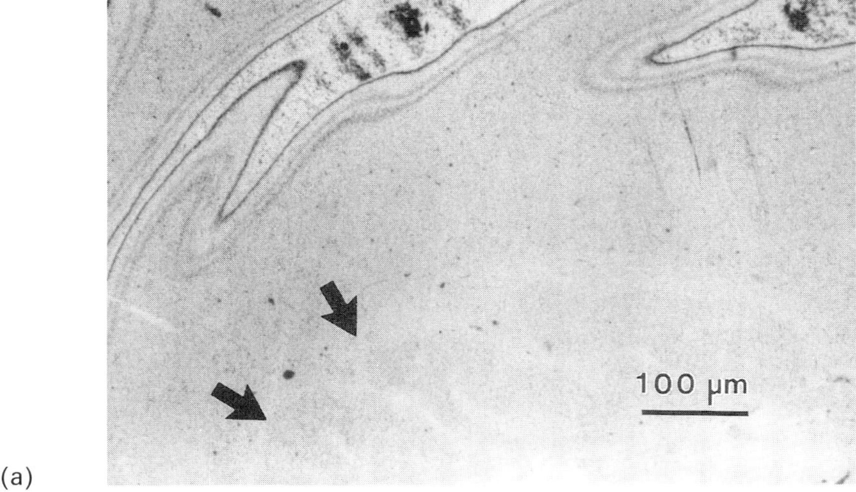



(a)

(b)

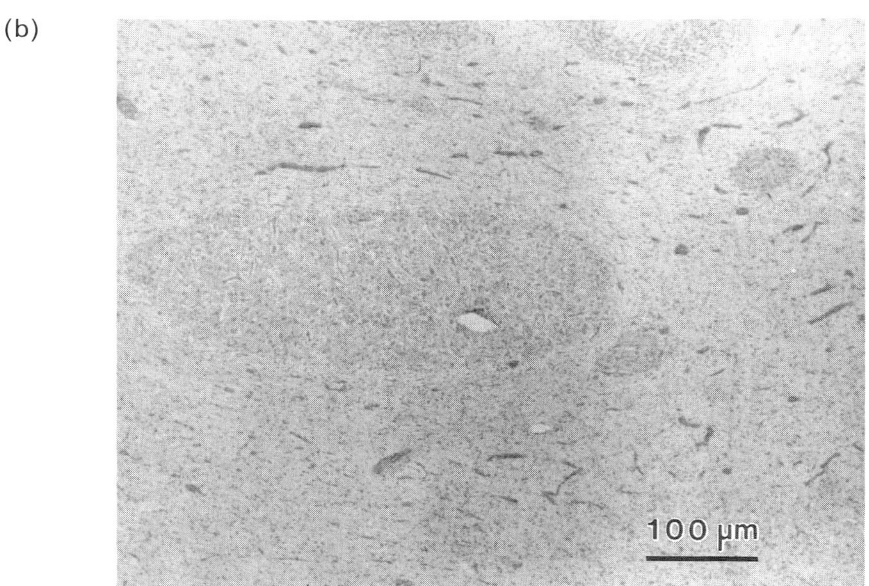



**Fig. 21.5** Dysplasia of cerebellar dentate nucleus. (*a*) Normal convoluted pattern of the cerebellar dentate nucleus (arrowed). × 40. (*b*) This dentate nucleus is abnormal, lacking the normal shape and showing multiple nuclear aggregates composed of homogeneous cells. × 40.

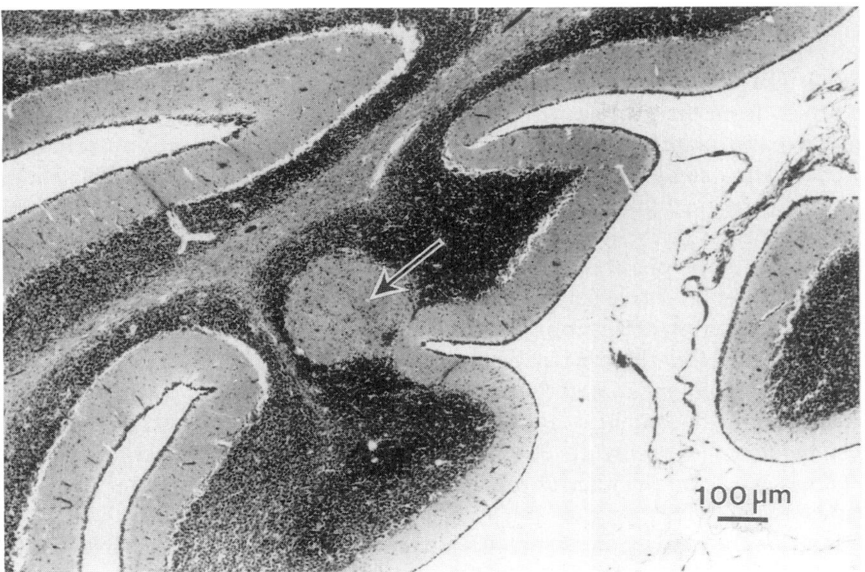

**Fig. 21.6** Cerebellar heterotopias. Section of a cerebellar hemisphere showing (arrowed) an ectopic aggregate of Purkinje cells and cells with molecular layer morphology located at the depth of a sulcus within the internal granular cell layer. This heterotopia includes cell types that are normally related to each other, but are displaced within the folia. × 19.

Thus, there is little information on the effects of exogenous retinoic acid on the development of the cerebellum and other rhombencephalic structures.

Do the affected rhombencephalic derivatives share an embryological origin or a developmental process that might explain their shared susceptibility to exogenously administered RA? The human cerebellum develops from the first rhombomere (R1) of the hindbrain. The first and second rhombomeres initially appear as a single rostral-most segment of the early hindbrain, and do not segregate until after the appearance of the more caudal segments of the rhombencephalon (Layer 1990). The cerebellum develops from neuroblasts that initially migrate laterally and dorsally to form the rhombic lips (the most lateral part of the alar plate). The alar plate is an area that is probably dorsal to the rhombomeric constrictions, which are evident ventrally. The neuroblasts of the thickened rhombic lips at the rostral end of the rhombencephalic roof then migrate medially to form the cerebellar plate. The first folia of the cerebellar vermis appear at 13–14 weeks gestational age (Lemire 1975). Similarly, the human pontine nuclei arise from neuroblasts near the junction of the rhombic lip and the most lateral part of the roof of the fourth ventricle caudal to the cerebellum (Essick 1912). These neuroblasts migrate ventrally, following a restricted pathway called the corpus pontobulbare, between the

7th and 8th cranial nerves, to the area of the pontine flexure. The precursors of the nuclei of the inferior olive originate from the rhombencephalon caudal to the origin of the pontine nuclei. The inferior olivary neuroblasts migrate ventrally from the entire width of the alar plate and, to some extent, from the lateral basal plate (Hanaway and Netsky 1971). Their migration probably proceeds in a caudal to rostral sequence (Hanaway and Netsky 1971). They arrive in the medulla beginning around stage 19 and form a recognizable nucleus medial to the hypoglossal nerve roots (Müller and O'Rahilly 1990). Thus, it is apparent that the cerebellum, pontine nuclei, and the inferior olivary nuclei all derive from neuroblasts migrating from the alar plate of the rhombencephalon at appropriate levels of the rostrocaudal axis. Tables 21.1 and 21.2 show that this area of the brain is susceptible to RA excess over a long period of development, and that a starting time as late as day 64 (case L) can cause major abnormality. It is also interesting to note that vermis agenesis resulted from exposure to RA at a wide range of developmental stages, whereas effects on the inferior olivary nuclei resulted only from later times of exposure.

It is not known why the alar plate-derived neuroblasts are susceptible to the effects of RA. Hanaway and Netsky previously reported two cases of the Dandy–Walker anomaly with malformations of the inferior olive, and referred to several other published cases where olivary heterotopia was illustrated or described. The authors concluded that heterotopic olivary neurons, or malformed inferior olivary nuclei, were a part of the Dandy–Walker anomaly. Our findings support this conclusion, but whereas Hanaway and Netsky found no other abnormal brainstem nuclei (including nuclei derived from alar plate neuroblasts), we found a number of cases with malformed pontine nuclei.

The alar plate-derived neuroblasts might be susceptible to RA because of the presence of cellular RA binding protein (CRABP) or RA receptors (RARs). CRABP is a cellular protein that specifically binds RA, but whose function in normal embryogenesis is poorly understood. Two types of CRABP have been found, CRABP I and II. Some indirect evidence suggests that CRABP I may control the free levels of RA available to bind to the nuclear receptors. Alternatively, CRABP may participate in the transfer of RA into the nucleus. The CRABPs may also have other functions that are yet to be delineated. Several groups have studied the temporal and segment-specific localization of CRABP in the brains of vertebrate embryos using immunohistochemical methods (Maden *et al.* 1989; Dencker *et al.* 1990; Maden *et al.* 1990; Momoi *et al.* 1990) or *in situ* hybridization (Vaessen *et al.* 1989; Vaessen *et al.* 1990; Ruberte *et al.* 1991). In general, CRABP localization within the embryonic vertebrate brain is restricted to the hindbrain and the midbrain roof. Because our report focuses on hindbrain malformations in humans, we will only refer to the localization of CRABP within that part of the brain. Dencker and co-workers (1990) reported CRABP localization with neurons of the mouse rhombencephalic floor, walls, and the thin roof of the 4th ventricle.

They also noted that the degree of immunostaining varied between some rhombomeres. Maden and co-workers (1990) have also localized CRABP to the cranial nerve ganglia of the hindbrain and to the entire thickness of the hindbrain neuroepithelium of the day 9 to day 12 mouse embryo. More recent findings relative to the apparent susceptibility of human alar plate-derived neuroblasts to RA, indicate that CRABP I is more specifically localized to alar plate and parts of the basal plate of the metencephalon and myelencephalon of day 12.5 mouse embryos (E. Ruberte, unpublished data).

Transcripts of RAR genes have also been localized within hindbrain areas (Ruberte *et al.* 1991). RAR-$\alpha$ and RAR-$\beta$ transcripts are expressed in the spinal cord and rostrally to specific levels of the hindbrain. RAR-$\alpha$ is expressed caudal to rhombomere 3 of day 9 (18–20 somite) mouse embryos, while RAR-$\beta$ is expressed from the rhombomere 6–7 boundary caudally along the closed portion of the neural tube. RAR-$\gamma$, in contrast, does not localize within rhombencephalic structures during any stage of mouse embryogenesis (Ruberte *et al.* 1990). Thus, while we do not yet know why derivatives of the rhombencephalic alar plate are specifically susceptible to RA, a potential role for CRABP, RAR-$\alpha$, or RAR-$\beta$ seems likely. Further study of an animal model of RA-induced Dandy–Walker anomalies may help to elucidate the pathogenesis of this malformation complex, which is one of the most common developmental anomalies of the cerebellum.

ACKNOWLEDGEMENTS

The Longitudinal Study of Infants Exposed to Isotretinoin *In Utero* is supported by a grant from Hoffman-La Roche, Inc. (Nutley). Special thanks to other members of the study group: Drs Lewis B. Holmes, Ailish Hayes, and Jane Adams; Ann Schunior, Barbara Battaglia, Dr Susan Tan-Torres, and Wendy Fishbeck. Jacqueline Ellis expertly crafted the manuscript.

We thank the families of all the children who are participating in our studies for their interest, cooperation, and commitment. The completion of this specific analysis was greatly aided by the cooperation of a number of pathologists who allowed us to review their autopsy materials and we also thank them.

REFERENCES

Adams, J. and Lammer, E. J. (1991). Relationship between dysmorphology and neuropsychological function in children exposed to isotretinoin 'in utero'. In *Functional neuroteratology*, (ed. T. Fujii). Teikyo University Press, Tokyo.
Bordarier, C. and Aicardi, J. (1990). Dandy-Walker syndrome and agenesis of the

cerebellar vermis: diagnostic problems and genetic counselling. *Developmental Medicine and Child Neurology*, **32**, 285–94.

Dencker, L., Annerwall, E., Busch, C., and Eriksson, U. (1990). Localization of specific retinoid-binding sites and expression of cellular retinoic acid-binding protein (CRABP) in the early mouse embryo. *Development*, **110**, 343–52.

Dermatologic Drugs Advisory Committee (1990). *Transcript of meeting, May* 12. Food and Drug Administration, Department of Health and Human Services, USA.

Dollé, P., Ruberte, E., Leroy, P., Morriss-Kay, G., and Chambon, P. (1990). Retinoic acid receptors and cellular retinoid binding proteins. I. A systematic study of their differential pattern of transcription during mouse organogenesis. *Development*, **110**, 1133–51.

Essick, C. R. (1912). The development of the nucleus pontis and the nucleus arcuatus in man. *American Journal of Anatomy*, **13**, 25–54.

Fantel, A. G., Shepard, T. H., Newell-Morris, L. L., and Moffett, B. C. (1976). Teratogenic effects of retinoic acid in pigtail monkeys (Macaca nemestrina). 1. General features. *Teratology*, **15**, 65–72.

Hanaway, J. and Netsky, M. G. (1971). Heterotopias of the inferior olive: relation to Dandy-Walker malformations. *Journal of Neuropathology and Experimental Neurology*, **30**, 380–9.

Hart, M. N., Malamud, N., and Ellis, W. G. (1972). The Dandy–Walker syndrome: a clinicopathological study based on 28 cases. *Neurology*, **22**, 771–80.

Hummler, H., Korte, R., and Hendrickx, A. G. (1990). Induction of malformations in the cynomolgus monkey with 13-cis-retinoic acid. *Teratology*, **42**, 263–72.

Irving, D. W., Willhite, C. C., and Burk, D. T. (1986). Morphogenesis of isotretinoin-induce microcephaly and micrognathia studied by scanning electron microscopy. *Teratology*, **34**, 141–53.

Kochhar, D. M. (1967). Teratogenic activity of retinoic acid. *Acta Pathologica et Microbiologica Scandinavica*, **70**, 398–404.

Lammer, E. J., Chen, D. T., Hoar, R. M., *et al.* (1985). Retinoic acid embryopathy. *New England Journal of Medicine*, **313**, 837–41.

Lammer, E. J., Hayes, A. M., Schunior, A., and Holmes, L. B. (1988). Unusually high risk for adverse outcomes of pregnancy following fetal isotretinoin exposure. *American Journal of Human Genetics*, **43**, A58.

Layer, P. G. (1990). Patterning of chick brain vesicles as revealed by peanut agglutinin and cholinesterases. *Development*, **109**, 613–24.

Lemire, R. J., Loeser, J. D., Leech, R. W., and Ellsworth, C. A. (1975). Cerebellum. In *Normal and abnormal development of the human central nervous system*, pp. 144–65. Harper and Row, Hagerstown, MD.

Maden, M., Ong, D. E., Summerbell, D., and Chytil. F. (1989). The role of retinoid-binding proteins in the generation of pattern in the developing limb, the regenerating limb and the nervous system. *Development*, (Suppl), 109–19.

Maden, M., Ong, D. E., and Chytil, F. (1990). Retinoid-binding protein distribution in the developing mammalian nervous system. *Development*, **109**, 75–80.

Momoi, M., Yamagata, T., Ichihashi, K., Yanagisawa, M., Yamakado, M., and Momoi, T. (1990). Expression of cellular retinoic acid binding protein in the developing nervous system of the mouse embryo. *Developmental Brain Research*, **54**, 161–7.

Müller, F., and O'Rahilly, R. (1990). The human rhombencephalon at the end of the embryonic period proper. *American Journal of Anatomy*, **189**, 127–45.

Ruberte, E., Dollé, P., Krust, A., Zelent, A., Morriss-Kay, G., and Chambon, P. (1990). Specific spatial and temporal distribution of retinoic acid receptor gamma transcripts during mouse embryogenesis. *Development*, **108**, 213–22.

Ruberte, E., Dollé, P., Chambon, P., and Morriss-Kay, G. (1991). Retinoic acid receptors and cellular retinoid binding proteins: II. Their differential pattern of transcription during early morphogenesis in mouse embryos. *Development*, **111**, 45–60.

Shenefelt, R. E. (1972). Morphogenesis of malformations in hamsters caused by retinoic acid: relation to dose and stage at treatment. *Teratology*, **5**, 103–18.

Vaessen, M.-J., Kootwijk, E., Mummery, C., Hilkens, J., Bootsma, D., and Van Kessel, A. G. (1989). Preferential expression of cellular retinoic acid binding protein in a subpopulation of neural cells in the developing mouse embryo. *Differentiation*, **40**, 99–105.

Vaessen, M.-J., Meijers, J. H. C., Bootsma, D., and Van Kessel, A. G. (1990). The cellular retinoic acid-binding protein is expressed in tissues associated with retinoic acid-induced malformations. *Development*, **110**, 371–8.

Webster, W. S., Johnston, M. C., Lammer, E. J., and Sulik, K. K. (1986). Isotretinoin embryopathy and the cranial neural crest: an *in vivo* and *in vitro* study. *Journal of Craniofacial Genetics and Developmental Biology*, **6**, 211–22.

# Index